Lightning Transients and Protection for Renewable Energy System

Qiuqin Sun · Li Zhang · Xiao Zhong · Feng Bin ·
Zhi Zheng

Lightning Transients and Protection for Renewable Energy System

Springer

Qiuqin Sun
College of Electrical and Information Engineering
Hunan University
Changsha, Hunan, China

Xiao Zhong
College of Electrical and Information Engineering
Hunan University
Changsha, Hunan, China

Zhi Zheng
College of Electrical and Information Engineering
Hunan University
Changsha, Hunan, China

Li Zhang
School of Electrical Engineering
Shandong University
Jinan, Shandong, China

Feng Bin
School of Physics and Electronic Science
Changsha University of Science and Technology
Changsha, Hunan, China

ISBN 978-981-97-7266-7 ISBN 978-981-97-7267-4 (eBook)
https://doi.org/10.1007/978-981-97-7267-4

This Springer imprint is published by the registered company Springer Nature Singapore Pte Ltd.
The registered company address is: 152 Beach Road, #21-01/04 Gateway East, Singapore 189721, Singapore

Preface

This book is dedicated to the lightning transients and protection for renewable energy systems, including both wind energy and solar energy. In addition to the formation mechanism of lightning transients, the practical engineering application has been also addressed. This book is comprehensive, and it includes the electromagnetic transient modeling methodology, electromagnetic coupling for PV system, lightning protection design, testing, and evaluation. This book is particularly suitable for readers who are interested in lightning phenomenon and renewable energy system. It would benefit researchers, engineers, and graduate students in the fields of electrical engineering, renewable energy, computational electromagnetism, etc.

Changsha, China — Qiuqin Sun
Jinan, China — Li Zhang
Changsha, China — Xiao Zhong
Changsha, China — Feng Bin
Changsha, China — Zhi Zheng

Acknowledgments

This book *Lightning Transients and Protection for Renewable Energy System* is a collaboration of the following four authors. The word count of the whole book is about 200 thousand.

Professor Qiuqin Sun has contributed to the contents of Chaps. 2–5 in Part I, the word count of which is about 90 thousand, and he is also responsible for the content revision, polishing, and editing of the whole book. Associate Professor Li Zhang is responsible for completing Chaps. 7–9 in Part II, the word count of which is about 40 thousand. Dr. Xiao Zhong has contributed to the contents of Chap. 6 in Part I. Dr. Feng Bin has contributed to the contents of Chap. 1 in Part I. Dr. Zhi Zheng is responsible for completing Chaps. 10–11 in Part II, the word count of which is about 30 thousand.

Meanwhile, the book publishment is supported by grants from the National Natural Science Foundation of China (No. 52277145) and Hunan Natural Science Foundation of China (No. 2023JJ10011). The authors express deep gratitude for their support.

Acknowledgments

Contents

About the Authors

Qiuqin Sun Professor, Hunan University, Changsha, China.

IEEE Senior Member, IET Chartered Engineer. He received degrees of B.Sc. (2006) and Ph.D. (2012) from Chongqing University, Chongqing, and Shandong University, Jinan, China, respectively, both in Electrical Engineering. He joined in Jiangsu Electrical Power Company Research Institute as a senior electrical engineer in 2012 and then came to Hunan University, Changsha, China, in 2014. From June 2014 to May 2015, he worked as an Honorary Research Fellow at the University of Liverpool. He is currently a professor at the College of Electrical and Information Engineering, Hunan University. He has authored or co-authored about 100 journal papers, including about 30 IEEE Transactions papers. He serves as editorial board member of journal *High Voltage Apparatus*, *Electric Power Engineering Technology*, and academic editor of *Mathematical Problems in Engineering*. He is the recipient of Distinguished Young Scholars of Hunan Province. He can be reached by email at sunqq@hnu.edu.cn.

Li Zhang Associate Professor, Shandong University, Jinan, China.

He received his degree of Ph.D. (2009) from the School of Electrical Engineering at Shandong University, Jinan, China. From July 2019 to October 2020, he served as a visiting scholar at the University of Florida in the USA. Currently, he is a Ph.D. candidate supervisor and the head of high voltage institute, Shandong University. His research primarily focuses on applied electromagnetics, high voltage, and insulation technology. He is a leader and principal investigator for four national natural science foundation projects. He can be reached by email at zhleee@sdu.edu.cn.

Xiao Zhong received his B.Sc. degree in electrical engineering from Hefei University of Technology, Hefei, China, in 2019. Currently, he is studying for the Ph.D. degree in electrical engineering at the School of Electrical and Information Engineering of Hunan University, Changsha, China. His research interests include lightning electromagnetic transients in power systems and lightning protection design of PV systems. He can be reached by email at zhongxiao@hnu.edu.cn.

Feng Bin Lecturer, Changsha University of Science and Technology, Changsha, China.

He received his degree of Ph.D. (2020) from School of Electrical and Information Engineering at Hunan University, Changsha, China. Currently, he is a master supervisor and the director of entrepreneurship and innovation center at School of Physics and Electronic Science, Changsha University of Science and Technology. His research primarily focuses on artificial intelligence-based object detection and insulation online monitoring for power equipment. He can be reached by email at dianzi8349@csust.edu.cn.

Zhi Zheng received the B.S. degree from Taiyuan University of Technology, Taiyuan, China, in 2017 and the M.S. degree from North China Electric Power University, Baoding, China, in 2020. He received Ph.D. (2024) from College of Electrical and Information Engineering at Hunan University, Changsha, China. His research interest is electromagnetic transient analysis and overvoltage protection for offshore wind power system. He can be reached by email at zhengzhi@hnu.edu.cn.

Part I
Lightning Electromagnetic (EM) Transients and Protection for Photovoltaic System

Chapter 1
Introduction to Photovoltaic System

1.1 Background

The photovoltaic (PV) power generation system is mainly composed of large-area PV panels, direct current (DC) combiner boxes, DC distribution cabinets, PV inverters, alternating current (AC) distribution cabinets, grid connected transformers, and connecting cables. Since the output voltage of single PV cell is very small, multiple PV cells are often connected in series through a foil-plated thin copper wire in order to obtain a higher output voltage [1]. The PV cell in series can be equivalent to a straight wire, whose two ends represent positive and negative electrodes, respectively. Both positive and negative output terminals of PV module are connected to the junction box in parallel with a bypass diode, which provides an alternative current path to mitigate the effect of shadows or flares. To prevent water penetration, the bottom of PV cell is filled with insulation material [1] (Fig. 1.1).

Generally, PV power generation systems are installed on the metal bracket with a tilt angle, and these brackets are placed in the wilderness or on the top of building. Besides, the bracket and frame of panel are connected to common ground. PV power generation systems have the characteristics of high installation density, large covering area, and high proportion of metal material. It is estimated that a 100 MW PV power station occupies nearly 20 km^2 [2]. Because the equipment is exposed to the open area for a long time, they are very easy to suffer from the lightning damage, including direct lightning overvoltage, induced lightning overvoltage, lightning intrusion wave and so on. The statistical results show that damage caused by lightning strikes accounts for 26% of PV array accidents [3], and the proportion is higher for areas with lots of lightning activity. There have been many reports about lightning strike accidents in PV power stations [4–6]. For example, in 2010, a PV power station in Xuzhou, China, undergone induced lightning intrusion, resulting in the destruction of control system of single-axis tracking unit. In 2016, a PV power generation system in Xizang, China, was stroked by lightning, leading to obvious lightning stripes on some of the PV panels. In 2017, a 30 MW PV power station was invaded by lightning overvoltage,

Q. Sun et al., *Lightning Transients and Protection for Renewable Energy System*,
https://doi.org/10.1007/978-981-97-7267-4_1

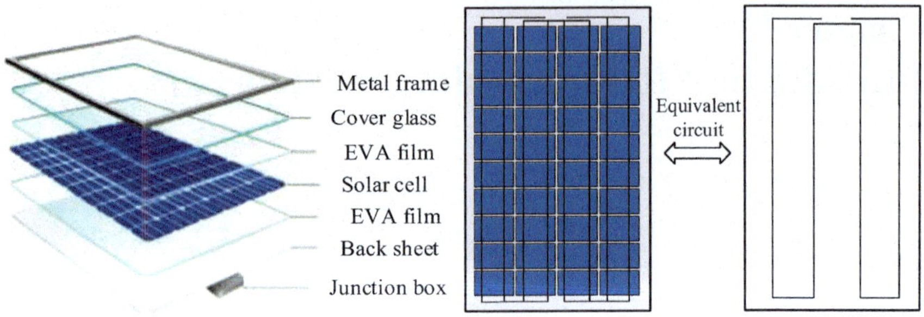

Fig. 1.1 Structure of PV module

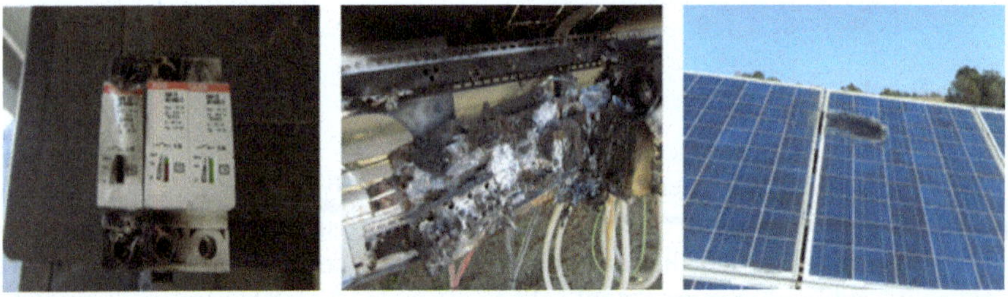

Fig. 1.2 Damage caused by lightning surge to PV system

which caused that plenty of bypass diodes was broken down, more than 200 junction boxes of PV module were bulged and even burned due to low open circuit voltage of diodes, and a number of PV-15A fuses were blown out.

The main hazards of lightning strikes to PV systems include that lightning may directly hit the PV panels, causing the permanent damage or ablation of equipment, or the formed electromagnetic (EM) pulse propagates into space, generating surges on nearby DC circuits. In extreme cases, the lightning-induced overvoltage in the photovoltaic series circuit may be as high as several thousand volts. The PV modules use a large amount of semiconductor material, such as silicon, with low insulation strength, and poor resistance to overvoltage and overcurrent, and thus sensitive to EM interference. Since external lightning protection devices introduce lightning currents into the earth, high voltage may also be generated on the grounding network, resulting in the damage of AC and DC sides secondary equipment such as inverters and controllers. When the bypass diode of PV module fails, a large amount of reverse current backflows, leading to the junction box to be burned. In addition, lightning surge can cause permanent degradation of PV cells, resulting in the decrease of conversion efficiency and output power. Hence, although the PV cell may still operate normally without apparent discharge or leakage currents on the surface, it cannot provide the maximum power. However, in the maximum power tracking control mode, the above-mentioned faulty batteries may turn into load and provoke reverse current, resulting in local overheating of PV array and even fire accident.

The book will present a comprehensive overview on the lightning transient effects in PV systems, the potential-induced degradation mechanism, and the lightning protection measures for PV arrays.

1.2 Lightning Electromagnetic Modeling of Photovoltaic System

The strong EM radiation generated around the lightning discharge channel will form EM coupling on the PV module interconnection cables, and then interfere with or even damage the electronic equipment in connection with these cables. Due to the low instalment height, there are little electrostatic induction component but strong EM induction component on the PV array. The induced overvoltage of PV array involves three aspects, i.e., modelling of lightning channel, calculation of lightning EM field, and coupling mechanism [7]. Based on the semi-analytical expression of the magnetic vector potential, a three-dimensional semi-analytical numerical calculation method is proposed to investigate the EM transients process caused by nearby lightning strikes in PV modules, and the common and differential mode voltage of PV arrays [8]. The University of Florida conducted field measurements for the transient effects of induced lightning strikes on PV arrays, using high-speed data loggers to monitor voltage, threshold voltage to trigger event recording, and transient voltage suppressors (TVS) to maintain voltage peaks on the DC side [9]. German scholars carried out lightning-induced overvoltage experiments to establish an inductive coupling mechanism between PV modules and lightning current [10]. The lighting surge generator was used to inject the impulse current into the grid-connected PV power station, and then the overcurrent and overvoltage on both DC cable and photovoltaic inverter terminal were measured [11, 12].

The PV array consists of DC cable, PV support bracket, component frame, and thin copper wire, all of which may be acted as the coupling channels of lightning EM fields. There are two methods, including transmission line model [14, 15] and full-wave model [16], to simulate the conductor structure in PV arrays [13]. The former assumes that the lightning EM wave is a transverse EM wave (TEM) which propagates along the conductor, and it is often used to calculate the EM transient response of transmission lines struck by lightning. In [17, 18], researchers from Beijing Jiaotong University proposed a method to calculate the parameters of large-scale bracket with horizontal, vertical, or inclined structure and grounding device, established the circuit model of bracket, and obtained the transient voltage of each node of bracket using EMTP software under the condition of direct lightning strike. In [19], PSCAD software was employed to build the transient model of PV module, power cable, combining manifold, and grid-connected inverter, and conduct the lightning-induced overvoltage simulation. Nevertheless, the induced current in the metal frame and PV bracket would affect the EM field within adjacent DC cable and thin copper wire, and

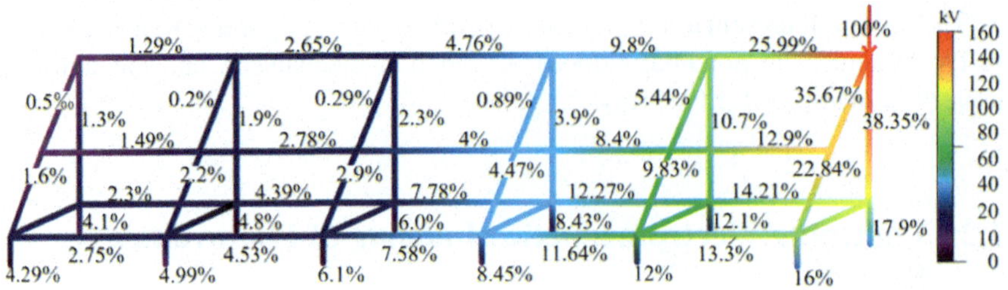

Fig. 1.3 Lighting transient distribution on PV bracket structure [17, 18]

thus the EM coupling mechanism among bracket, wire, and cable cannot be ignored (Fig. 1.3).

As a numerical method, full-wave model is utilized to solve the integral equation of EM field, and it mainly contains partial element equivalent circuit (PEEC) method, generalized modified mesh current (MMC) method and finite difference time domain (FDTD) method. Researchers from Hong Kong Polytechnic University proposed a PEEC based on vector fitting technology which takes into account the frequency-dependent effect of PV bracket and frame. According to above technique, the amplitude-frequency characteristics curve of conductor impedance was obtained at different frequencies [20, 21]. Moreover, the lightning-induced overvoltage on PV modules and bypass diodes was calculated under different installation methods, and the generation and damage mechanism of lightning-induced overvoltage in PV modules and arrays was analyzed [20]. The vector fitting-based PEEC method was applied to modelling an actual PV power station located in a transmission line corridor and studying the transient processes of PV power station when overhead lines were struck by lightning [22]. It was found that the lightning current would dissipate into the earth after PV system was hit by lightning [22]. Several scholars utilized MMC method to solve the lighting-induced time-domain transient process of PV arrays on lossy soil [23]. This method introduces radial branches into original circuit topology and employs node charges as additional state variables, but it is only suitable for simulating the EM coupling mechanism of single wire on the soil or in the air [23]. Zhang et al. [24] used finite element method (FEM) to analyze the lightning strike transient characteristics of PV brackets, DC cables and grounding grids. Despite of considering the dispersion effect of soil, the thin wire structure in the PV module was ignored. Besides, the induced overvoltage on DC cables at different positions in the PV array with different grounding modes was evaluated, and it was discovered that the equipotential bonding of adjacent brackets could improve the lightning protection capability of PV system. Presently, the research on lightning strike characteristics of PV arrays mainly focuses on roof or flat terrain conditions. In actual, complex terrain will change the distribution of EM fields around lightning strike channels. Nonetheless, there is no accurate experimental measurements and reliable models regarding the transient coupling mechanism of PV arrays on complex terrain so far.

1.3 Lightning Damage to Photovoltaic System

PV modules, consisting of a PN junction, face the problem of degradation such as light-induced degradation (LID) and potential-induced degradation (PID). In the absence of light, ideal PV cell is in cut-off state. However, the potential difference and leakage current between crystalline component and glass, as well as bracket and frame, causes Na + ions in the sodium calcium silicate glass of PV modules to accumulate on the silicon surface or enter its interior, thus affecting the active region of semiconductor material. Consequently, polarization and degradation occur in PN junctions, resulting in the decrease of open-circuit voltage and output power, and the form of PID. Additionally, when PV cell is completely shaded, reverse current will leak between ethylene vinyl acetate (EVA) and polyethylene terephthalate (PET) backplane, between EVA and glass cover, and between EVA and metal frame [25]. Specially, the first leakage has the greatest influence on the leakage path and is an important cause of PID. This phenomenon is particularly obvious under high pressure, high temperature and high humidity (Figs. 1.4 and 1.5).

The mechanism of PID of PV modules is highly complex and easily influenced by various factors such as anti-reflection layer, substrate material, packaging material, component structure, grounding mode, and system architecture. Field experience shows that the grounding of metal frame leads to a negative bias voltage between the frame and the component near the negative electrode of inverter, but a positive bias voltage between the frame and the component near the positive electrode. It

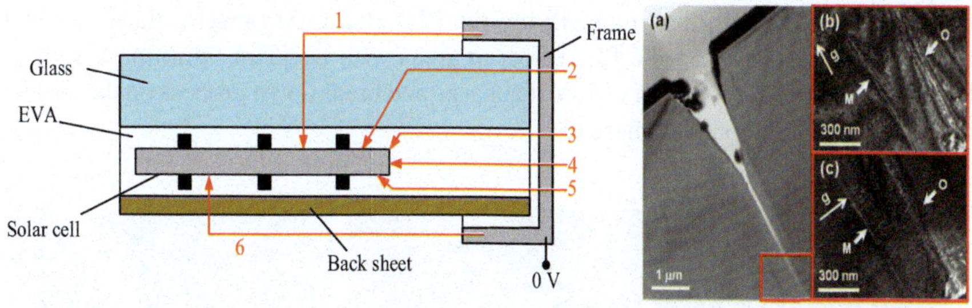

Fig. 1.4 Leakage current path in PV modules

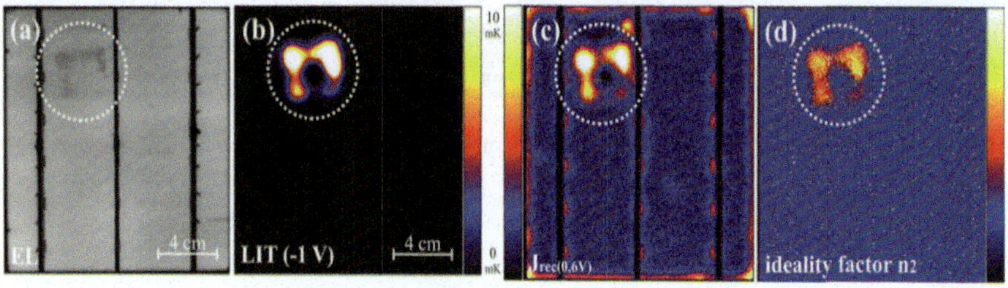

Fig. 1.5 PV module defect detection based on EL

indicates that components closer to the negative terminal of inverter are more severely affected by PID effect, otherwise the opposite [26]. In addition, substrate material is one important factor for PID. Low doping can cause the depletion region of PN junction to widen and uneven doping in PV cells can lead to uneven sheet resistance, thereby reducing the PID effect of silicon wafers. Research institutions in United States, Japan, and Germany conducted PID experiments on PV modules by applying reverse DC bias voltage (typically between 250 and 1000 V) to the junction box, and then successfully measured *I–V* characteristics curve and output power of PV cells under different conditions [27–29]. SunPower company analyzed the quantum efficiency of PV cells in the wavelength range of 300 to 1100 nm, and used Halm PV cell and module *I-V* tester to measure the open circuit voltage (V_{oc}), short-circuit current (I_{sc}), fill factor (FF), and conversion efficiency [30]. By testing the electroluminescent characteristics of PV cells, selecting different bright and dark regions (i.e. normal and abnormal regions) in the EL image, and conducting scanning electron microscopy-energy dispersive X-Ray spectrometry (SEM–EDS) test on the surface of PV cells, the identification of PID was realized. Jilin University established an equivalent circuit of PV components using resistors and capacitors, revealed the distribution of leakage current, and simulated the output power of PV cells [31].

Under the condition of lightning strike, high surge voltage is induced in the series DC cable and PV modules. When the surge voltage exceeds certain threshold, the resistance between PV cell and insulation material in the panel decreases, and the leakage current increases. In severe cases, this may result in discharge and cause bubbles or breakdown within the battery. Additionally, the electric field generated by the lightning surge may accelerate the migration of metal ions such as Na + on the surface of PV cell, thus intensifying the PID effect. At present, there is a lack of systematic research on the PID effect in areas with frequent lightning activities. Figure 1.6 illustrates the observed arc discharge and breakdown process on the surface of PV cell by high-speed camera.

Fig. 1.6 Results of lightning strike experiments on PV cells

1.4 Lightning Protection for Photovoltaic System

The existing lightning protection measures of PV arrays mainly derives from relevant experience in the construction and power industries. The commonly used direct lightning flash protection devices consist of lightning arrester, down lead, and grounding device. The lightning arresters installed on PV arrays are divided into independent and non-independent types [32]. To prevent the adverse impact of heat spot, the lightning rod should adopt retractable structures (Fig. 1.7). During frequent thunderstorms, the height of lightning rod can be properly increased to expand the protection range; in autumn and winter, it can be appropriately reduced or even lowered to zero. The protection area of lightning rod can be determined by the protection angle and rolling ball methods [32]. However, excessively high lightning rod may generate flare effects, reduce the power generation efficiency of PV cells, and decrease the anti-interference ability for lightning EM pulses [33]. When lightning strikes the lightning arrester in regions with high soil resistivity, the ground potential rises and the current flows backwards, that is, the current enters the cable and bracket through surge protective device (SPD) and ground down lead. Therefore, nether lightning arrester nor dedicated grounding grids are installed in PV power stations [34]. By adopting double or multiple rods, the height and number of lightning rods can be greatly reduced [35]. In addition, the lightning protection capability of PV arrays can be maximized by optimizing the structure of PV brackets, increasing the number of grounding points, and dissipating lightning currents [21]. Furthermore, the adjacent PV bracket and frame can be connected by using equal potential, forming an M-shaped grid structure, to avoid excessive potential difference between the conductors.

Currently, the lightning surge protection of PV power stations mainly relies on SPD. When the voltage exceeds trigger threshold, SPD will enter conductive state and lightning pulse will be discharged to the ground, thereby being able to effectively confining the lightning surge within the tolerance range of safeguarded equipment. According to the lightning risk and importance, equipment protection is divided into

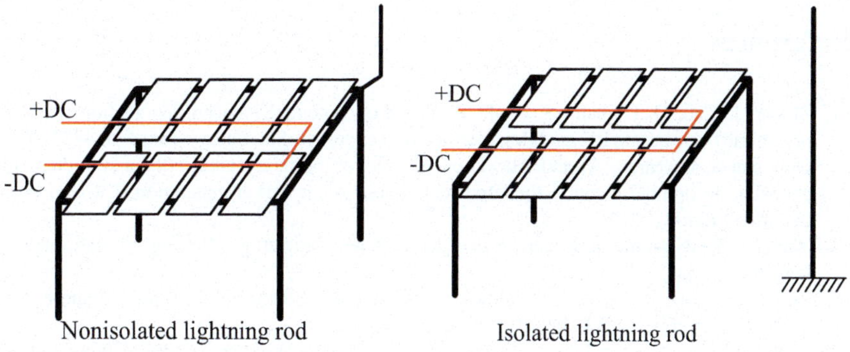

Fig. 1.7 Lightning arrester of PV array

four grades [9]. In Grade I, SPD is installed in the high-voltage side of transformer and combining manifold to prevent the lightning from invading the power grid side and protect the PV modules and their junction boxes. In Grade II, SPD is installed in the controller to protect the control module of PV system. In Grade III, SPD is installed in the inverter. In Grade VI, SPD is installed at the front of load as a backup protection [36]. Due to the limited protective range of SPD, it can only ensure that the overvoltage amplitude at both ends of the installed equipment remains within a safe threshold. Usually, the volume and energy withstand of SPD in weak electricity systems are small. Hence, SPD will be destroyed when encountering large lightning surge.

Other measures to suppress lightning surge include installing EM shielding devices on DC cables [37, 38], installing Faraday cage on the surface of components, and improving the wiring method for PV modules. Faraday cage itself constitutes an equal potential body, and the potential in the cage equal to zero. Since the charge is uniformly distributed on the surface of the cage without internal electric field, the cage can not only achieve lightning protection effect by blocking the EM field, but also play a role in flow shunting and equalization. Besides, because the lightning current symmetrically flows through the metal shielding layer of Faraday cage into the ground, the EM field in the cage weakens each other and interference intensity is reduced. Meng et al. [35] studied the influence of cable layout scheme and proposed an optimized wiring method according to the conclusion that the larger the area of wiring loop, the greater the amplitude of lightning surge. It was found that for redundant DC cable, the "8" type winding layout could facilitate the generation of opposite potentials on its two sides, thereby decreasing the induced voltage [39]. Furthermore, the lightning surge could be suppressed through changing the installation method and number of bypass diodes. Compared with the forward installation, the reverse installation of bypass diodes could greatly reduce the amplitude and duration of lightning overcurrent [40]. Increasing the number of bypass diodes in series could significantly enhance the overall ability to withstand surges and prevent them from being penetrated.

References

1. Zhang, J. Cao, D. Diaham, S. et al.: Research on potential induced degradation (PID) of polymeric backsheet in PV modules after salt-mist exposure. Sol. Energy. 188, 475–482 (2019).
2. Ogbonnaya, C. Turan, A. and Abeykoon, C.: Novel thermodynamic efficiency indices for choosing an optimal location for large-scale photovoltaic power generation. J. Clean Prod. 249(10), 119405 (2020).
3. Wang, Y. Study on lightning transients of large photovoltaic arrays. Beijing: Beijing Jiaotong University, 2021.
4. Zhong, T. Kang, K. Li, H.: Breakdown Accident analysis of large area photovoltaic bypass diode. Sol. Energy. 12, 53–60 (2018).
5. Sheng, C. Yang, S. Shen, D. et al.: Safety and Effectiveness analysis of lightning protection for mountain photovoltaic power station. Sol. Energy. 04, 26–28 (2016).

6. Yang, S. Zhang, T. Shen, D. et al.: Technical analysis of lightning protection system on DC side of photovoltaic Power Station. Sol. Energy. 05, 53–57 (2016).
7. Formisano, A. Hernández, J. C. Petrarca, C. et al.: Modeling of PV module and DC/DC converter assembly for the analysis of induced transient response due to nearby lightning strike. Electronics. 10 (120), 01–25 (2021).
8. Formisano, A. Petrarca, C. and Hernandez, J. C.: Assessment of induced voltages in common and differential-mode for a PV module due to nearby lightning strikes. IET Renew. Power Gener. 13, 1369–1378 (2019).
9. Hernandez, J. C. Vidal, P. G. and Jurado, F.: Lightning and surge protection in photovoltaic installations. IEEE Trans. Power Deliv. 23(4), 1961–1971 (2008).
10. Stern, H. Karner, H. C. "Lightning induced EMC phenomena in photovoltaic modules," Proceedings of International Symposium on Electromagnetic Compatibility, Dallas, TX, USA, 442–446 (1993).
11. Sakai, K. Yamamoto, K. "Lightning protection of photovoltaic power generation system," Proceedings of International Symposium on Lightning Protection (XII SIPDA), Belo Horizonte, Brazil, 335–339 (2013).
12. Pretorius, P. H. "On in situ lightning current impulse testing in a large free field PV plant," Proceedings of 34th International Conference on Lightning Protection, Rzeszow, Poland, 1–8 (2018).
13. Sun, Q. Zhong, X. et al.: Three-dimensional modeling on lightning induced overvoltage for photovoltaic arrays installed on mountain. J. Clean Prod. 288, 125084 (2021).
14. Oufi, E. Alfuhaid, A. and Saied, M.: Transient analysis of lossless single phase nonuniform transmission lines. IEEE Trans. Power Deliv. 9(3), 1694–1700 (1994).
15. Zou, J. Lee, J. Ji, Y. et al.: Transient simulation model for a lightning protection system using the approach of a coupled transmission line network. IEEE Trans. Electromagn. Compat. 49(3), 614–622 (2007).
16. Davidson, D. "A review of important recent developments in full-wave CEM for RF and microwave engineering," Proceedings of 3th International Conference on Computational Electromagnetics and Its Applications, Beijing, PS/1–PS/4 (2004).
17. Wang, Y. Zhang, X. and Tao, S.: Modeling of lightning transients in photovoltaic bracket systems. IEEE Access. 7, 12262–12271 (2019).
18. Wang, Y. Zhang, X. Tao, S. et al.: Research on lightning transient of photovoltaic support system. Acta Energiae Solaris Sinica. 39(10), 2788–2796 (2018).
19. Demirel, E. Dolgun, G. K. and Keçebas, A.: Comprehensive transient analysis on control system in a photovoltaic power plant under lightning strike. Sol. Energy. 233, 142–152 (2022).
20. Zhang, Y. Chen, H. and Du, Y.: Lightning protection design of solar photovoltaic systems: methodology and guidelines. Electr. Power Syst. Res. 174, 105877 (2019).
21. Zhang, Y. Chen, H. Du, Y. et al.: Considerations of photovoltaic system structure design for effective lightning protection. IEEE Trans. Electromagn. Compat. 62, 1333–1341 (2020).
22. Zhang, Y. Chen, H. Du, Y. et al.: Transients in solar photovoltaic systems during lightning strikes to a transmission line. Int. J. Electr. Power Energy Syst. 234, 106885 (2021).
23. Tu, Y. Zhang, C. Hu, J. et al.: Research on lightning over voltages of solar arrays in a rooftop photovoltaic power system. Electr. Power Syst. Res. 94, 10–15 (2013).
24. Zhang, Y. Li, B. Du, Y. et al.: Effective grounding of the photovoltaic power plant protected by lightning rods. IEEE Trans. Electromagn Compat. 63(4), 1128–1136 (2021).
25. Dhere, N. G. Shiradkar, N. S. and Schneller, E.: Evolution of leakage current paths in MC-Si PV modulesfrom leading manufacturers undergoing high-voltage bias testing. IEEE J. Photovolt. 4(2), 654–658 (2014).
26. Hacke, P. Smith, R. Terwilliger, K. et a1.: Testing and analysis for lifetime prediction of crystalline silicon PV modules undergoing degradation by system voltage stress. IEEE J. Photovolt. 3(1), 246–253 (2012).
27. Breitenstein, O. Langenkamp, M. Lang, O. et a1.: Shunts due to laser scribing of solar cells evaluated by highly sensitive lock-in the mography. Sol. Energy Mater. Sol. Cells. 65(23), 55–62 (2001).

28. Araujo, S. V. Zacharias, P. and Mallwitz, R.: Highly efficient single-phase transformerless inverters for grid-connected photovoltaic systems. IEEE Trans. Ind. Electron. 57(9), 1118–1128 (2010).
29. Kolar, J. W. Wolbank T. M. and Schrodl, M. "Analytical calculation of the RMS current stress on the DC link capacitor of voltage DC link PWM converter systems," Proceedings of Ninth International Conference on Electrical Machines and Drives (Conf. Publ. No. 468), Canterbury, UK, 81–89 (1999).
30. Zhang, J. W. Cao, D. K. Cui, Y. C. et al.: Investigation of the influence of electron avalanche on the crystallinity of backsheet in solar photovoltaic system for sustainable energy. J. Clean Prod. 189, 169–175 (2018).
31. Liu, F. Jiang, L. Yang, S. Y.: Ultra-violet degradation behavior of polymeric backsheets for photovoltaic modules. Sol. Energy. 108, 88–100 (2014).
32. Charalambous, C.A. Kokkinos, N.D. and Christofides, N.: External lightning protection and grounding in large-scale photovoltaic applications. IEEE Trans. Electromagn. Compat. 56, 427–434 (2014).
33. Ahmad, N. I. Ab-Kadir, M. Z. A. Izadi, M. et al.: Lightning protection on photovoltaic systems: a review on current and recommended practices. Renew. Sust. Energ. Rev. 82, 1611–1619 (2018).
34. Sabihaa, N. A. Alsharefa, M. Metwaly, M. K. et al.: Sustaining electrification service from photovoltaic power plants during backflow lightning overvoltages. Electr. Power Syst. Res. 186, 106386 (2020).
35. Meng, D. Chen, Z. Yan, G.: Research on meteorological disaster risk assessment of photovoltaic power plant-taking Hubei Province as an example. Acta Energiae Solaris Sinica. 41(5), 359–364 (2020).
36. Ding, M. Wang, W. Wang, X. et al.: Overview of the impact of large-scale photovoltaic power generation on power systems. Proceedings of CSEE. 34(1), 1–14 (2014).
37. Stapleton, G. Neill, S. Wang Y. et al.: Solar photovoltaic grid-connected power generation system. China: China Machine Industry Press, 2014.
38. Uman, M. A. All about lightning. Toronto: Dover, 1986.
39. Coetzer, K. M. Wiid, P. G. and Rix, A. J. "PV installation design influencing the risk of induced currents from nearby lightning strikes," Proceedings of International Conference on Clean Electrical Power (ICCEP), Otranto, Italy, 204–213 (2019).
40. Chen, W. Ai, X. Wu, T. et al.: Review on the impact of grid-connected photovoltaic power generation system on power grid. Electric Power Automation Equipment. 33(2), 26–32 (2013).

Chapter 2
Mechanism of Lightning Electromagnetic Coupling for Photovoltaic Array

The metal frame is [1] generally grounded to reduce the induced overvoltage caused by direct and indirect lightning electromagnetic fields. The PV panel is installed on a bracket structure with a slant angle, as depicted in Fig. 2.1. The PV array is boosted up to 35 kV by a DC/AC inverter, and then collected by a cable to substation. It would be further enhanced to 220 kV to power grid. While a lightning strikes nearby ground, the current will produce a strong electromagnetic radiation and a voltage will be induced on the wiring loop of PV array [2]. The associated electronic device like the bypass diode [3] and the inverter [4] may be damaged. The schematic diagram of PV array under nearby lightning strike [5] is depicted in Fig. 2.1.

2.1 Induced Voltage of Photovoltaic Array on Flat Ground

The mathematical model for the induced voltage involves three major steps, i.e., preprocessing, the magnetic-field computation and the induced voltage calculation. The flowchart is given in Fig. 2.2.

2.1.1 Lightning Channel Model

Supposing a lightning channel of length H with current I, as illustrated in Fig. 2.3, according to the Biot-Savart's law, the magnetic induction for a point at a distance R_1 to the channel satisfies [6],

$$B = \frac{\mu_0}{4\pi} \int_H \frac{I \times \vec{e}_R}{R_1^2} dl \tag{2.1}$$

Q. Sun et al., *Lightning Transients and Protection for Renewable Energy System*,
https://doi.org/10.1007/978-981-97-7267-4_2

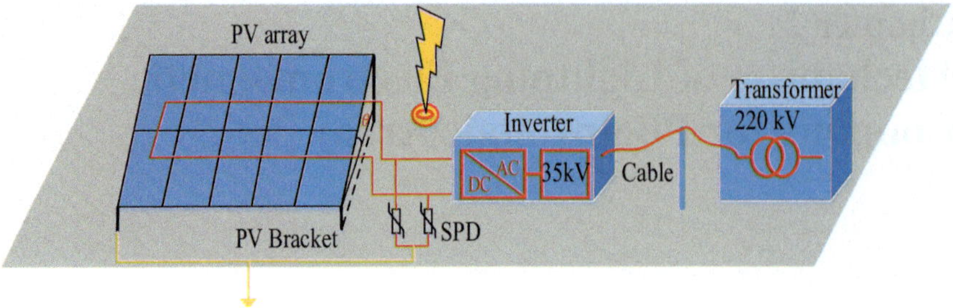

Fig. 2.1 Schematic diagram of PV system

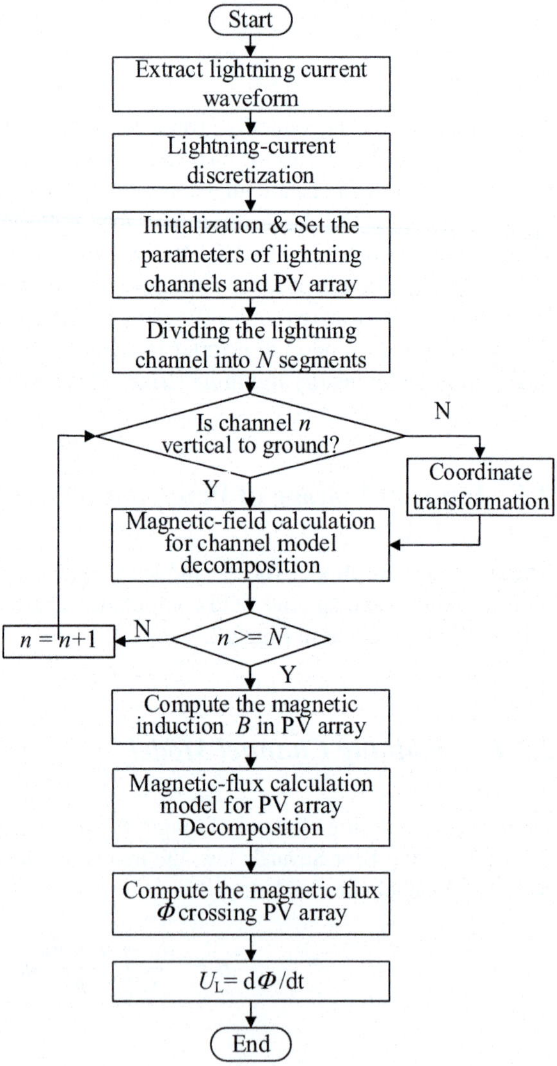

Fig. 2.2 Flowchart of proposed model for the induced voltage

Fig. 2.3 Magnetic-field model for lightning channel

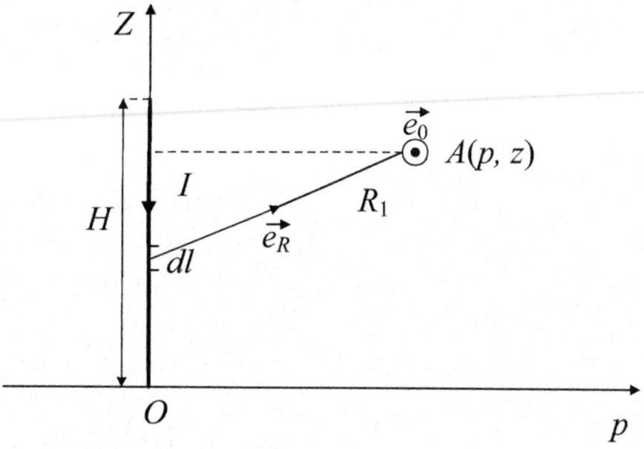

where μ_0 is the permeability in air, *dl* is the vector of line element. Many mathematical expressions have been proposed to represent the lightning current based on long-term observation. The Heidler function is employed herein and the current can be described by Eq. (2.2) [7],

$$i(t) = \frac{I}{\eta}\frac{(t/\tau_1)^n}{1+(t/\tau_1)^n}\exp\left(-\frac{t}{\tau_2}\right) \tag{2.2}$$

where I_0 is the peak value of the current, η is the correction factor, τ_1 and τ_2 are time constants which determine the rise time and decay time, respectively.

The propagation of stepped leader from cloud to the ground plays a vitally important role in the shape of lightning channel. The streamer would split into many branches in the front of leader tip. Generally, the branch channel of lightning leader above a height of 500 m has little impact on the PV panel since the distance is extraordinarily large. Thus, it is represented by a vertical downward conductor. The stepped leader model is employed [8–10]. The conical region for the leader tip is 120°, and the step length is assumed to be 5 m in this paper. The stepped leader in the branch area is assumed to propagate along the following routes: (i) the stepped leader is considered to propagate within the streamer conical region in the front of leader tip; (ii) one of these leader channels propagates toward the ground within 60° with respect to the vertical direction; (iii) the propagating leader channel would not intersect with other branched channels due to the repulsive force of charges. The schematic for the propagation of lightning stepped leader is depicted in Fig. 2.4.

2.1.2 *Magnetic Flux for PV Array on Plate Ground*

The interconnection of solar cells in PV panel forms a loop in geometry. The induced voltage is essentially determined by the magnetic flux passing through the loop. The magnetic flux is significantly influenced by the oblique angle between the panel and

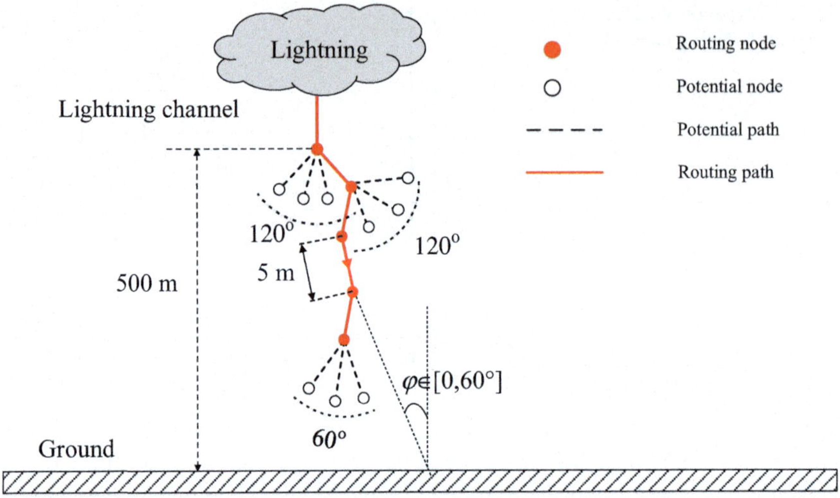

Fig. 2.4 Schematic diagram of lightning channel propagation [11]

the ground. A PV panel with angle θ is given in Fig. 2.5. The total flux can be obtained by integrating the magnetic field intensity over surface S_{pv},

$$\Phi = \iint_{\Sigma} B\,ds_{pv} = \frac{\mu_0 I \sin\alpha_0}{4\pi} \iint_{\Sigma} \frac{1}{\sqrt{x^2+y^2}} [\frac{z}{\sqrt{x^2+y^2+z^2}} - \frac{z-L}{\sqrt{x^2+y^2+(z-L)^2}}]\,ds_{pv}$$

$$= \frac{\mu_0 I \sin\alpha_0}{4\pi} \sqrt{1+\left(\frac{k_x}{k_z}\right)^2+\left(\frac{k_y}{k_z}\right)^2}$$

$$\cdot \int_{y_1}^{y_2}\int_{x_1}^{x_2} \frac{1}{\sqrt{x^2+y^2}} [\frac{z(x,y)}{\sqrt{x^2+y^2+z(x,y)^2}} - \frac{z(x,y)-L}{\sqrt{x^2+y^2+(z(x,y)-L)^2}} dxdy \tag{2.3}$$

where $\boldsymbol{K}$ is the normal vector of S_{pv}, $z(x, y)$ represents the expression of S_{pv}, D_{pv} is the projection of S_{pv} on O_{XY}, and α_0 represents the angle between lightning channel and $\boldsymbol{K}$,

$$\cos(\alpha_0) = \frac{\vec{K}\cdot\vec{I}}{|\vec{K}||\vec{I}|} \tag{2.4}$$

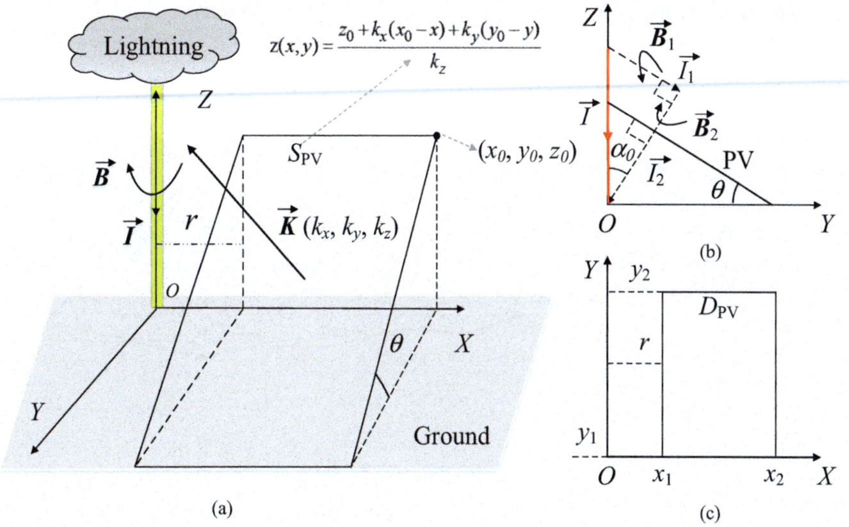

Fig. 2.5 Plane Projection of PV panel. **a** Magnetic-flux model. **b** Lightning current decomposition. (c) Projection of PV surface

2.1.3 Magnetic Flux for PV Array Installed on Mountain

Suppose the PV array is installed on the mountain, as depicted in Fig. 2.6a, it needs to integrate the conical surface to acquire magnetic flux passing through PV array [12], as given in Eq. (2.5).

$$\Phi = \iint\limits_{\Sigma} \vec{B}_x(x, y, z)dydz + \vec{B}_y(x, y, z)dxdz + \vec{B}_z(x, y, z)dxdy \tag{2.5}$$

Considering the angle δ between lightning channel and mountain peak, the surface is illustrated in Fig. 2.6b. The magnetic induction $\boldsymbol{B}$ of PV array installed on mountain is,

$$\begin{aligned} \vec{B} = & \frac{\mu_0 I \sin\delta}{4\pi\sqrt{x^2+y^2}} \\ & \left[\frac{z-h_m}{\sqrt{x^2+y^2+(z-h_m)^2}} - \frac{z-h_m-H}{\sqrt{x^2+y^2+(z-h_m-H)^2}}\right](\vec{e}_y+\vec{e}_z) \end{aligned} \tag{2.6}$$

where $\boldsymbol{e}_{\mathbf{y}}$ and $\boldsymbol{e}_{\mathbf{z}}$ are unit vectors in y-axis and z-axis direction, respectively; h_{m} is the altitude of mountain peak, β represents the hillside angle, and H is lightning channel length. The conic surface can be expressed as,

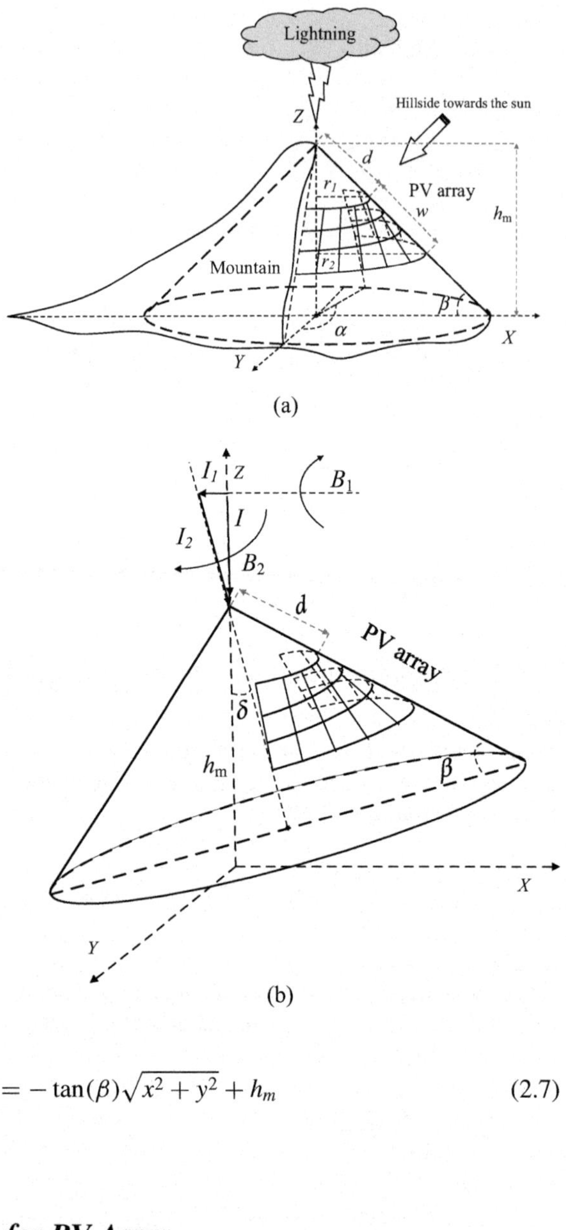

Fig. 2.6 Model of PV array installed on mountain. **a** Geometry of PV array. **b** Coordinate conversion [11]

$$z = -\tan(\beta)\sqrt{x^2 + y^2} + h_m \tag{2.7}$$

2.1.4 Induced Voltage for PV Array

After acquiring the magnetic flux crossing internal loop, the induced voltage U_L can be obtained,

$$U_L = -\frac{d\Phi}{dt} \tag{2.8}$$

If the lightning channel is not vertical to the ground, a coordinate transformation is required at each segment [13, 14], with the lightning channel as the z-axis, the end of the lightning channel as the origin, and the coordinates of the endpoints of PV array can be determined as [15],

$$[(x, y, z) \cdot \vec{e}_x,\ (x, y, z) \cdot \vec{e}_y,\ (x, y, z) \cdot \vec{e}_z] \tag{2.9}$$

where (x, y, z) is the initial point coordinate, and $\boldsymbol{e}_\mathbf{x}$ is unit vectors in x-axis direction. If the origin $\boldsymbol{O}$ (0, 0, 0) has changed, before bringing into Eq. (2.9), (x, y, z) needs to be subtracted from the original origin $\boldsymbol{O}$.

2.2 Electromagnetic Transient Model for Photovoltaic Module

The proposed circuit model of PV module includes the transient model of PV-cell string, the equivalent circuit of PV frame, the electromagnetic coupling between metal frame and PV-cell string, etc., as illustrated in Fig. 2.7. The details of the circuit model are presented as follows.

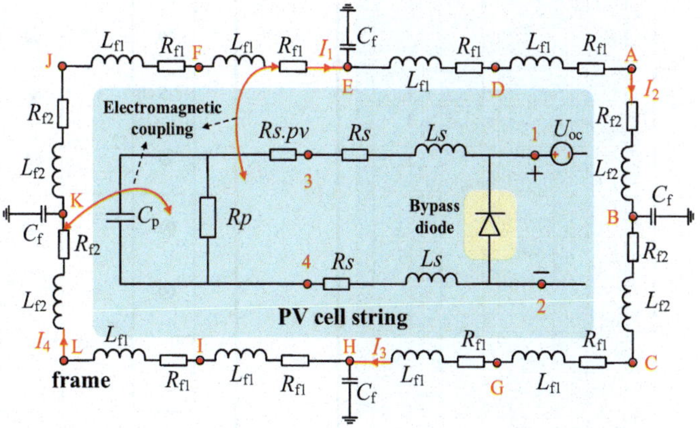

Fig. 2.7 Proposed circuit model of PV module [16]

2.2.1 Electromagnetic Coupling Between Frame and Solar-Cell String

The lightning current flowing in PV frame would generate an impulse magnetic field, and the PV-cell wiring tend to interact with this field. The mutual inductances are used to quantify the relationship between the impulse current and the induced voltage of the module. The cells inside a PV module are connected in series by means of a straight wire, as shown in Fig. 2.8.

Supposing a straight conductor of length l with current I, e.g., L_1, L_2, L_3 and L_4, the magnetic field strength for a point at a distance R to the conductor can be obtained by,

$$\begin{aligned} \boldsymbol{B} &= \frac{\mu_0 I}{4\pi}\int_{-l/2}^{l/2}\frac{r\cdot\boldsymbol{e}}{\left[r^2+(z-y)^2\right]^{3/2}}dy \\ &= \frac{\mu_0 I\cdot\boldsymbol{e}}{4\pi}\left[\frac{z+l/2}{\sqrt{r^2+(z+l/2)^2}}-\frac{z-l/2}{\sqrt{r^2+(z-l/2)^2}}\right] \end{aligned} \quad (2.10)$$

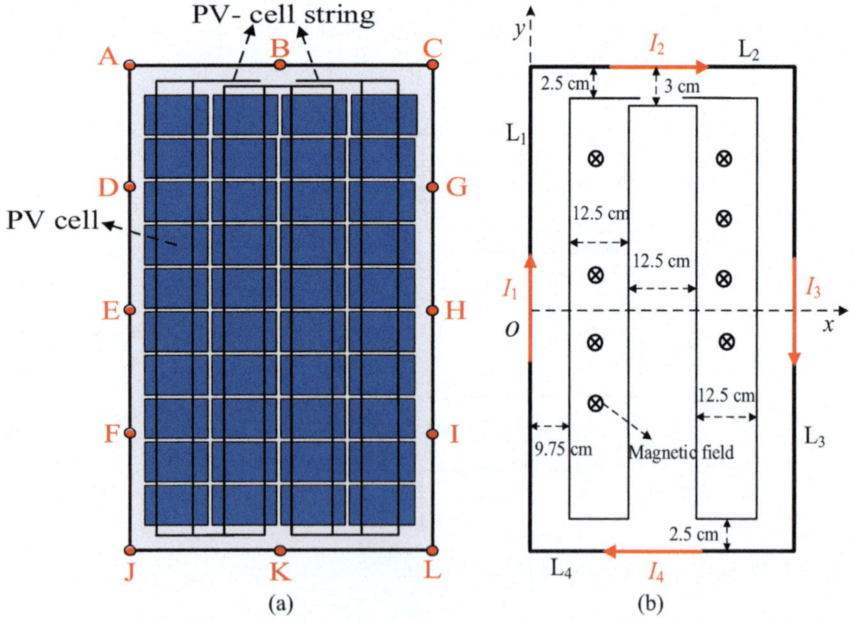

Fig. 2.8 PV module. **a** Schematic diagram of PV module. **b** Equivalent loop of conventional PV-cell wiring. The symbol ⊗ denotes the magnetic field and the direction is inward [16]

where μ_0 represents the permeability in air, and it is 1.25×10^{-6} H/m; $\boldsymbol{e}$ is the unit vector, which is inward as shown in Fig. 2.8b; (r, z) is the point in cylindrical coordinates. Considering the loop formed by thin conductor in Fig. 2.8, the flux Φ can be determined by integrating the magnetic field over the surface of loop.

$$\begin{aligned}\Phi &= \oint_S \boldsymbol{B}ds \\ &= \oint_S \frac{\mu_0 I}{4\pi x}\left[\frac{y+l/2}{\sqrt{x^2+(y+l/2)^2}} - \frac{y-l/2}{\sqrt{x^2+(y-l/2)^2}}\right]ds\end{aligned} \tag{2.11}$$

According to the Faraday's law, the induced voltage U_{OC} between the positive and negative terminals of a PV module can be calculated,

$$\begin{aligned}U_{OC} &= -\frac{d\Phi}{dt} \\ &= \frac{\mu_0}{4\pi}\oint_S \frac{dI}{xdt}\left[\frac{y+l/2}{\sqrt{x^2+(y+l/2)^2}} - \frac{y-l/2}{\sqrt{x^2+(y-l/2)^2}}\right]ds\end{aligned} \tag{2.12}$$

Therefore, the mutual inductance M_i between the straight conductor L_i of the frame and the wiring loop can be derived,

$$M_i = \oint_S \frac{\mu_0}{4\pi x}\left[\frac{y+l/2}{\sqrt{x^2+(y+l/2)^2}} - \frac{y-l/2}{\sqrt{x^2+(y-l/2)^2}}\right]ds \tag{2.13}$$

where S is the area of wiring loop, and it can be divided into several parts. The frame is made up of four straight conductors, and the induced voltage U_{OC} is equal to,

$$U_{OC} = -\sum_{i=1}^{4} M_i \frac{dI_i}{dt} \tag{2.14}$$

Once the mutual inductances are determined, the induced voltage U_{OC} can be calculated by substituting the lightning currents on each side of the frame (such as I_1, I_2, I_3 and I_4) into Eq. (2.14). I_1, I_2, I_3 and I_4 are obtained by solving the equivalent circuit of the frame in Fig. 2.7. The failure of the bypass diode (BPD) can be predicted from the induced voltage between the positive and negative terminals of the BPDs. The BPDs of a PV panel may be breakdown when U_{OC} exceeds the maximum repetitive reverse voltage for BPDs.

2.2.2 Dynamic Model of Solar-Cell String

The PV-cell string is represented by a transient circuit in the proposed model, as shown in Fig. 2.7 Since the lightning current has a large number of high-frequency components and its amplitude is much greater than that of photocurrents, the current source of cells is ignored. The transient model of a PV-cell string, as depicted in Fig. 2.7, can be equivalent to the capacitance C_p, which includes the junction capacitance and diffusion capacitance of P–N junction [17]. The PV-cell string, as illustrated in Fig. 2.8a, includes 40 cells connected in series. The capacitance of a PV-cell string ranges from 1.4 to 10.2 μF according to different operating conditions [17, 18]. The parallel resistor R_p and series resistor $R_{s.pv}$ are the sum of the resistances of the PV cells forming a cell-string [16]. The impedance Z_{pv} of PV-cell string is [17],

$$Z_{pv} = \left[R_{s.pv} + \frac{R_p}{(2\pi f R_p C_p)^2 + 1}\right] - j\left[\frac{2\pi f R_p^2 C_p}{(2\pi f R_p C_p)^2 + 1}\right] \tag{2.15}$$

According to Eq. (2.15), each of the three equivalent circuit parameters of PV-cell string (R_p, $R_{s.pv}$ and C_p) can be almost determined independently by frequency analysis. When the frequency f is close to zero, the term "$2\pi f R_p C_p$" is almost equal to zero, and the impedance Z_{PV} has only the resistance. Equation (2.15) can be approximated as,

$$Z_{pv} = R_{s.pv} + R_p \tag{2.16}$$

Therefore, Z_{PV} is purely resistive and its magnitude is equal to $R_p + R_{s.pv}$. In contrast, the term "$2\pi f R_p C_p$" in Eq. (2.15) is almost infinite when the frequency f is extremely high. The imaginary part of the impedance Z_{PV} is also equal to zero, and it has only the resistance. Equation (2.15) can be approximated as,

$$Z_{pv} = R_{s.pv} \tag{2.17}$$

In principle, the branch C_p is almost short-circuited in the transient circuit of PV-cell string in the case of extremely high f, the magnitude of the impedance for the PV-cell string is approximately equal to $R_{s.pv}$. When the PV panel operates at the maximum power point, is 35.6 and is 0.8 Ω [16]. Given the impedance Z_{PV} of PV-cell string at a mid-frequency f, substituting $R_{s.pv}$ and R_p into Eq. (2.15), C_p can be determined. The magnitude of the impedance for the PV-cell string versus frequency is plotted in Fig. 2.9.

The copper wire connecting PV cells is very thin, and its resistance and inductance are nearly frequency independent [19]. The copper wire can be expressed by the DC resistance R_s and DC inductance L_s in series,

$$R_s = \frac{l_s}{\gamma_s S} \tag{2.18}$$

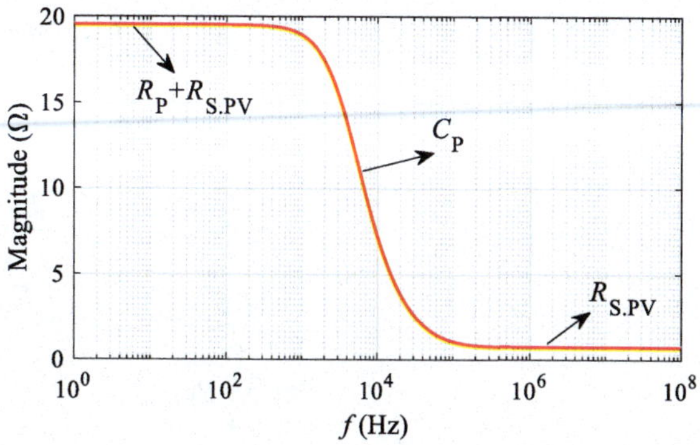

Fig. 2.9 Magnitude of bode plot of PV-cell string impedance [16]

$$L_s = \frac{\mu_0 l_s}{2\pi}\left(\ln\frac{2l_s}{r_s} - 1\right) \tag{2.19}$$

where l_s is the length of copper wire, γ_s is the conductivity of copper; S is the cross-sectional area of copper wire, μ_O is the permeability in the air and r_s represents the radius of copper wire. According to the measured parameters of the wire inside a PV module in [19], the resistance R_s and inductance L_s are taken as 75 mΩ and 1.15 μH in our study, respectively.

The BPD, as illustrated in Fig. 2.7, is connected in parallel with the PV-cell string at the output of PV panel in the reverse direction. The BPD provides a low resistance path for the current go around a series of PV cells that have been shaded. In order to protect each PV-cell string of PV panel, several BPDs are often connected in series inside a junction box, and they are equivalent to a BPD in our study. Under normal operation, the BPD is opened. As the lightning transient takes a very short interval, the BPD is treated as an opened switch in the proposed model.

2.2.3 Equivalent Circuit of Metal Frame

The linear-lumped elements are used in the circuit of frame and bracket. The top and bottom edges of the frame are equally divided into four segments, respectively. Their lengths are the same, which are represented by resistance R_{f1} and inductance L_{f1} in series. Since the length of the left and right edges are inconsistent with that of the top and bottom edges, they are divided into two segments, which are represented by resistance R_{f2} and inductance L_{f2}.

The supporting bracket of PV array generally consists of several bars and four grounding legs, and the equivalent circuit is illustrated in Fig. 2.10 The entire circuit of the bracket contains tens of elements, and hence only part of them are marked due to the symmetrical structure.

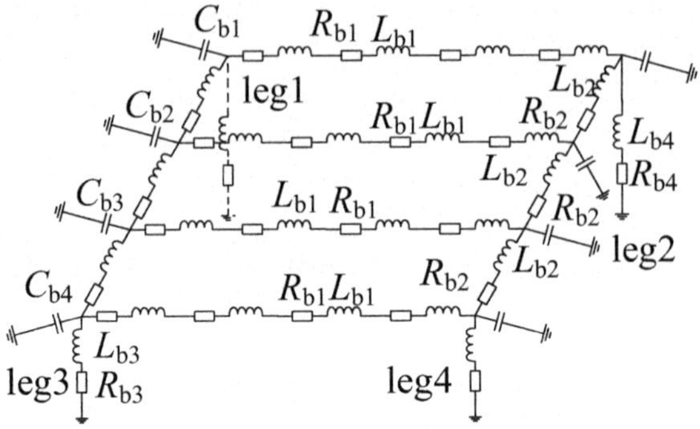

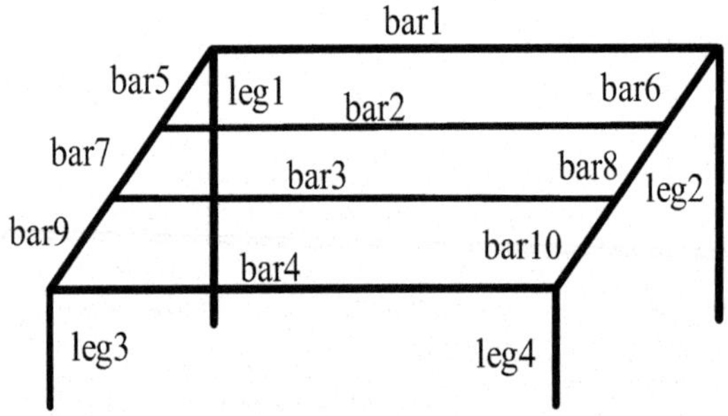

Fig. 2.10 PV supporting bracket. **a** Equivalent circuit. **b** Schematic diagram [16]

The PV panel frame as well as the supporting bracket is C-shaped. The resistance and internal inductance of the C-shaped steel are calculated using the cylindrical model [20]. Considering the high frequency displacement current, the resistance of the frame can be obtained,

$$R_f = \frac{2\pi l}{\gamma\left[\arccos\left(\frac{r-\delta}{r}\right)r^2 - 2\pi r(r-\delta)\sin\left(\arccos\left(\frac{r-\delta}{r}\right)\right)\right]} \tag{2.20}$$

where l is the length of segment, γ is the conductivity of frame, which is set to be 3.4×10^7 S/m; μ is the permeability of frame; δ represents the penetration depth of high frequency current. It is determined by the cut-off frequency f_e of lightning current,

$$\delta = \sqrt{2\pi f_e \mu \gamma} \tag{2.21}$$

The inductance of the frame also depends on the f_e [21],

$$L_f = \frac{\mu l}{2\pi}\left(\ln\frac{2l}{r} - 1\right) + \frac{\mu l}{8\pi}\xi \tag{2.22}$$

The quantity ξ is defined as,

$$\xi = \frac{4}{kr}\frac{\mathrm{Re}[J_1(kr)]\mathrm{Re}[J_1(kr)]' + \mathrm{Im}[J_1(kr)]\mathrm{Im}[J_1(kr)]'}{\mathrm{Re}[J_1(kr)]'^2 + \mathrm{Im}[J_1(kr)]'^2} \tag{2.23}$$

where k is equal to $\sqrt{2\pi f_e \mu \gamma}$, J_1 is the Bessel function of the first kind; ε_0 is the permittivity of vacuum and it is 8.85×10^{-12} F/m. Assuming that the angle between the frame and ground is θ. The height for the lowest branch of metal frame is h_0. If not mentioned, θ and h_0 are 30° and 1 m, respectively. The frame capacitance to ground is given as,

$$C_f = \int_l \frac{2\varepsilon_0 \pi}{\ln\left[\frac{2(h_0 + l\sin\theta)}{r}\right]} dl \tag{2.24}$$

The glass, ethylene vinyl acetate (EVA) film, PV cells and back sheet are encapsulated together to form a PV panel. There are a number of leakage current pathways from PV cells to frame, as plotted in Fig. 2.11a. Since the bulk resistances of the back sheet and EVA are significantly larger than the surface resistance of the glass, the major leakage current pathway is through the glass and EVA, then crosses through the surface of the front glass [22]. The leakage currents flowing through other pathways are almost negligible. The PV module surface insulation is dependent on surface conditions, and the surface resistance of the front glass $R_{\text{s-glass}}$ generally ranges from 100 kΩ to 4.6 MΩ [23]. The bulk resistances of the glass and EVA (R_{glass} and R_{EVA}) range from 70 MΩ to 3.8 GΩ, and their capacitances C_{glass} and C_{EVA} range from 1.4 to 79 nF [23]. The model of the insulation layers in PV modules are given in Fig. 2.11b, and the glass insulation layer is represented by R_{glass} and C_{glass} whereas the EVA insulation layer is represented by R_{EVA} and C_{EVA}.

The equivalent circuit of a PV module, as illustrated in Fig. 2.12, is built in the environment of electromagnetic transient program (EMTP). It is a powerful tool to solve the transients in electric circuit [24]. The entire circuit is huge and the elements are thus packaged in two components of Fig. 2.12. The equivalent circuit of the frame, including the 12 RL branches (such as R_{f1}, L_{f1} and R_{f2}, L_{f2}, etc.) representing the four edges, is integrated in the gray component. This component has 12 terminals

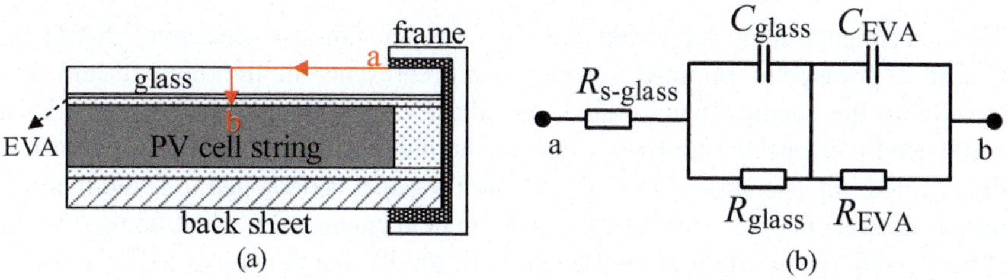

Fig. 2.11 Cross-section of PV module. **a** Leakage current pathways. **b** Model of insulation layers

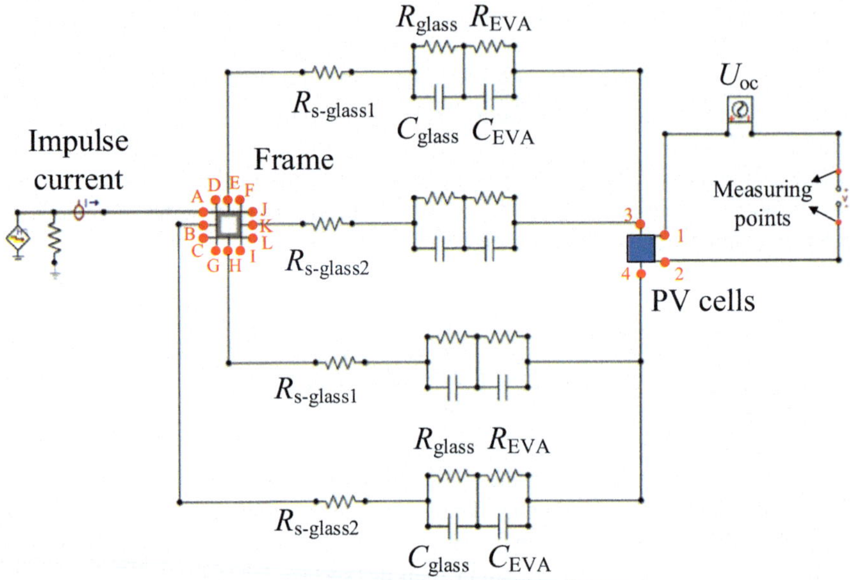

Fig. 2.12 Proposed model of PV module in EMTP [16]

(A, B, C…K, L), which correspond to the nodes of four edges in the frame (A, B, C…K, L), as marked in Figs. 2.7 and 2.8. The blue component is the dynamic model of PV-cell string. Its four terminals are Node 1, Node 2, Node 3 and Node 4, as shown in Fig. 2.7. Node 3 and 4 are connected to the model of the insulation layers for the PV module whereas the output terminals of the PV module are Node 1 and 2.

2.3 Experiment on Indirect Lightning Strikes at Photovoltaic Module

2.3.1 Experiment Setup

The experimental setup is presented in Fig. 2.13. The impulse generator (ES-5110A) is used to produce an impulse current, which represents the lightning current. It is injected to the conductor in parallel with the PV panel with a distance of 0.25 m and its end is connected to the negative of the impulse generator. The waveform of the injected impulse current is 8/20 μs. The PV panel is darkened to avoid the light influence, and only the effect of the magnetic field generated by the impulse current is presented. The open-circuit voltage U_{oc} of the PV panel is recorded by a digital oscilloscope (MDO30245).

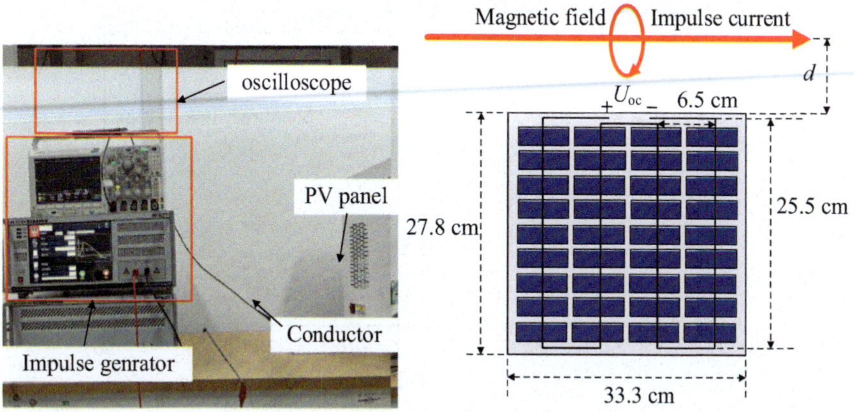

Fig. 2.13 Schematic diagram of experiment

2.3.2 *Validation of Calculated Model*

The comparison between the calculation values and the experimental results is summarized in Table 2.1. The deviations for the measured voltages and the calculated values are relatively small and reasonable, and their waveforms are close to each other, as shown in Fig. 2.14. The calculation model in our work can be validated by the experimental results.

Since the connection wires between the conductor and the outputs of the impulse generator are not vertical to the PV panel, the current on the wires produces the EM interference and the original magnetic field generated by the current on the conductor is weakened. Therefore, most of the measured results are slightly lower than the calculated results.

Table 2.1 Comparison with experimental result (V)

Amplitude of impulse current (kA)	Measured	Model in our work
0.75	0.52	0.50
1.00	0.64	0.66
1.50	0.96	0.99
2.00	1.24	1.36
2.50	1.60	1.70

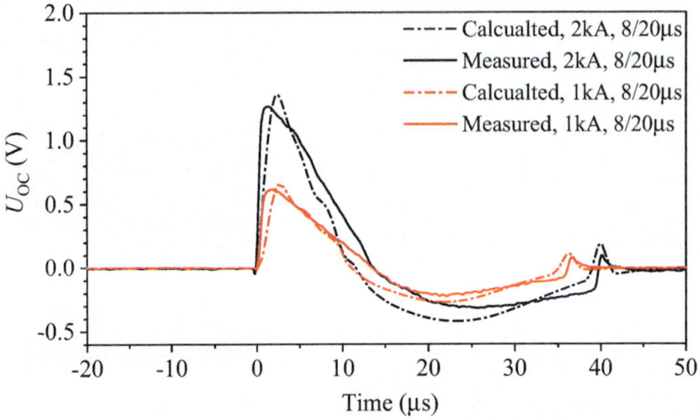

Fig. 2.14 Comparison with experimental induced voltage of PV panel

2.3.3 Results and Discussion

The proposed model is applied to evaluate the lightning overvoltage within a PV array formed by 8 series connected modules. The effect of PV-cell wiring scheme on the induced voltage is analysed. The influence of grounding structures of supporting brackets is also discussed.

The lightning transient in the DC side of a PV system is studied, including DC cable, PV modules and the bracket, as shown in Fig. 2.15 The equivalent circuit of the bracket for the PV array shown in Fig. 2.15 is presented in Fig. 2.10 Similar to the equivalent circuit of the frame for PV panel, the parameters of equivalent circuit of the bracket can be calculated by:

$$\begin{cases} R_{bi} = \dfrac{2\pi\, l_{bi}}{\gamma\left[\arccos\left(\frac{r_{bi}-\delta}{r_{bi}}\right) r_{bi}^2 - 2\pi\,(r_{bi}-\delta) r_{bi} \sin\left(\arccos\left(\frac{r_{bi}-\delta}{r_{bi}}\right)\right)\right]} \\ L_{bi} = \dfrac{\mu\, l_{bi}}{2\pi}\left(\ln\dfrac{2\, l_{bi}}{r_{bi}} - 1\right) + \dfrac{\mu\, l_{bi}}{8\pi}\xi \\ C_{bi} = \displaystyle\int_{l_{bi}} \dfrac{2\varepsilon_0\pi}{\ln\left[\frac{2(h_{bi}+l_{bi}\sin\theta)}{r_{bi}}\right]} dl \end{cases} \quad (2.25)$$

where the penetration depth of high frequency current δ can be obtained by Eq. (2.21), and the quantity ξ is determined by Eq. (2.23). l_{bi} is the length of branch i in the bracket, such as 0.6 and 0.66 m in Fig. 2.15 h_{bi} is the height for the lowest side of branch i, such as 1, 1.33, 1.66 and 2 m. r_{bi} is the equivalent radius of each branch for the bracket with the C-shaped structure.

According to Eq. (2.25) and the dimensions of the bracket, the parameters of the equivalent circuit for the bracket can be determined, as listed in Table 2.2. The

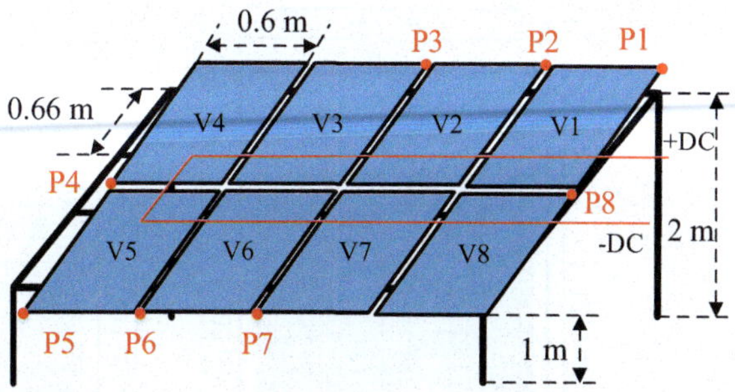

Fig. 2.15 Schematic diagram of PV array

PV array consists of eight PV modules connected in series (V1, V2, …, V7, V8), as illustrated in Fig. 2.15. Each PV panel in the PV array are modelled by the proposed model shown in Fig. 2.7, which is built in EMTP, as shown in Fig. 2.12. The parameters of the equivalent circuit for the PV panel are listed in Table 2.2. The maximum repetitive reverse voltage V_{RRM} of single BPD is 35 V, and its forward voltage V_F is 0.55 V. The different points of the PV array are set to be the lightning strike points, such as P1, P2, …, P7, P8 in Fig. 2.15.

In terms of the conventional PV-cell wiring, the cells are connected in series along the vertical direction, as shown in Fig. 2.16a. A new wiring scheme is proposed in this article, and cells are connected in series along the horizontal direction, as illustrated in Fig. 2.16b. The equivalent loop is distinct from the conventional one. The output voltage, current, and power are unchanged, and the sizes of PV panels are the same. The two PV-cell wiring schemes are compared in this section.

The mutual inductance between the PV-cell string and the frame highly depends on the loop of PV module. The wiring scheme has a great impact on the electromagnetic coupling. The lightning striking points are irregular and random. For a more comprehensive investigation about the effect of the two wiring on the induced voltage, the typical lightning striking points are considered, such as P1, P2, P3 and P7 in Fig. 2.15.

The induced voltages between the outputs of PV panels are calculated, and the peak voltages without the polarity are presented in Fig. 2.17 and Table 2.3. The induced voltages within a PV array for abovementioned two wirings are close to each other in most cases. For the PV array with new wiring, the maximum voltage

Table 2.2 Parameters of equivalent circuit for bracket [16]

Quantity	Value	Parameter	Value
R_{b1}	190.4 mΩ	L_{b3}	0.8 μH
R_{b2}	209.4 mΩ	L_{b4}	1.8 μH
R_{b3}	7.7 mΩ	C_{b1}	15.8 pF
R_{b4}	15.4 mΩ	C_{b2}	19.8 pF
L_{b1}	29.0 μH	C_{b3}	20.7 pF
L_{b2}	32.8 μH	C_{b4}	17.9 pF

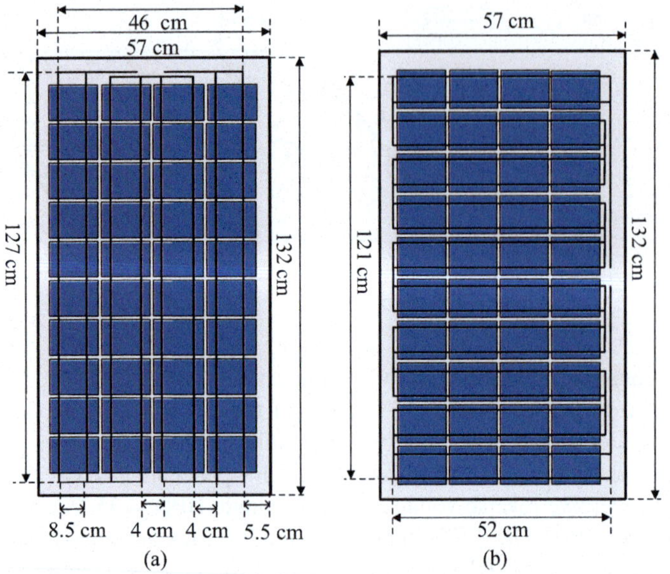

Fig. 2.16 Two PV-cell wirings. **a** Conventional PV-cell wiring. **b** New PV-cell wiring

can almost reach up to 1.3 kV. The peak voltage changes from 1.3 to 1.1 kV as the new wiring becomes the conventional wiring, as shown in Fig. 2.17, and the reduction is almost 15%. Table 2.3 shows that the induced voltages for most of PV panels with the conventional wiring are lower than that of the new wiring.

However, not all PV modules with the conventional wiring have a smaller induced voltage. The induced voltage of PV module is influenced by bracket and frame. The bracket changes the lightning-current distribution within a PV array and hence the magnetic field. If the direction of magnetic field generated by the current flowing in the bracket is opposite to that generated by the current in metal frame, the resultant magnetic field is partially cancelled out and the induced voltage would decline.

When the induced voltage is negative, the BPDs are at the forward bias, and the failure of the BPDs caused by lightning surge will not occur. In contrast, the BPDs are exposed to the reverse voltage when the induced voltage is positive, and the BPDs may be breakdown if the induced voltage magnitude is larger than $3V_{\mathrm{RRM}}$. For instance, the induce voltages of PV panels with the conventional wiring (V2, V3, V4 and V7) could fail when lightning strikes at P1.

The lightning strike point could affect the induced voltage as well. It ranges from 5.5 to 2.0 kV for the PV array with conventional wiring under different strike points, as shown in Fig. 2.18. Similar result is also observed in the PV array with new wiring. The distribution of induced voltage within a PV array can be summarized as: the lightning strikes on the rear side (e.g., P1, P2, and P3) of PV array would result in a greater lightning surge, especially for the middle point (P3 and P7). The smaller

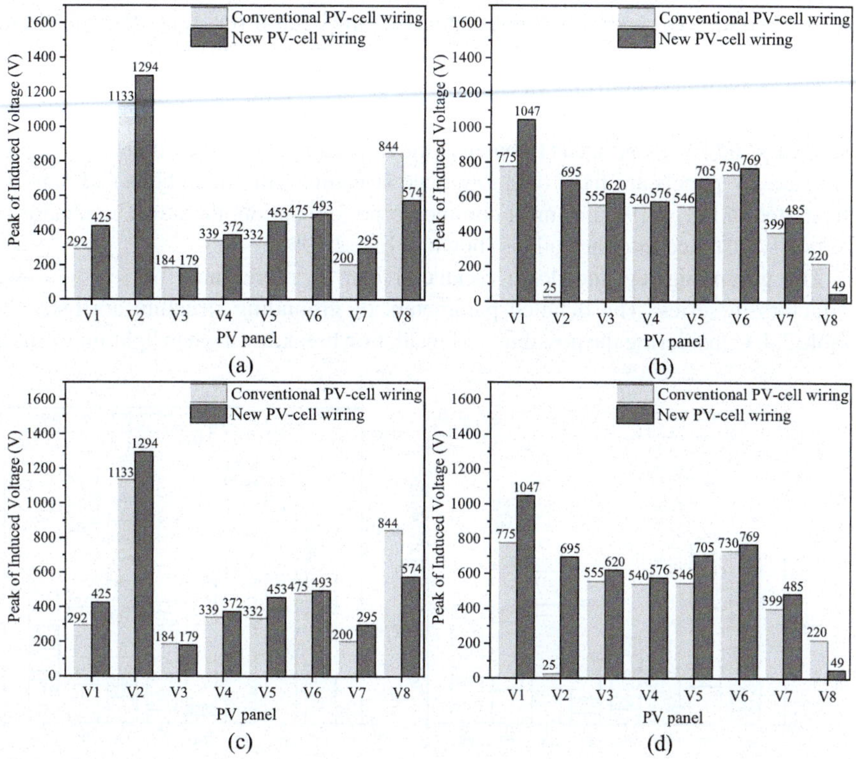

Fig. 2.17 Induced voltage of PV module with the two PV-cell wirings. **a** Lightning strike at P1. **b** Lightning strike at P2. **c** Lightning strike at P3. **d** Lightning strike at P7 [16]

Table 2.3 Induced voltage of PV module with different wirings (V) [16]

PV	Strike at P1		Strike at P2		Strike at P3		Strike at P7	
	CW	NW	CW	NW	CW	NW	CW	NW
V1	292	425	775	1047	68	22	702	803
V2	1133	1294	25	695	1009	1176	521	559
V3	184	179	555	620	680	60	666	692
V4	339	372	540	576	1222	1332	215	325
V5	332	453	546	705	791	989	22	196
V6	475	493	730	769	699	770	1175	1367
V7	200	295	399	485	839	897	490	164
V8	844	574	220	49	80	136	899	1150

Note CW is conventional PV-cell wiring, and NW is new PV-cell wiring induced voltage occurs as the lightning strikes the central row (e.g., P4 and P8) of PV array.

The grounding system plays a vital role for dissipating the energy originating from lightning channel to the earth. The design of grounding system for the application of large-scale PV plant needs to balance the installation cost, the efficiency and the lifetime. PV panels are fixed on a grounded structure with metal bolts and nuts, as illustrated in Fig. 2.19a. For most common type foundation, the down conductor is vertically installed into the soil, as shown in Fig. 2.19b.

The grounding leg and down conductor can be represented by resistors and inductors in series. The detailed parameters of grounding structure are listed in Table 2.4. Considering the possibility of insulation breakdown when lightning strikes

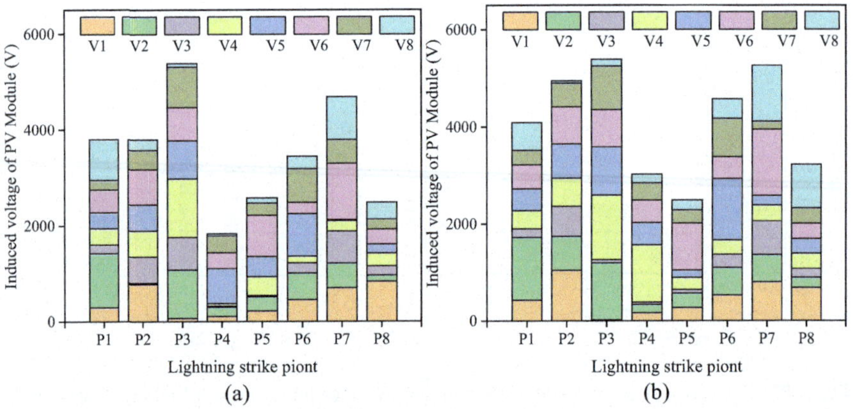

Fig. 2.18 Induced voltage of PV module under different lightning strikes. **a** Conventional PV-cell wiring. **b** New PV-cell wiring [16]

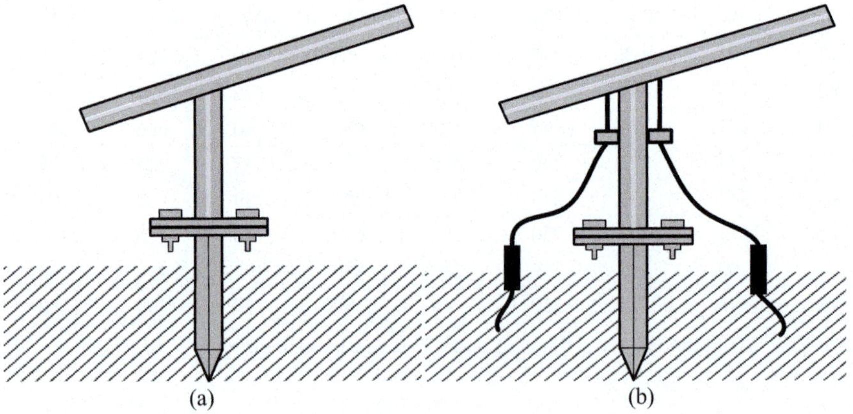

Fig. 2.19 Grounding structure for PV array. **a** Grounding leg. **b** Grounding leg with additional down conductors

Table 2.4 Parameters of grounding leg [16]

Grounding structure	Material	Height
Grounding leg	Aluminum	1, 2 m
Grounding leg	Steel	0.25, 1.25 m
Grounding leg	Aluminum	0.25, 1.25 m
Down conductor	Copper	0.25, 1.25 m

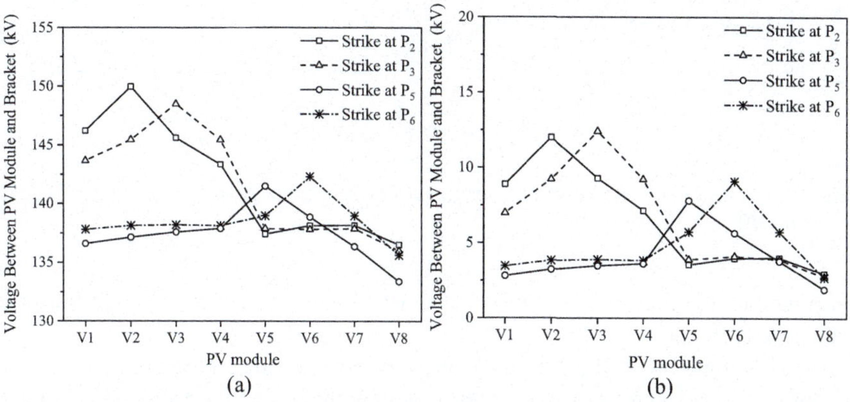

Fig. 2.20 Voltage between PV module and bracket with steel grounding legs. **a** Without additional down conductor. **b** With additional down conductor [16]

the bracket, the voltage between PV module and bracket is calculated. The grounding structure, including the grounding leg with different materials and additional down conductors, are studied. Firstly, the steel grounding leg and aluminium grounding leg are discussed. Supposing the heights of the front and the rear grounding legs are 0.25 and 1.25 m, respectively. The results are given in Figs. 2.20, and 2.21.

Given the mounting height, the peak voltage with steel grounding legs is several times that of aluminium, as shown in Figs. 2.20a and 2.21a. In terms of the former, it can even reach up to 120 kV, which can directly break down the PV modules. The impedance of steel grounding legs are tens of times higher than that of aluminium legs owing to its high permeability ($\mu = 75\mu_0$). It means that the steel grounding legs cannot immediately conduct the lightning current into the ground, and most of PV modules on supporting bracket may be damaged.

Two measures are recommended to decrease the voltage between PV module and supporting bracket. One is to use aluminium grounding legs, as plotted in Fig. 2.21a, the peak voltage is only 12 kV. It can ensure that the voltage is at a safe level. In addition to the lightning overvoltage, the choice of grounding leg material should also consider the mechanical strength and the anti-corrosion performance. Presently, most of grounding legs in large-scale PV systems still employ steel [25, 26]. The other measure is the use of additional down conductors, as shown in Fig. 2.20b.

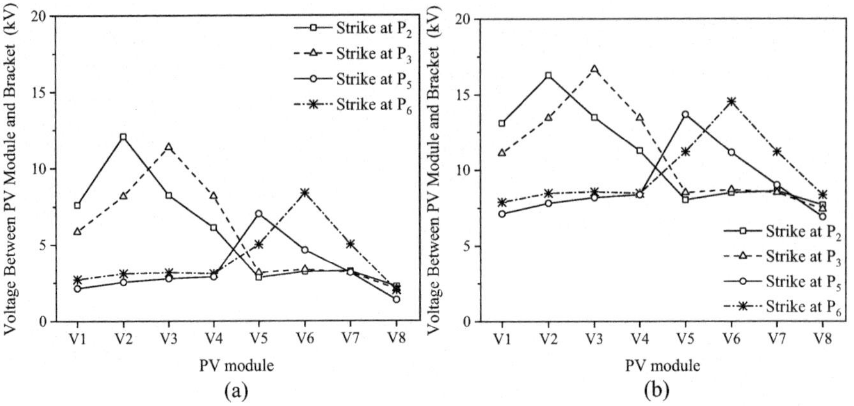

Fig. 2.21 Voltage between PV module and bracket with aluminium grounding legs. **a** 0.25, 1.25 m. **b** 1, 2 m [16]

In principle, the presence of additional down conductors benefits the diversion of lightning current.

The lightning overvoltage for a PV array varies in a wide range. The area near the striking point has a greater value. In terms of the zone away from the striking point or near the earth, the voltage of metal frame and supporting bracket is relatively small. Furthermore, the influence of grounding leg height is studied. Clearly, the higher the grounding leg, the larger the impedance and hence the greater the voltage.

2.4 Conclusion

A novel transient model of a PV module is developed for calculating the lightning induced voltage. The metal frame is divided into segments, and the mutual inductance between the PV-cell string and each segment of frame is derived. Since the electromagnetic coupling of each branch can be studied, it is more suitable for lightning transient analysis. The proposed model is validated, and it is more efficient and comprehensive than previous models.

The induced voltages of PV modules are determined by the magnetic field generated by the lightning current in metal frame and bracket. The PV-cell wiring can change the loop areas inside PV modules, and it has a great influence on the induced voltage. It is also closely related to the lightning strike point. The impulse magnetic field is partly cancelled out when the lightning strikes the central row of a PV array. Furthermore, the lightning overvoltage between PV modules and brackets is reduced by the additional down conductors. It facilitates the diversion of lightning current and the grounding resistance of the bracket becomes smaller.

References

1. Ahmad, N. I. Ab-Kadir, M. Z. A. Izadi, M. et al.: Lightning protection on photovoltaic systems: a review on current and recommended practices. Renew. Sust. Energ. Rev. 82, 1611–1619 (2018).
2. Hernandez, J. C. Vidal, P. G. and Jurado, F.: Lightning and surge protection in photovoltaic installations. IEEE Trans. Power Deliv. 23(4), 1961–1971 (2008).
3. Jiang, T. Grzybowski, S. "Electrical degradation of photovoltaic modules caused by lightning induced voltage," Proceedings of Proceedings of the IEEE Electrical Insulation Conference, Philadelphia, PA, USA, 107–110 (2014).
4. Yang, H. Liu, X. "Design of PV charge and discharge controller in insulator monitoring system," Proceedings of Proceedings of the 2th International Conference on Artificial Intelligence, Management Science and Electronic Commerce, Dengleng, 2039–2042 (2011).
5. Baba, Y. Rakov, V. A.: Voltages induced on an overhead wire by lightning strikes to a nearby tall grounded object. IEEE Trans. Electromagn. Compat. 48, 212–224 (2006).
6. Stratton, J. Electromagnetic Theory. New York: McGraw-Hill, 1941.
7. Heidler, F. Cvetic, J. Stanic, B.: Calculation of lightning current parameters. IEEE Trans. Power Deliv. 14, 399–404 (1999).
8. Biagi, C. J. Uman, M. A. et al.: Observations of stepping mechanisms in a rocket-and-wire triggered lightning flash. J. Geophys. Res. 115, D16117 (2010).
9. Hill, J.D. Uman, M.A. Jordan, D.M.: High-speed video observations of a lightning stepped leader. J. Geophys. Res. 116, D23215 (2011).
10. Shi, W. Li, Q. M. Zhang, L.: A stepped leader model for lightning including charge distribution in branched channels. J. Appl. Phys. 116, 103303 (2014).
11. Sun Q. Q, Zhong, X, et al.: Three-dimensional modeling on lightning induced overvoltage for photovoltaic arrays installed on mountain. J. Clean. Prod., vol. 288, 125084 (2021).
12. Davis, P. J. Rabinowits, P. Methods of Numerical Integration. New York: Dover Publications, 1975.
13. Andreotti, A. Petrarca, C. Rakov, V. A. et al.: Calculation of voltages induced on overhead conductors by nonvertical lightning channels. IEEE Trans. Electromagn. Compat. 54, 860–870 (2012).
14. Sakakibara, A.: Calculation of induced voltages on overhead lines caused by inclined lightning studies. IEEE Trans. Power Deliv. 4, 683–693 (1989).
15. Spigel, M. R. Mathematical Handbook of Formulas and Tables. New York: McGraw-Hill, 1968.
16. Zhong, X, Sun Q. Q, et al.: A Novel Transient Model of Photovoltaic Module Considering Electromagnetic Coupling. IEEE J. Photovolt., vol. 13, 122–132 (2023).
17. Kim, K. A. Krein, P. T. and Seo, G. et al. "Photovoltaic AC parameter characterization for dynamic partial shading and hot spot detection," Proceedings of 2013 Twenty-Eighth Annual IEEE Applied Power Electronics Conference and Exposition (APEC), Long Beach, CA, USA, 109–115 (2013).
18. Formisano, A. Petrarca, C. and Hernandez, J. C.: Assessment of induced voltages in common and differential-mode for a PV module due to nearby lightning strikes. IET Renew. Power Gener. 13, 1369–1378 (2019).
19. Zhang, Y. Chen, H. and Du, Y.: Lightning protection design of solar photovoltaic systems: methodology and guidelines. Electr. Power Syst. Res. 174, 105877 (2019).
20. Wang, Y. Zhang, X. and Tao, S.: Modeling of lightning transients in photovoltaic bracket systems. IEEE Access. 7, 12262–12271 (2019).
21. Ruehli, A. E.: Inductance calculations in a complex integrated circuit environment. IBM J. Res. Develop. 16, 470–481 (1972).
22. Dhere, N. G. Shiradkar, N. S. and Schneller, E.: Evolution of leakage current paths in MC-Si PV modules from leading manufacturers undergoing high-voltage bias testing. IEEE J. Photovoltaics. 4(2), 654–658 (2014).
23. Hernandez, J. C. Vidal, P. G. and Medina, A.: Characterization of the insulation and leakage currents of PV generators: Relevance for human safety. Renew. Energy. 35, 593–601 (2010).

24. Can-Am EMTP User Group, Alternative Transients Program (ATP) Rule Book. Portland, OR, 1987–1998.
25. Zhang, Y. Chen, H. Du, Y. et al.: Considerations of photovoltaic system structure design for effective lightning protection. IEEE Trans. Electromagn. Compat. 62, 1333–1341 (2020).
26. Lorenzo, G. D. Araneo, R. Mitolo, M. et al.: Review of O&M practices in PV plants: failures, solutions, remote control, and monitoring tools. IEEE J. Photovolt. 10(4), 914–926 (2020).

Chapter 3
Lightning Induced Voltage of Photovoltaic Array on Complex Terrains

3.1 Induced Voltage of Photovoltaic Array on Flat Ground

3.1.1 Influence of Oblique Angle for Photovoltaic Array

Suppose a PV array installed on flat ground, twenty PV individual panels are connected into a group, and eight groups are wired together and then linked to the inverter, as shown in Fig. 3.1a. Considering the lightning channel is vertical and perpendicular to the ground, the horizontal distance R, from PV array to the lightning channel, is 5 m. The detailed dimension of single PV panel is depicted in Fig. 3.1b.

Many mathematical expressions have been proposed to represent the lightning current based on long-term observation. The Heidler function is employed herein and the current can be described by,

$$i(t) = \frac{I}{\eta}\frac{(t/\tau_1)^n}{1+(t/\tau_1)^n}\exp\left(-\frac{t}{\tau_2}\right) \tag{3.1}$$

where I_0 is the peak value of the current and set to be 10 kA in this section; η is the correction factor, which is generally 0.98; τ_1 and τ_2 are time constants which determine the rise time and decay time, respectively, they are set to be 10 and 350 μs in this section, respectively.

The induced voltage is closely related to the oblique angle θ and the results are plotted in Fig. 3.2 For instance, the voltage amplitude reaches almost 3.85 kV when θ is 45°. The surge would transmit from DC side into the inverter, and such a high surge can easily damage the bypass diode and the electronic device. Suppose the oblique angle θ is 90° and the panel is vertical to the ground, the magnetic field due to nearby lightning would vertically passes through the internal loop, resulting in the largest magnetic flux as well as the induced voltage. Hence, the higher the oblique angle θ the more magnetic flux and the larger the induced voltage. The selection of an appropriate θ is necessary, if permitted. The surge protection devices (SPD) shall

Q. Sun et al., *Lightning Transients and Protection for Renewable Energy System*,
https://doi.org/10.1007/978-981-97-7267-4_3

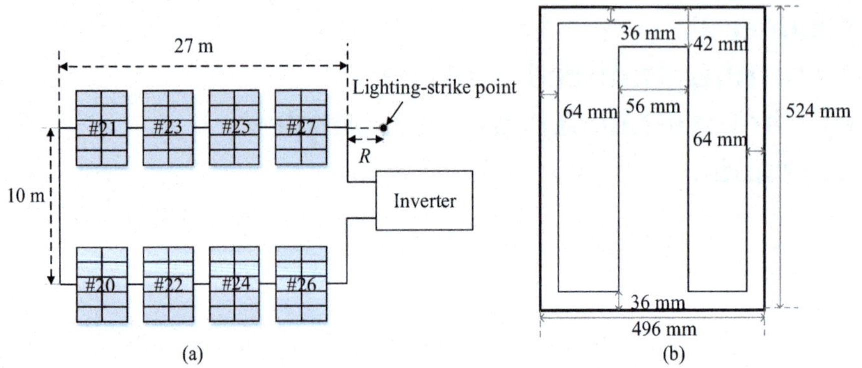

Fig. 3.1 Schematic diagram of PV array. **a** Wiring scheme of PV array. **b** Dimension of single PV panel [1]

be installed at the entry as close as possible to the protected equipment and withstand the expected discharge capacity.

The lightning channel length also has an impact on the induced voltage. Since the higher the lightning channel, the lower the magnetic field strength within the loop. Once the lightning channel exceeds a certain level, the magnetic flux within the entire loop is negligible. In this case, when the lightning channel length is greater than 200 m, such influence is quite small and almost negligible. If not mentioned, the length H and oblique angle θ are set to be 500 m and 30°, respectively in the following analysis.

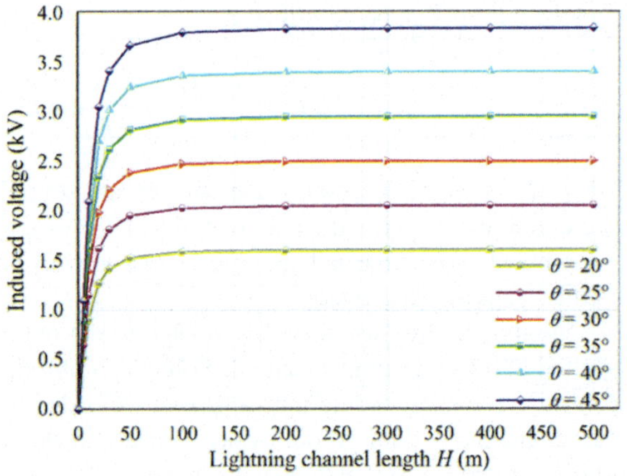

Fig. 3.2 Influence of lightning channel length on induced voltage [1]

3.1.2 Influence of Lightning Location

Given the distance between the strike point and the center of PV array, as illustrated in Fig. 3.3, the open-circuit voltage is computed. Table 3.1 lists the induced voltage on the terminal of PV array for five lightning strike points. The point S_i ($i = 1, 2, 3, 4$) is characterized by the direction angle φ and the radius R_l. The relationship between the induced voltage and the direction angle φ is plotted in Fig. 3.4.

It can be obtained that the lightning channel generates a large induced voltage on left and right sides. In contrast, the voltages on the upper (S_4, 90°) and lower points (S_2, 270°) is relatively small, and that of the lower point (S_2, 270°) is the smallest. Obviously, the magnitude of induced voltage is inversely proportional to the distance R_l.

Given the distance between lightning channel and PV center, different lightning strike points would yield different electromagnetic field. Theoretically, the lightning current can be regarded as a radiation source. To facilitate the analysis, the electromagnetic field outside radiation radius R_0 is neglected and the points S_1 and S_4 are supposed to be the center of circle, respectively. The interaction area of internal loop for S_1, as illustrated in Fig. 3.5, is larger than that of S_4. Providing the lightning current and hence the electromagnetic intensity, the magnetic flux flowing the loop of point S_1 is greater than that of S_4. Therefore, it generates a higher induced

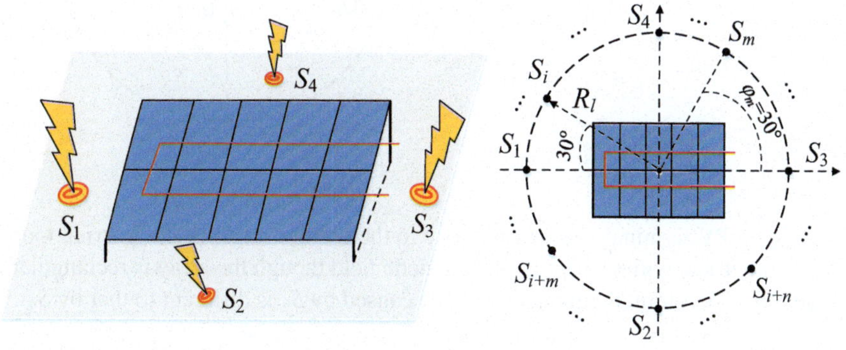

Fig. 3.3 Relative position between lightning strike spots and PV array [1]

Table 3.1 Induced voltage at different lightning-strike point [1]

R_l (m)	S_1 (kV)	S_2 (kV)	S_i (kV)	S_m (kV)	S_4 (kV)
15	4.29	2.30	3.32	2.56	2.40
17.5	2.98	1.98	2.60	2.17	2.05
20	2.33	1.74	2.07	1.87	1.79
25	1.66	1.40	1.59	1.47	1.42
30	1.30	1.16	1.27	1.20	1.18

Fig. 3.4 Variation of induced voltage versus lightning strike point [1]

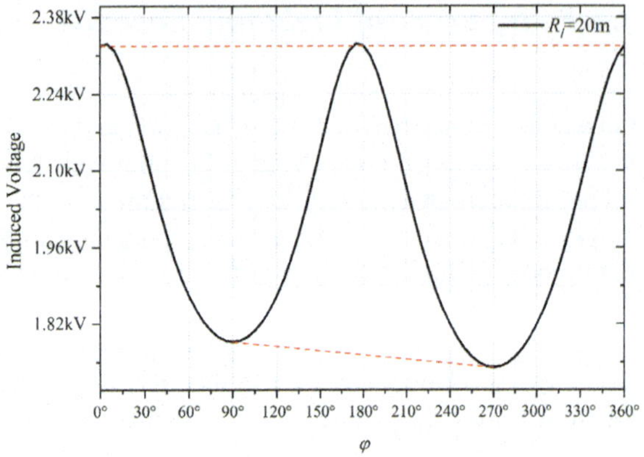

Fig. 3.5 Comparison of magnetic flux area centered on S_1 and S_4

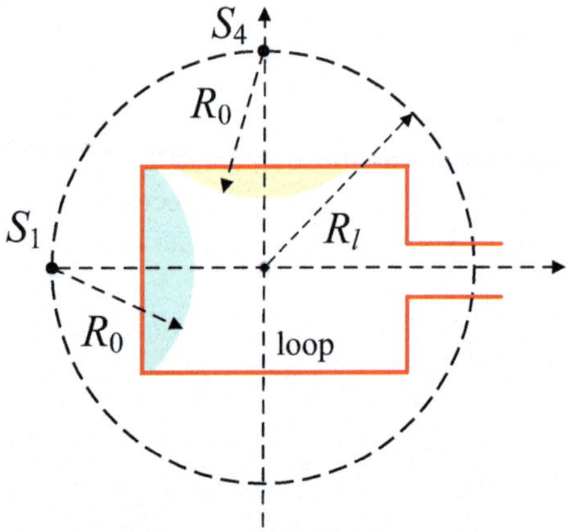

voltage on PV terminal. Meanwhile, due to the oblique angle θ, the internal loop is not symmetrical with respect to the magnetic field though the array is rectangular in geometry. Therefore, the induced voltage caused by S_2 is different to that by S_4.

3.1.3 *Influence of Lightning Channel Shape*

Lightning is a rather stochastic natural phenomenon. The lightning channel path is complex and it is a key factor for the induced voltage. Five lightning channels are randomly produced using the stepped leader model presented in Chap. 2. The peak of lightning current is 10 kA, and the horizontal distance R is set to be 5 m. The channel path and the induced voltage are given in Fig. 3.6. Clearly, the vertical channel has the smallest value amongst them. Previous analysis suggests that the vertical channel produces the highest induced voltage when θ is 90° since all the magnetic flux crosses

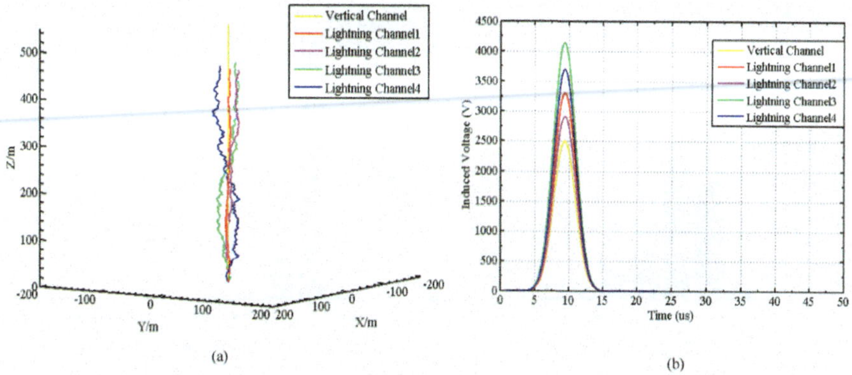

Fig. 3.6 Induced voltages by different lightning channels. **a** Random lightning channel path; **b** Induced voltage [1]

the loop. In practice, θ is often around 30°, and the induced voltage caused by the vertical channel is generally lower than that of others.

In addition, the direction of lightning channel also plays a role. The voltage would be relatively higher if the channel is prone towards PV array. Figure 3.7 illustrates the statistical distribution of induced voltage generated by 100 random lightning discharges, and the details are listed in Table 3.2. Clearly, the induced voltage distribution is jointly determined by the distance R and θ. It has a wide range and approximately normal distributed where most values cluster in the middle of the range and the rest taper off symmetrically toward either extreme. A comparison is performed in Fig. 3.7. It can be found that the majority of the induced voltages is shifted from [3–4 kV] to [4–5 kV] as θ increases from 20° to 40°. Despite the low probability of occurrence, several impulsive voltages reach around 10 kV and even up to 11 kV. As the distance R increases from 5 to 15 m, the mean value declines; the standard deviation is also smaller since the voltage becomes more concentrated in distribution.

The bypass diodes used in most PV array have a repetitive peak reverse voltage of 1–2 kV and they would suffer the permanent breakdown when the impulse voltage exceeds 3 kV. Consider the critical value is 3 kV and the lightning failure rate for this system is evaluated, as given in Table 3.2. Clearly, the severity is inversely proportional to the distance from the striking point. The influences of those occur 50 m beyond is small.

3.1.4 Effect of Metal Frame

Supposing the lightning current is 10 kA and in very close proximity to the PV panel with a horizontal distance R is 1 m, a circulating current would be induced across the metal frame. The peak of the current gets to 80 A, as depicted in Fig. 3.8a. It would

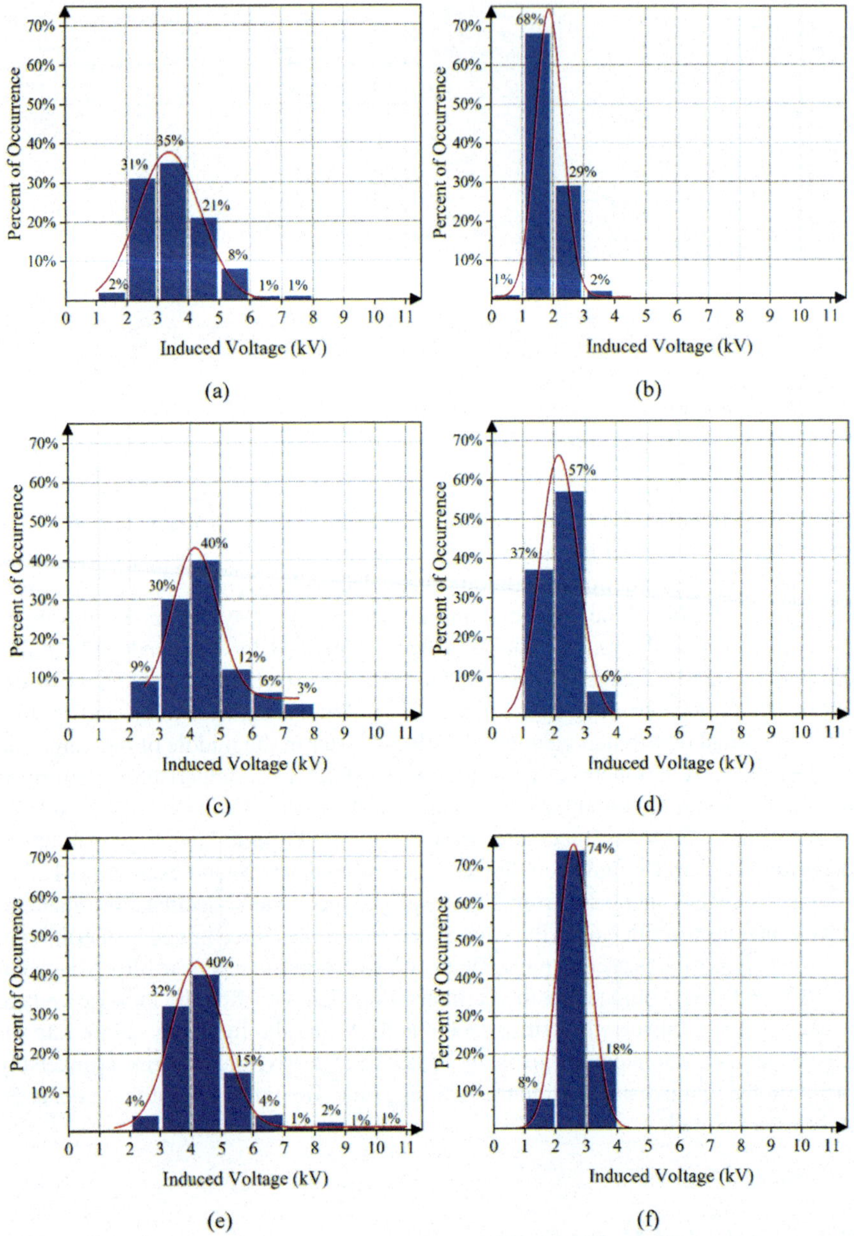

Fig. 3.7 Induced voltage distribution of PV arrays by 100 random lightning discharges. **a** $\theta = 20°$, $R = 5$ m; **b** $\theta = 20°$, $R = 15$ m; **c** $\theta = 30°$, $R = 5$ m; **d** $\theta = 30°$, $R = 15$ m; **e** $\theta = 40°$, $R = 5$ m; **f** $\theta = 40°$, $R = 15$ m [1]

Table 3.2 Statistics of induced voltage for random lightning channels [1]

R	5 m	10 m	50 m
Average (kV)	4.243	2.743	0.866
Max (kV)	7.698	4.517	1.195
Min (kV)	2.321	1.761	0.549
Std. Dev	1.112	0.577	0.123
Failure rate	0.91	0.3	0

create a reverse magnetic field with respect to the lightning current. Consequently, the current in frame partially offset the field and the induced voltage of internal loop. The frame effect of PV panel is elucidated in Fig. 3.8b, and the details are listed in Table 3.3. The induced voltage of internal loop with frame is lower than that without frame. The frame reduction factor R_F, defined as the ratio between the maximum voltage without and with frame, depends on the PV module type and the oblique angle. It can attain up to 2.58 and only one third is remained.

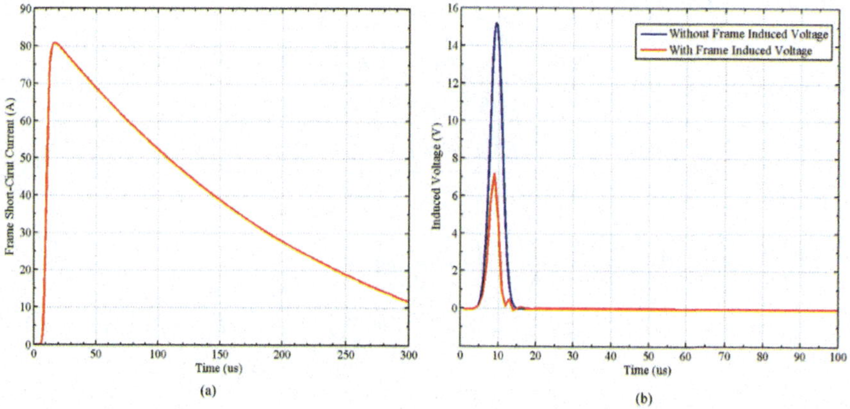

Fig. 3.8 Frame effect of PV panel. **a** Induced current along the aluminum frame; **b** Induced voltage of single PV panel [1]

Table 3.3 Induced voltage of PV panel considering frame (V) [1]

R(m)	20°			30°			40°		
	Frame	No frame	R_F	Frame	No frame	R_F	Frame	No frame	R_F
0.5	7.43	17.17	2.31	13.83	26.54	1.92	15.69	35.89	2.29
1.0	4.03	10.10	2.50	7.21	15.27	2.11	9.52	20.24	2.13
1.5	2.75	7.12	2.58	4.80	10.66	2.22	6.18	14.01	2.27
2.0	2.26	5.50	2.43	3.49	8.18	2.34	4.80	10.70	2.23

3.2 Induced Voltage of Photovoltaic Array on Irregular Ground

3.2.1 *Finite-Difference Time-Domain (FDTD) Method*

The alternatively Yee cell [2] approach derived for cylindrical coordinates is used, which can represent a three dimensional (3D) problem into a 2D finite-difference scheme. Assuming that the permittivity ε and permeability μ in air are constants, the lightning electromagnetic field (LEMF) can be regarded as transverse electric (TE) field in cylindrical coordinates. In TE field, the electric fields are transverse to the direction of propagation, it can be represented by two components $\boldsymbol{E}_{\mathbf{r}}$ and $\boldsymbol{E}_{\mathbf{z}}$ in cylindrical coordinates. While the magnetic fields are normal to the direction of propagation, which has a component $\boldsymbol{H}_{\phi}$. The finite difference equations for LEMF in cylindrical coordinates can be determined as,

$$H_{\phi}^{n+1/2}\left(i+\frac{1}{2},j+\frac{1}{2}\right)=\frac{2\mu-\sigma_m\Delta t}{2\mu+\sigma_m\Delta t}H_{\phi}^{n-1/2}\left(i+\frac{1}{2},j+\frac{1}{2}\right)+\frac{2\Delta t}{2\mu+\sigma_m\Delta t}\left[\frac{E_r^n\left(i+\frac{1}{2},j+1\right)-E_r^n\left(i+\frac{1}{2},j\right)}{\Delta z}-\frac{E_z^n\left(i+1,j+\frac{1}{2}\right)-E_z^n\left(i,j+\frac{1}{2}\right)}{\Delta r}\right] \tag{3.2}$$

$$E_r^{n+1}\left(i+\frac{1}{2},j\right)=\frac{2\varepsilon-\sigma\Delta t}{2\varepsilon+\sigma\Delta t}E_r^n\left(i+\frac{1}{2},j\right)+\frac{2\Delta t}{2\varepsilon+\sigma\Delta t}\cdot\frac{H_{\phi}^{n+1/2}\left(i+\frac{1}{2},j+\frac{1}{2}\right)-H_{\phi}^{n+1/2}\left(i+\frac{1}{2},j-\frac{1}{2}\right)}{\Delta z} \tag{3.3}$$

$$E_z^{n+1}\left(i,j+\frac{1}{2}\right)=\frac{2\varepsilon-\sigma\Delta t}{2\varepsilon+\sigma\Delta t}E_z^n\left(i,j+\frac{1}{2}\right)-\frac{2\Delta t}{2\varepsilon+\sigma\Delta t}\cdot\frac{H_{\phi}^{n+1/2}\left(i+\frac{1}{2},j+\frac{1}{2}\right)-H_{\phi}^{n+1/2}\left(i-\frac{1}{2},j+\frac{1}{2}\right)}{\Delta r} \tag{3.4}$$

where σ and σ_{m} are the electric conductivity and magnetic conductivity, respectively. Δr and Δz are the space increment in r and z, respectively; Δt is the time increment, as shown in Fig. 3.9a.

In order to reduce the computational cost, it is necessary to choose the absorbing boundary condition. The most used boundaries are Mur's absorbing boundary [3] and Liao's absorbing boundary [4]. They are deduced from the outgoing wave equation. Table 3.4 lists the wave reflection coefficients for Mur's absorbing boundary and Liao's absorbing boundary, respectively. The results show that Mur's absorbing

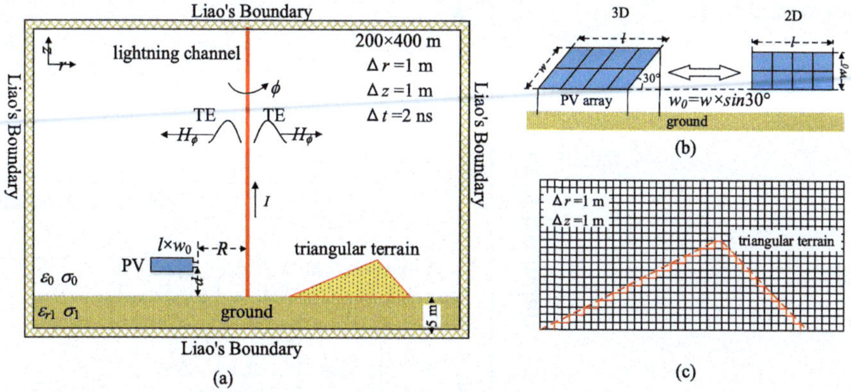

Fig. 3.9 Schematic diagram of 2D FDTD model. **a** TE wave propagation in cylindrical coordinates. **b** Conversion between 3D PV arrays and 2D PV arrays. **c** Staircase technique [6]

boundary has a relatively larger error when the waves are not incident perpendicular to absorbing boundary, e.g., the corner of object which has a high reflection coefficient. On the contrary, Liao's absorbing boundary keeps a low reflection coefficient for these cases. It is thus selected in this article. As illustrated in Fig. 3.10, Liao's second-order absorbing boundary for $\boldsymbol{Ez}$ can be expressed as,

$$\begin{aligned}
E_z^{n+1}\left(0, j+\frac{1}{2}\right) &= (1-S)(2-S)E_z^n\left(0, j+\frac{1}{2}\right) \\
&\quad + 2S(2-S)E_z^n\left(\Delta r, j+\frac{1}{2}\right) + S(S-1) \\
&E_z^n\left(2\Delta r, j+\frac{1}{2}\right) - \frac{(1-S)^2(2-S)^2}{4}E_z^{n-1}\left(0, j+\frac{1}{2}\right) - S(1-S)(2-S)^2 \\
&E_z^{n-1}\left(\Delta r, j+\frac{1}{2}\right) - \left[S^2(2-S)^2 - \frac{S(2-S)(1-S)^2}{2}\right]E_z^{n-1}\left(2\Delta r, j+\frac{1}{2}\right) \\
&- S^2(S-1)(2-S)E_z^{n-1}(3\Delta r, j+\frac{1}{2}) - \frac{S^2}{4}(S-1)^2E_z^{n-1}\left(4\Delta r, j+\frac{1}{2}\right)
\end{aligned} \quad (3.5)$$

Here, $S = v\Delta t/\Delta r$ and v is the propagation speed of electromagnetic wave. The modified transmission line model with exponential decay (MTLE) is utilized to represent the lightning return stroke current. Since the sum of two functions are

Table 3.4 Reflection coefficient at different absorbing boundary conditions [6]

Incidence angle	Mur's (%)	Liao's (%)
0	6.7	5.9
26.6°	9.1	6.8
45°	13.8	7.4

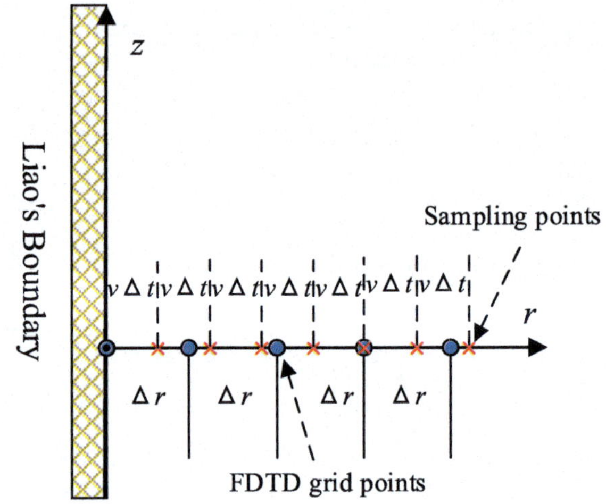

Fig. 3.10 Schematic diagram of Liao's absorbing boundary [6]

particularly suitable for approximating the front of the channel-base current (CBC) and similar to the return stroke current observed for natural lightning, it is used to express the CBC [5],

$$I(0,t) = \frac{I_{BD}[1-\exp(-\frac{t}{\tau_1})]^2 \exp(-\frac{t}{\tau_2})}{\eta} + I_C[\exp(-\frac{t}{a}) - \exp(-\frac{t}{b})] \tag{3.6}$$

where I_{BD} is 28.51 kA and I_{C} is 22.36 kA in this article; τ_1 and τ_2 are set to be 0.1232 and 3.0695 μs, respectively; a and b are 48.73675 and 6.3317 μs, respectively; η can be described by Eq. (3.7).

$$\eta = \left(\frac{2\tau_2}{\tau_1 + 2\tau_2}\right)^2 \left(\frac{\tau_1}{\tau_1 + 2\tau_2}\right)^{\frac{\tau_1}{\tau_2}} \tag{3.7}$$

The dipole technique is used to calculate the H-field component $\boldsymbol{H}_{\phi 0}$ around the lightning channel, as line excitation source of FDTD model. Therefore, the formula of $\boldsymbol{H}_{\phi}$ at the location of lightning channel needs to be modified.

$$\begin{aligned} H_{\phi}^{n+1/2}\left(i+\frac{1}{2},j+\frac{1}{2}\right) &= \frac{2\mu - \sigma_m \Delta t}{2\mu + \sigma_m \Delta t} H_{\phi}^{n-1/2}\left(i+\frac{1}{2},j+\frac{1}{2}\right) \\ &+ \frac{2\Delta t}{2\mu + \sigma_m \Delta t}\left[\frac{E_r^n(i+\frac{1}{2},j+1) - E_r^n(i+\frac{1}{2},j)}{\Delta z}\right. \\ &\left. - \frac{E_z^n(i+1,j+\frac{1}{2}) - E_z^n(i,j+\frac{1}{2})}{\Delta r}\right] \\ &+ H_{\phi 0}^{n+1/2}\left(i+\frac{1}{2},j+\frac{1}{2}\right) \end{aligned} \tag{3.8}$$

Due to the angle θ between PV array and ground, the equivalent area of magnetic flux crossing is not the actual area of PV array. According to the vector decomposition of the magnetic field $\boldsymbol{H}_{\phi}$, the surface crossed by the vertical vector is the equivalent 2D surface, as depicted in Fig. 3.9b. To simulate the irregular geometry of complex terrain, the staircase technique is utilized to approximate the curved surface (triangle, trapezoid, etc.). In order to avoid possible errors in surface discretization, the minimized space step is required in staircase approximation, as depicted in Fig. 3.9c. The flowchart of FDTD method for induced voltage on PV array is given in Fig. 3.11. Firstly, the magnetic field $\boldsymbol{H}_{\phi 0}$ around the lightning channel is calculated by the dipole technique. After building the 2D Yee cell, the LEMF propagation is acquired and hence the LEMF distribution is determined. Finally, the magnetic field $\boldsymbol{H}_{\phi}$ of Yee cells in PV array is summed, and the induced voltage U_{OC} can be obtained based on the Faraday's law.

3.2.2 Photovoltaic Array on Lossy Soil

Supposing the PV array installed on flat ground with the thickness of 5 m, and the horizontal distance R is set to be 10 m, as illustrated in Fig. 3.12a. In practice, the electrical properties of ground are complex and vary at different depths. Thus, two soil models are adopted, i.e., the monolayer model and the stratified model. The latter represents the inhomogeneous characteristics of soil. The soil thickness of each layer is 1 m, as depicted in Fig. 3.12a. The electromagnetic wave propagation is influenced not only by the lossy ground, but also the obstacles such as earth mounds, hillocks, etc. The shape of obstacle is taken as an isosceles trapezoid, as illustrated in Fig. 3.12b. The impact of obstacle location on the induced voltage of PV array is discussed.

The induced voltages for PV array on lossy flat ground are analyzed in detail, as shown in Fig. 3.13. It is clear that the induced voltage is heavily dependent on the ground conductivity. The peak value becomes higher and the waveform becomes a little wider as the conductivity decreases, such trend is similar to that on overhead lines.

If the earth is regarded as an infinite conducting plane, less magnetic strength occurs within ground. Consequently, the electromagnetic waves would be reflected against the surface and partially cancel out the magnetic field above the ground; the induced voltage would drop, as described in Fig. 3.14. As the mounting height d increases, the effect of reflected waves is weakened and the induced voltage increases (see Fig. 3.13). The results also indicate that the peak value of induced voltage rises as the ground permittivity increases. The waveform becomes steeper and the rise time is lower, which implies that the high frequency component is produced due to the earth capacitance.

The electrical characteristics of soil are complex and influenced by many factors, such as depth and frequency, etc. For comparison purpose, the stratified model with

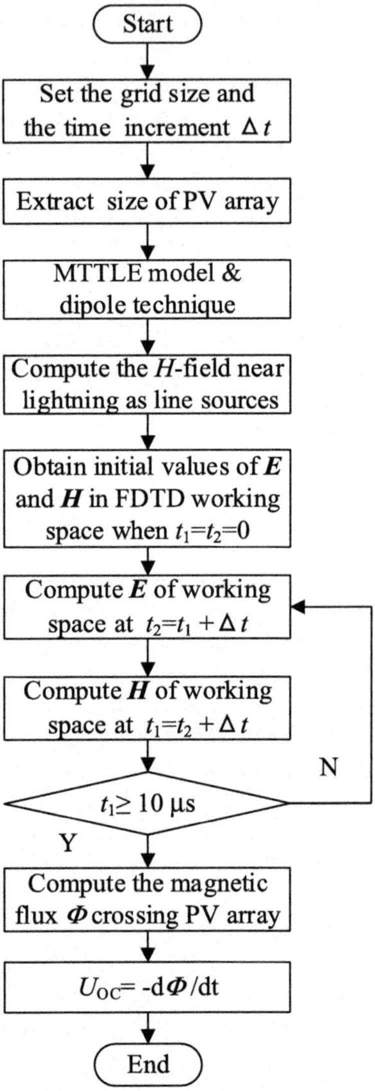

Fig. 3.11 Flowchart of FDTD method for induced voltage on PV array

five layers of soil is also investigated. Table 3.5 lists the maximum open-loop voltage V_{OCmax} in layered soil. It suggests that the lossy soil within 2 m below the ground plays a vital role in LEMF. If the conductivity of surface soil is very small, the impact of deeper soil cannot be simply neglected. Since the surface soil with low conductivity cannot effectively mitigate the reflected wave from the deeper soil, the LEMF above the ground is severely distorted. The open-loop voltage is thereby minimized. Compared with the conductivity, the soil permittivity has a minor and similar effect.

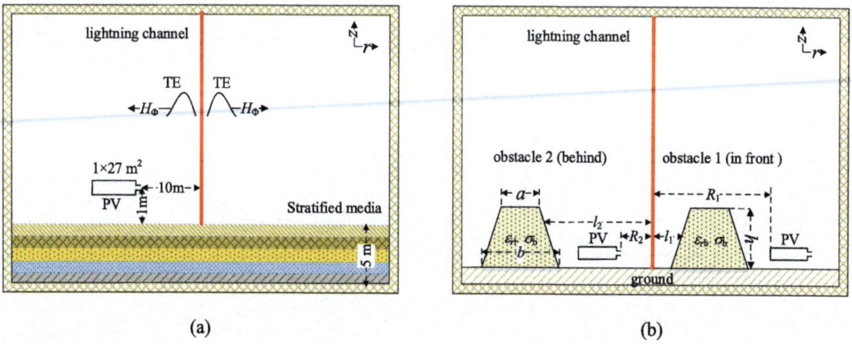

Fig. 3.12 PV array installed on flat ground. **a** Stratified media model of earth. **b** Schematic diagram of obstacle [6]

The ground conductivity and relative permittivity are set to be 0.5 mS/m and 10, respectively. The results for the case that the obstacle is in front of PV array are listed in Table 3.6. It can be found that the obstacle in front of PV array could reduce the induced voltage, the effect becomes more obvious as the conductivity and permittivity of obstacle increase. Concerning the electromagnetic field, part of waves passes through the obstacle, and the lossy medium consumes the energy. The wave entering into the PV array loop becomes weaker. The shape of obstacle also impacts the propagation of electromagnetic wave. The path of electromagnetic wave in triangular obstacle is shorter than in trapezoidal obstacle with the same height. The narrower the upper base of trapezoid, the closer it is to triangle, the lower the energy loss. Therefore, the induced voltage in the narrower trapezoid is larger than the wider one, as shown in Table 3.6.

From Table 3.7, it can be seen that the obstacle behind PV array affects the induced voltage less whether a great or small conductivity and permittivity. The distance between the obstacle behind PV array and lightning channel is farther. The greater the distance, the weaker the magnetic strength, despite the presence of the obstacle. Such influence is small compared to the distance and it is almost negligible.

3.2.3 Photovoltaic Array on Mountainous

The photograph of PV system installed on mountain is located in Shanxi, China. As shown in Fig. 3.15, three typical irregular terrains are studied in this article. Here, T_1 is the conical terrain, which roughly represents geographical condition of hilltop; T_2 can be expressed by trapezoidal terrain, which represents the hill ridge; T_3, which is V-shaped, generally represents the area between two hillsides. The schematic of T_1, T_2 and T_3 are illustrated in Fig. 3.16a–c, respectively. The height of the mountain in Fig. 3.15 is around several tens of meters.

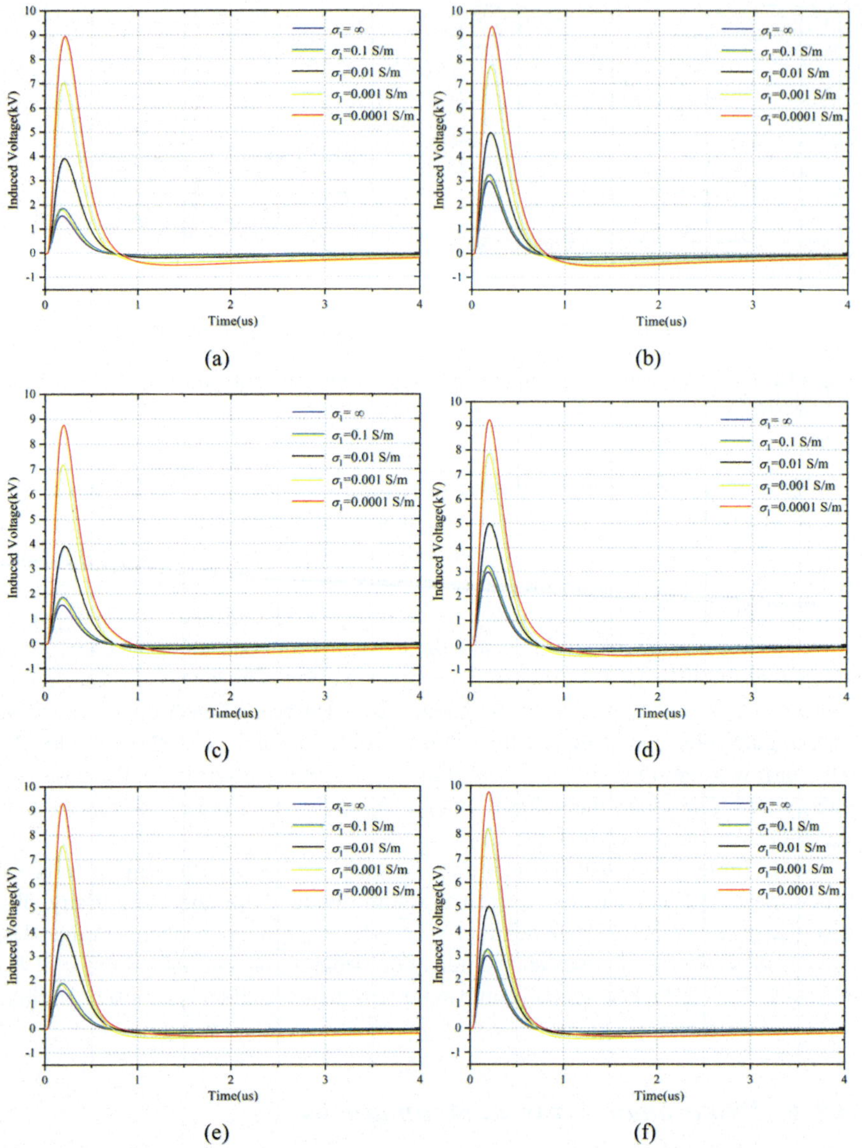

Fig. 3.13 Induced voltages of PV array under single layer soil. **a** $\varepsilon_{r1} = 1$, $d = 1$ m. **b** $\varepsilon_{r1} = 1$, $d = 2$ m. **c** $\varepsilon_{r1} = 10$, $d = 1$ m. **d** $\varepsilon_{r1} = 10$, $d = 2$ m. **e** $\varepsilon_{r1} = 25$, $d = 1$ m. **f** $\varepsilon_{r1} = 25$, $d = 2$ m [6]

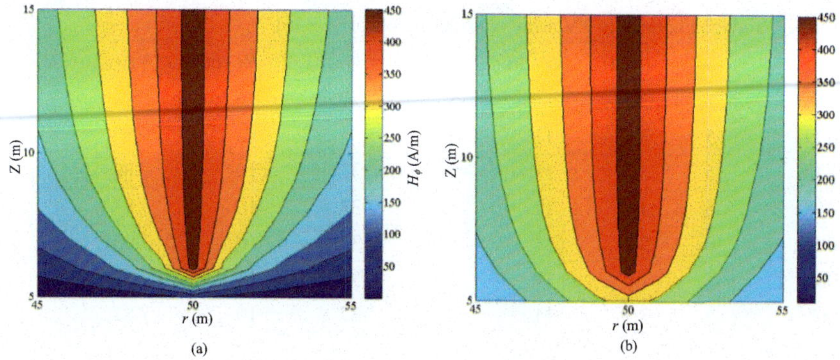

Fig. 3.14 ***H***-field over flat ground. **a** $\sigma_1 = 0.1$ S/m. **b** $\sigma_1 = 0.0005$ S/m [6]

Table 3.5 Induced voltage of PV array under stratified soil [6]

$[\sigma_1, \sigma_2, \sigma_3, \sigma_4, \sigma_5]$ (mS/m)	$[\varepsilon_{r1}, \varepsilon_{r2}, \varepsilon_{r3}, \varepsilon_{r4}, \varepsilon_{r5}]$	V_{OCmax} (kV)
[1,2,3,4,5]	[10,10,10,10,10]	6.91
[1,2,5,4,3]	[10,10,10,10,10]	6.88
[1,5,2,3,4]	[10,10,10,10,10]	6.50
[5,1,2,3,4]	[10,10,10,10,10]	5.19
[0.1,0.1,10,10,10]	[10,10,10,10,10]	7.14
[0.1,0.1,1,1,1]	[10,10,10,10,10]	8.07
[0.1,0.1,0.1,0.1,0.1]	[10,20,25,30,35]	9.22
[0.1,0.1,0.1,0.1,0.1]	[35,10,20,25,30]	9.28

Note $[\sigma_1, \sigma_2, \sigma_3, \sigma_4, \sigma_5]$ represents the conductivity of five layers of soil, respectively; $[\varepsilon_{r1}, \varepsilon_{r2}, \varepsilon_{r3}, \varepsilon_{r4}, \varepsilon_{r5}]$ represents the permittivity of five layers of soil, respectively

Table 3.6 Induced voltage of PV array considering obstacle 1 [6]

σ_b (S/m)	ε_{rb}	a (m)	b (m)	h (m)	l_1 (m)	R_1 (m)	V_{OCmax} (kV)
0	1	3	11	5	6	18	4.30
0.0001	5	3	11	5	6	18	4.03
0.001	5	3	11	5	6	18	3.31
0.01	5	3	11	5	6	18	2.55
0.001	30	3	11	5	6	18	3.17
0.001	5	1	9	5	6	18	3.52
0.001	5	3	11	3	6	18	3.73
0.001	5	3	11	5	6	20	3.01
0.001	5	3	11	5	7	20	3.00
0.001	5	3	11	5	8	20	2.99

Table 3.7 Induced voltage of PV array considering obstacle 2 [6]

σ_b (S/m)	ε_{rb}	a (m)	b (m)	h (m)	l_2 (m)	R_2 (m)	V_{OCmax} (kV)
0	1	3	11	5	37	10	7.85
0.0001	5	3	11	5	37	10	7.90
0.001	5	3	11	5	37	10	7.88
0.01	5	3	11	5	37	10	7.83
0.001	30	3	11	5	37	10	7.92
0.001	5	1	9	5	37	10	7.88
0.001	5	3	11	3	37	10	7.87

Fig. 3.15 Photograph of PV system installed on mountain [6]

The induced voltage of PV array on mountain with the height of 50 and 100 m, are 4.19 and 4.20 kV, respectively. The mountain height is set to be 50 m in following study. The lightning channel is assumed vertical and placed at most likely striking points of mountain terrain, as shown in Fig. 3.16. The LEMF in PV array is mainly generated by the lightning current element within low height, and the height of lightning channel is thus set to be 200 m. The working space of 200 m × 400 m is composed of 1 × 1 m cell in dimension. For convergence purpose, the time increment Δt is set to be 2 ns to satisfy the space increment requirement.

The LEMF on mountain surface is complex and distorted more severely than that on flat ground. The common mountain geography can be divided as triangular, trapezoidal and V-shaped terrains, as illustrated in Fig. 3.16. Suppose the relative permittivity and conductivity of ground are 10 and 0.5 mS/m, respectively. Figure 3.17 shows the induced voltage for PV array installed on triangular terrain. The mountain conductivity plays a crucial role and the induced voltage is inversely proportional to it. By contrast, the permittivity has a minor impact and the voltage changes little with it when the mountain conductivity is high. The LEMF on the mountain surface is unevenly distributed, due to the dramatic differences in electric characteristics between mountain and air. The LEMF far from mountain surface is more uniformly

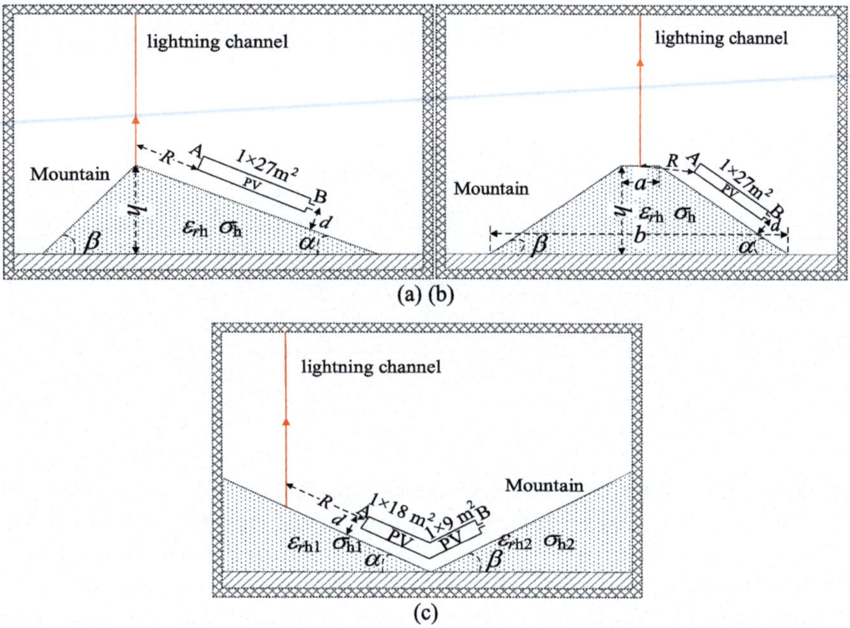

Fig. 3.16 PV array installed on three irregular terrains. **a** Triangular terrain. **b** Trapezoidal terrain. **c** V-shaped terrain [6]

distributed; therefore, the induced voltage is higher when the mounting height d is greater.

The triangular terrain has a slope, and the electromagnetic field is inevitably influenced by it. Figure 3.17 also indicates that the PV array installed on triangular terrain with steeper slope has a lower induced voltage. The intensity of reflected wave is enhanced and LEMF distortion is severe on the surface with steeper slope. Correspondingly, the LEMF near PV array with steeper mountain slope is weaker.

Fig. 3.17 Induced voltages of PV array installed on triangular terrain at distance $R = 9.5$ m [6]

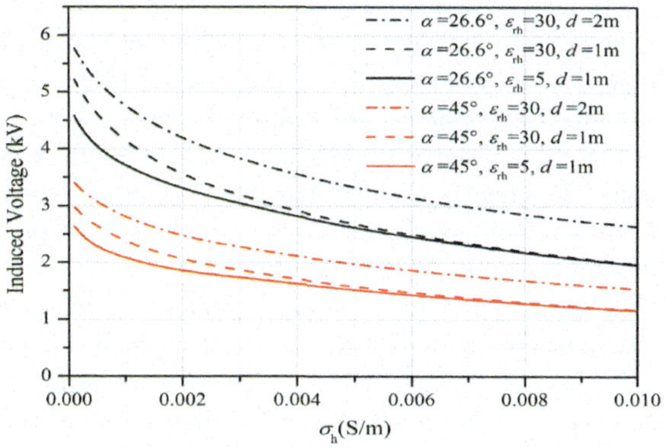

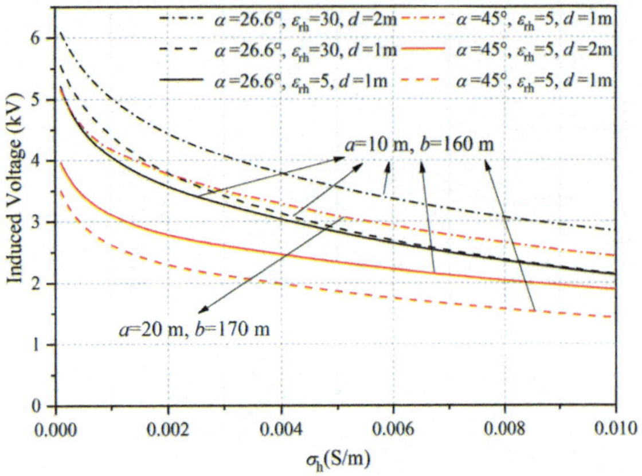

Fig. 3.18 Induced voltages of PV array installed on triangular terrain at distance $R = 10.0$ m [6]

In addition, the mountain with steeper slope means that the wave propagation path in mountain becomes longer at the same distance R, and more energy is dissipated.

The induced voltages for PV array installed on trapezoidal terrain are given in Fig. 3.18. It has the same trend with that in triangular terrain. A smooth slope, a large resistivity and a high permittivity would result in a greater induced voltage. Given the distance R between lightning channel and PV array, the induced voltage on trapezoidal terrain is greater than that on triangular terrain. The paths of electromagnetic waves can be divided into two routes, one is in air and the other is in the mountain.

The propagation paths in the mountain are different for diverse mountain terrains, and the energy consumption of waves propagating on triangular mountain is greater than that on trapezoidal mountain. Physically, the intensity of reflected waves caused on triangular terrain is greater than that on trapezoidal terrain. Some electromagnetic waves propagate through the upper base of the trapezoidal, which is similar to flat ground and less reflected waves are generated. Hence, the PV array on the trapezoidal terrain with wider upper base would have a greater voltage. If the distance R is sufficiently far, the voltages of these two different terrains are very close, and the topography influence becomes small.

As far as the slope is concerned, the hillside near the lightning strike has lower voltage when the slope is steeper, such trend is similar to that in triangular terrain. However, the trend reverses for the other hillside owing to the relatively close distance from lightning channel to PV array on steeper slope. The result is further confirmed in Fig. 3.20. Clearly, the magnetic field strength on V-shaped terrain as well as the induced voltage is stronger than that of other two terrains, since the average distance from the lightning channel to the area S on V-shaped terrain is shorter than that on triangular terrain, as shown in Fig. 3.21.

Regarding V-shaped terrain, the induced voltages are given in Fig. 3.19. The high mountain conductivity also tends to reduce the voltage. As the mountain permittivity increases, the voltage rises to some extent. Both the mountain conductivity and permittivity could change the reflected wave intensity at the interface. If the mountain conductivity is small, the effect of permittivity becomes obvious and even greater than that of mounting height. Table 3.8 shows that the property of hillside near the lightning strike is a major factor not only because of the close distance but also a larger occupied area.

As far as the slope is concerned, the hillside near the lightning strike has lower voltage when the slope is steeper, such trend is similar to that in triangular terrain. However, the trend reverses for the other hillside owing to the relative close distance from lightning channel to PV array on steeper slope. The result is further confirmed

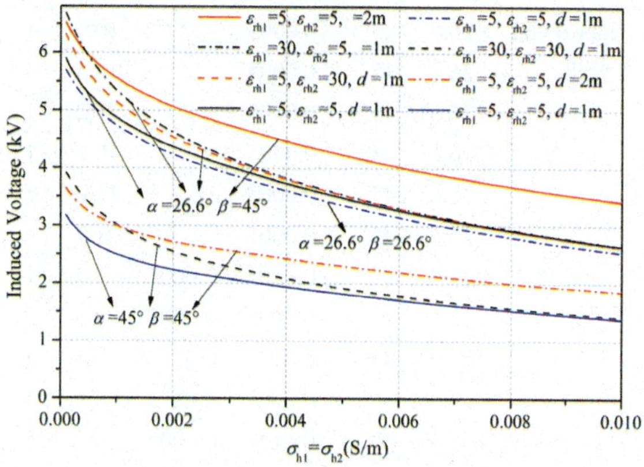

Fig. 3.19 Induced voltages of PV array installed on V-shaped terrain [6]

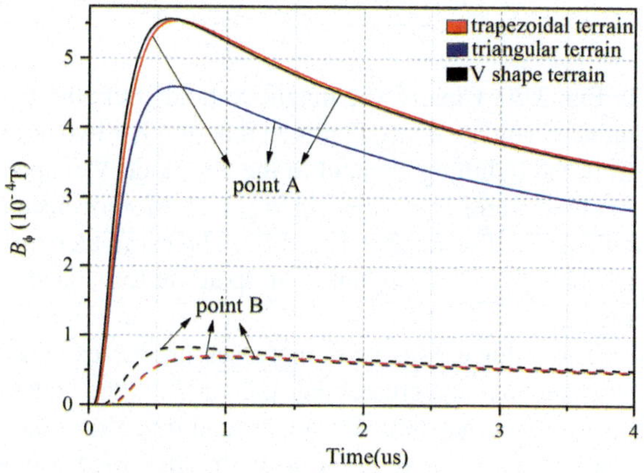

Fig. 3.20 Magnetic Field $\boldsymbol{B}_\phi$ over three typical mountain terrains [6]

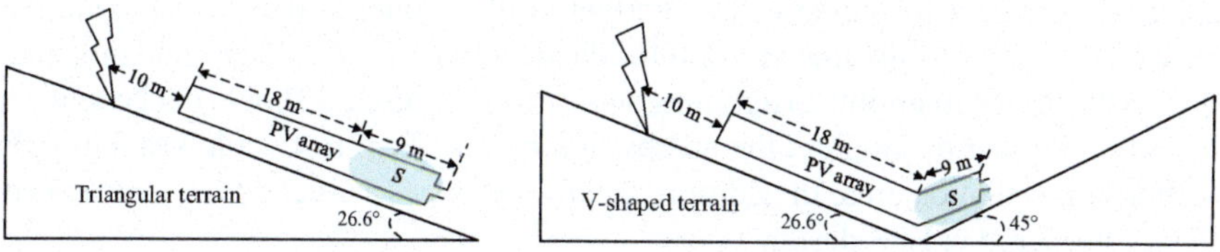

Fig. 3.21 Distances from lightning channel to PV array on different mountain terrains [6]

Table 3.8 Induced voltages of PV array installed on mountain [6]

Mountainous terrains	α	β	σ_{h1} (S/m)	σ_{h2} (S/m)	R (m)	d (m)	a (m)	b (m)	V_{OCmax} (kV)
Triangular terrain	26.6°	45°	0.0005	–	9.5	1	0	150	4.19
	26.6°	45°	0.0005	–	9.5	2	0	150	4.80
	26.6°	45°	0.0005	–	13.9	1	0	150	2.90
	26.6°	45°	0.0005	–	20.6	1	0	150	1.92
	26.6°	63.4°	0.0005	–	9.5	1	0	150	4.20
Trapezoidal terrain	26.6°	45°	0.0005	–	10.0	1	10	160	4.58
	26.6°	45°	0.0005	–	14.5	1	10	160	3.07
	26.6°	45°	0.0005	–	21.2	1	10	160	1.98
	26.6°	45°	0.0005	–	10.0	1	20	170	5.57
	26.6°	45°	0.0005	–	15.0	1	20	170	3.32
	26.6°	45°	0.0005	–	21.7	1	20	170	2.09
V-shaped terrain	26.6°	45°	0.0005	0.0005	9.5	1	–	–	5.59
	26.6°	45°	0.0005	0.0005	9.5	2	–	–	6.40
	26.6°	45°	0.01	0.01	9.5	1	–	–	2.65
	26.6°	45°	0.0005	0.01	9.5	1	–	–	4.81
	26.6°	45°	0.01	0.0005	9.5	1	–	–	3.09
	26.6°	63.4°	0.0005	0.0005	9.5	1	–	–	6.26
	45°	45°	0.0005	0.0005	9.2	1	–	–	2.94

in Fig. 3.20. Clearly, the magnetic field strength on V-shaped terrain as well as the induced voltage is stronger than that of other two terrains, since the average distance from the lightning channel to the area S on V-shaped terrain is shorter than that on triangular terrain, as shown in Fig. 3.21. Moreover, considering PV arrays installed on hillside, as illustrated in Fig. 3.22. The parameters of mountain terrain are presented in Table 3.9. The influence of span angle α on the induced voltage is plotted in Fig. 3.23.

The voltage rises as α increases until it exceeds 160°. For a PV array with a constant area, the greater the span angle α, the wider in horizontal direction. On the contrary, the narrower in the vertical direction, the shorter the equivalent distance from lightning strike point to the internal loop. Consequently, the greater the span angle would result in the higher the induced voltage. However, the PV array forms a curved surface as α increases to a certain level, leading to part of magnetic flux cancellation and it is the main reason for the decrease of the voltage. The mountain peak is the most susceptible to lightning strike, the span angle α should not be selected too large in order to suppress the nearby lightning strike effect while providing the sufficient installation area. In addition, the distance d, as a crucial factor, also needs to be considered in LPS design.

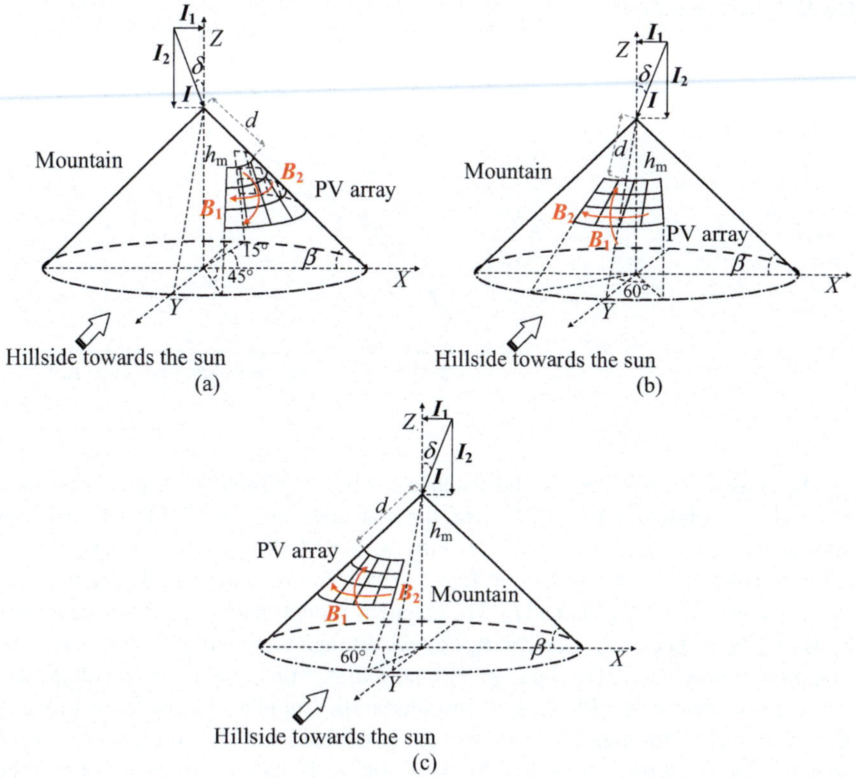

Fig. 3.22 Three installation modes of PV array. **a** Installed along hill ridge; **b** Installed in the middle of hillside; **c** Installed against hillside [1]

Table 3.9 Parameters of PV system on mountain terrain

Quantity	Symbol	Value
Hillside angle	β	30°
PV array area	S_{pv}	1080 m^2
Span angle for PV array	α	0–200°
Angle between lightning channel and mountain peak	δ	30°
Altitude of hill peak	h_m	200 m

There are many options for PV array arrangement. The geographic location of PV array installed on hillside, as depicted in Fig. 3.22, can be categorized into three modes: (i) Mode a, installed along hill ridge; (ii) Mode b, installed in the middle of hillside; (iii) Mode c, installed against hillside. As far as Mode a concerned, the span angle towards the sun is set to be 45° and the span angle on the other side is 15°.

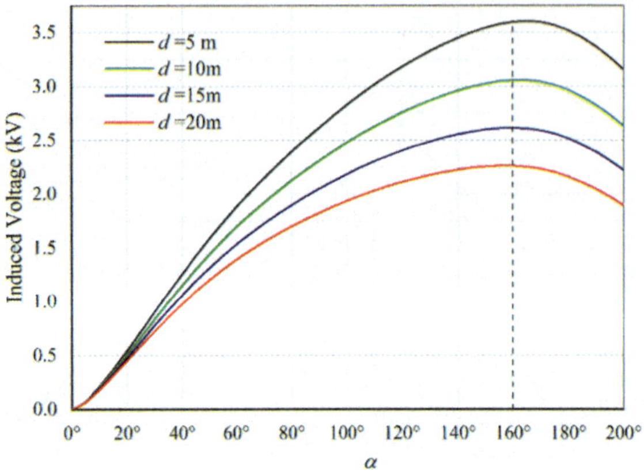

Fig. 3.23 Induced voltage for PV system installed on mountain [1]

The induced voltages versus the distance d is listed in Table 3.10, and the impacts of δ and β are plotted in Fig. 3.24. Here, the span angle α is 60°. Table 3.10 suggests that the induced voltage of Mode b is the largest, then Mode c, and the lowest for Mode a. In principle, the lightning current $\boldsymbol{I}$, as a vector, can be orthogonally decomposed into two components, i.e., $\boldsymbol{I}_1$ and $\boldsymbol{I}_2$. The magnetic inductions $\boldsymbol{B}_1$ and $\boldsymbol{B}_2$ are generated by $\boldsymbol{I}_1$ and $\boldsymbol{I}_2$, respectively. Clearly, $\boldsymbol{B}_1$ crosses through the internal loop, whereas $\boldsymbol{B}_2$ is approximately parallel to the loop. $\boldsymbol{I}_2$ contributes little to the induced voltage and it can be omitted. Since PV array is installed in the middle of hillside for Mode b, $\boldsymbol{B}_1$ almost passes through PV array surface in vertical, and the magnetic flux is the greatest amongst them. Regarding Mode a, part of $\boldsymbol{B}_1$ crosses from the side of the ridge to the other, and the magnetic flux is partially offset. Consequently, it has a relatively small value.

In addition, the hillside angle β and the angle between lightning channel and mountain peak δ also strongly impact the induced voltage. As β increases, the voltage decays drastically. For example, when β is less than 20°, the voltage hits 8 kV; if β exceeds 70°, the threat of lighting strike becomes quite small. As β is close to 90°, the curved surface of PV array is perpendicular to $\boldsymbol{I}_1$, and the magnetic field $\boldsymbol{B}_1$ generated by $\boldsymbol{I}_1$ will not cross the surface, as elucidated in Fig. 3.22. Hence, the induced voltage would be significantly minimized. On the contrary, the induced voltage is directly proportional to δ. As δ approaches to 0, $\boldsymbol{I}_2$ is almost equal to $\boldsymbol{I}$, whereas $\boldsymbol{I}_1$ tends to 0. Physically, the magnetic flux flow through PV loop declines.

Table 3.10 Induced voltages for three modes on mountain terrain [1]

d (m)	Mode a (kV)	Mode b (kV)	Mode c (kV)
5	1.04	2.80	1.88
10	0.94	2.53	1.70
15	0.85	2.30	1.53
20	0.77	2.08	1.40
25	0.70	1.90	1.27

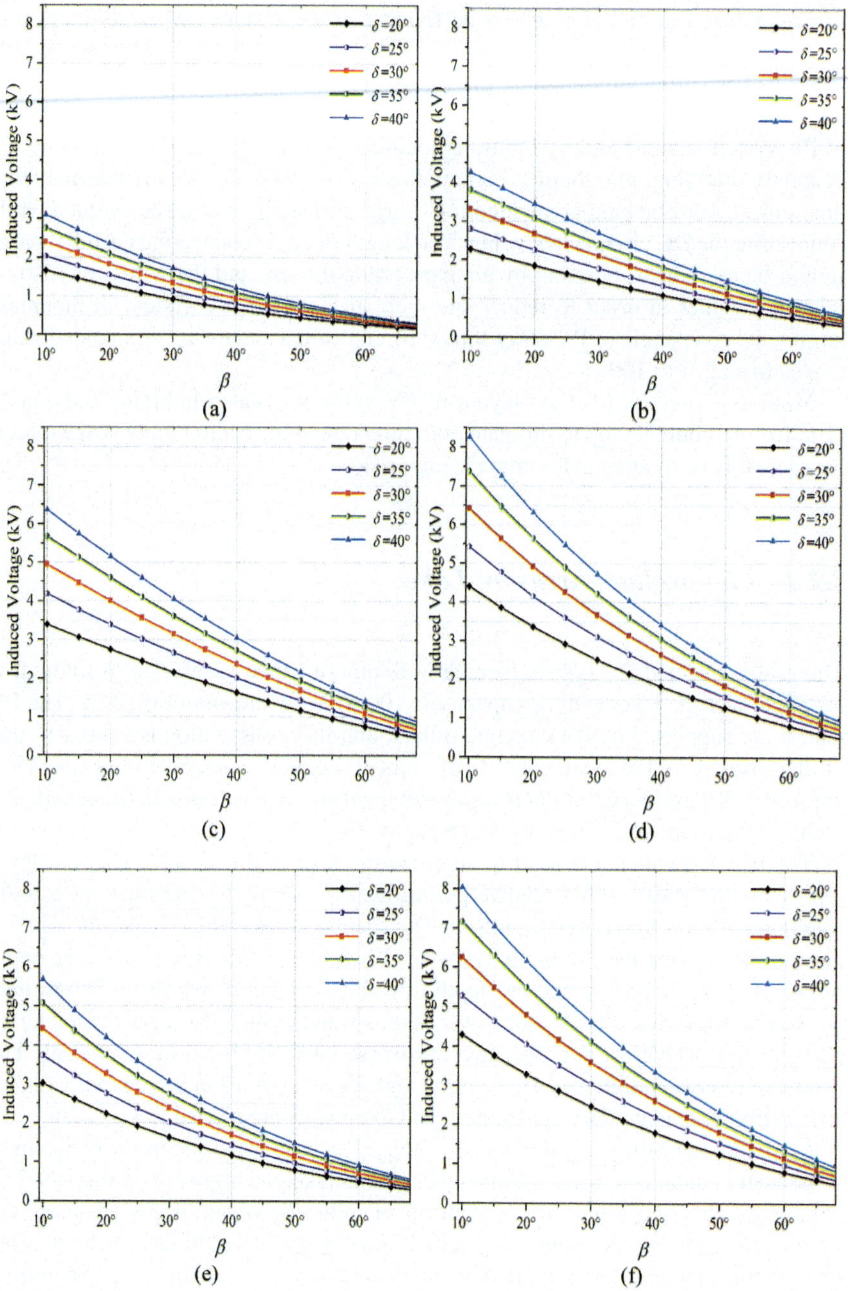

Fig. 3.24 Relationship between induced voltage and the hillside angle β. **a** Mode a, $\alpha = 80°$; **b** Mode a, $\alpha = 160°$; **c** Mode b, $\alpha = 80°$; **d** Mode b, $\alpha = 160°$; **e** Mode c, $\alpha = 80°$; **f** Mode c, $\alpha = 160°$ [1]

Comparing Fig. 3.24d, f, it can be found that the result of Mode b is close to that of Mode c when $\alpha = 160°$. The greater the span angle α, the wider area for PV array. If α is sufficiently enough, the layouts for Mode b and c are approximately the same, and hence the induced voltages are very close. The protection measures of PV system against nearby lightning includes grounding, equipotential bonding, magnetic shielding, line routing, and application of SPDs. Beside the internal loop, line routing must be paid great attention. Large surface loop must be avoided while connecting the DC circuits to a string. The length of cable connecting each PV panel should be as short as possible by an appropriate design, and the wires can also be twisted together in order to reduce the loop surface. For PV system in mountain terrain, the span angle of PV array should be controlled and the recommended value ranges from 20° to 100°.

Mode c is preferred for the layout of PV system; isolated lightning rod can be installed on mountain top to mitigate both direct and indirect lightning strike effects for special regions at an extraordinary high risk.

3.2.4 Photovoltaic Array on Lake

The grid-connected PV system located in Southern China is selected to study. It is installed over the lake, and the capacity is 100 MW, as shown in Fig. 3.25. The PV panels are supported by the concrete pillars, and its configuration is similar to that on flat ground. A PV array consists of eight PV panels connected in series. They are further linked to a common inverter and a wiring loop is formed. The length and width of the loop are 27 m and 1 m, respectively.

The two-layer media is used to simulate the lake in this article. The first layer represents the water with a relative permittivity ε_{r1} of 80–81 and the conductivity σ_1 ranges from 5.6 S/m to 0.05 mS/m. The second layer is the ground; the relative permittivity ε_{r2} and the conductivity σ_2 are 10 and 5 mS/m, respectively. The depth of lake d_1 is set to be 4 m and the depth of soil d_2 is 1 m. In contrast to flat ground, the medium for PV array installed over lake is water, which has a good conducting performance. The electric properties for diverse lakes are listed in Table 3.11. The distance R between lightning strike point and PV array is 10 m. Five kinds of lakes with different conductivity are studied, and the results are plotted in Fig. 3.26.

Similar to land, the peak of induced voltage rises as the lake conductivity declines. If the water contains fewer impurities like glacial lake, the higher the purity, the less conductive the lake water. The waveform of induced voltage becomes sharp and narrow. Simultaneously, a large negative voltage is produced. Recently, the floating PV system is increasingly popular which allows PV panels to be installed on unused areas of water and enable more daylight to pass through the structure. The installation height is approximately zero, and the induced voltage is analyzed in Fig. 3.26a. The voltage is generally lower than that of common PV array on lake. The induced voltage for floating PV array installed on saltwater is extremely small and nearly negligible.

(a)

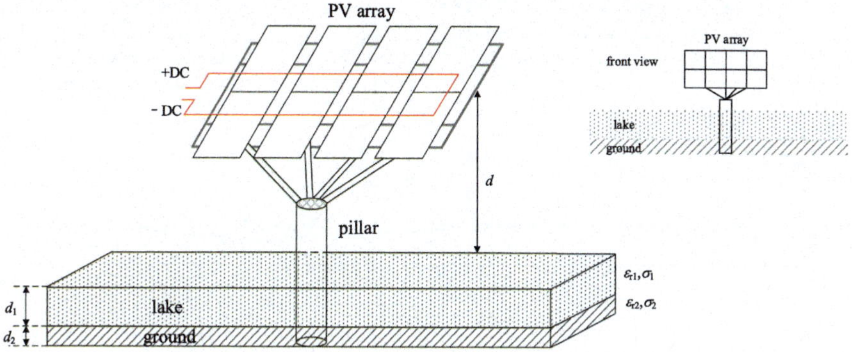

(b)

Fig. 3.25 PV array installed on lake, **a** Photograph of PV array on lake. **b** Schematic of PV array structure

Table 3.11 Conductivity and permittivity of lake [6]

Lake type	Relative permittivity	Conductivity (μS/m)
Glacial lakes	80.1–81	50–300
Fluvial lakes	80.1–81	5000–150,000
Fisheries lakes	80.1–81	15,000–50,000
Saltwater lake	80.1–81	5,600,000

Principally, the resistivity of saltwater lake is quite small, the lake surface can be approximately treated as an infinitely conducting plane. According to Eqs. (3.3) and (3.4), the electric field $\boldsymbol{E}_\mathbf{z}$ and $\boldsymbol{E}_\mathbf{r}$ in the lake surface are quite small due to the high conductivity of lake. The LEMF inside the saltwater can be neglected. The induced voltage is small when the PV array is close to its surface.

The waves are reflected not only on the boundary between lake and air, but also on that between lake and soil. According to the reflection law of electromagnetic wave,

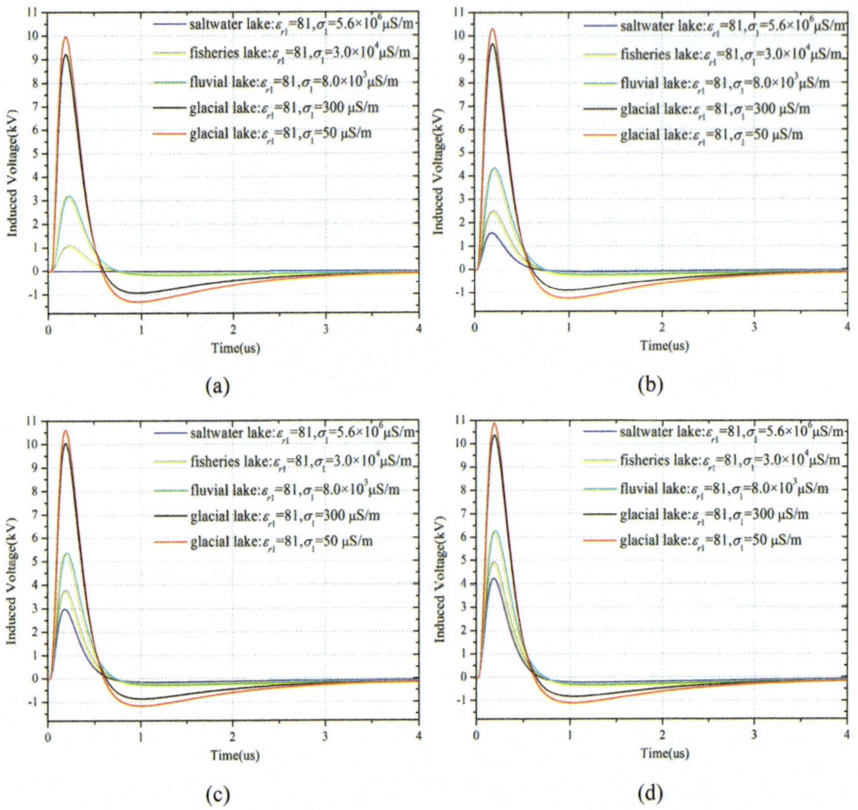

Fig. 3.26 Induced voltages of PV array installed on lake. **a** $d = 0$. **b** $d = 1$ m. **c** $d = 2$ m. **d** $d = 3$ m [6]

the higher the conductivity of lake, the greater the reflection coefficient and a lower voltage would be yielded. For the lake with a small conductivity, the reflected wave from the interface between lake and soil is strong. Since the wave speed in lake is slow, the reflected wave energy return to air is still sufficient. However, the incident magnetic strength has already attenuated, giving rise to the reverse impulse voltage.

Additionally, the lake water composition could change with the surrounding environment. Table 3.12 lists the seasonal variation of fluvial lake conductivity. The lightning induced voltage versus season is given in Fig. 3.27. Given the lightning and PV system parameter, the seasonal variation is evident. The voltages rise in winter and reach peak in January whereas it drops in summer and has the lowest value in June. For example, $d_1 = 3$ m, the induced voltage ranges from 3.82 to 5.22 kV during one year. If the lake is shallow, the voltage is quite high. The difference can reach 1.3 kV due to the lake depth.

Table 3.12 Seasonal variation in lake conductivity (nS/m) [6]

d_1 (m)	Jan	Feb	Mar	Apr	May	Jun	Jul	Aug	Sep	Oct	Nov	Dec
0.5	3.2	5.3	5.6	6.0	6.7	8.7	7.1	7.7	6.8	5.2	4.0	4.4
1.5	3.5	5.6	5.9	6.3	7.0	9.0	7.4	8.0	7.1	5.5	4.3	4.7
2.5	5.2	7.3	7.6	8.0	8.7	10.7	9.1	9.7	8.8	7.2	6.0	6.4
3.5	6.0	8.1	8.4	8.8	9.5	11.5	9.9	10.5	9.6	8.0	6.8	7.2

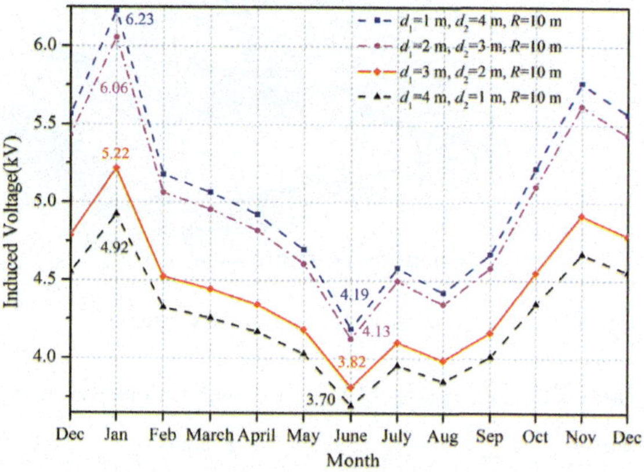

Fig. 3.27 Seasonal variation in induced voltages at different lake depths [6]

3.3 Induced Voltage of Rooftop Photovoltaic Array

3.3.1 *Structure of Rooftop PV Array*

There are two kinds of PV systems, one is the concentrated PV system and the other is the distributed PV system. Rooftop PV systems are distributed electricity generation options, which can be used to meet the energy consumption of building or provide electricity into power grid in certain situations. As illustrated in Fig. 3.28, a common rooftop PV system consists of PV modules, DC cables, PV inverter and mounting structure, as well as battery.

DC/AC inverters are often used in rooftop PV system. Three typical inverters are listed in Table 3.13. Depending on the maximum DC voltage of the inverter, PV modules have three common connections, as shown in Fig. 3.29. The maximum voltage rises as the number of PV modules connected in series increases.

The inverters are vulnerable to lightning surges. Once the voltage at the output of DC cable exceeds the maximum allowable voltage for the inverter, the capacitor at DC side would be broken-down and the PV inverter would fail. According to IEC 62,894 [7], the maximum allowable DC voltage for PV inverter of rooftop PV system is 1 kV.

The mounting system for rooftop PV arrays is often made of aluminum alloy or low-carbon steel. The legs and crossbars of the mounting system are generally

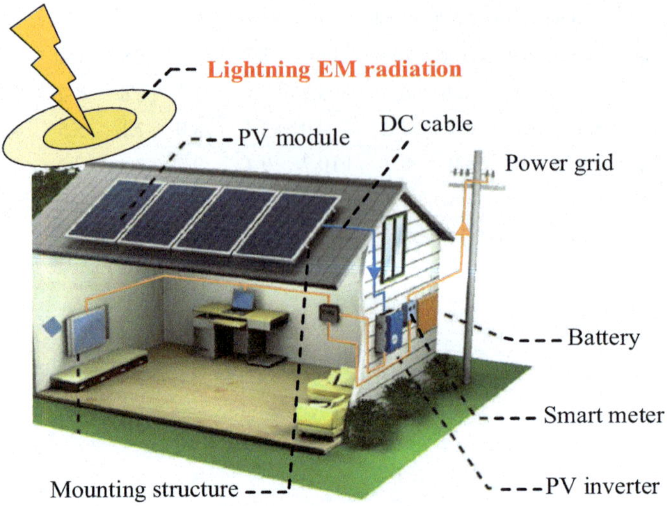

Fig. 3.28 Schematic diagram of rooftop PV system

Table 3.13 Parameters of PV inverter with various powers [8]

Power	275 kW	550 kW	1160 kW
Maximum input voltage	880 V DC	880 V DC	880 V DC
Range of MPPT input voltage	480–820 V	480–820 V	450–820 V
Peak efficiency	97.5%	98.5%	98.5%
Rated AC output power	250 kW	500 kW	1000 kW
Rated AC output current	534 A	1070 A	2138 A
Rated AC output voltage	270 V AC	270 V AC	270 V AC

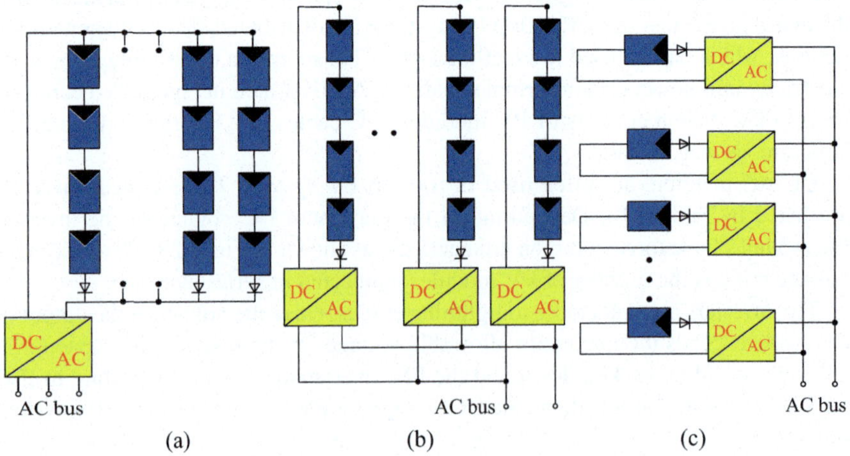

Fig. 3.29 Connection for PV inverter. **a** Type I. **b** Type II. **c** Type III [8]

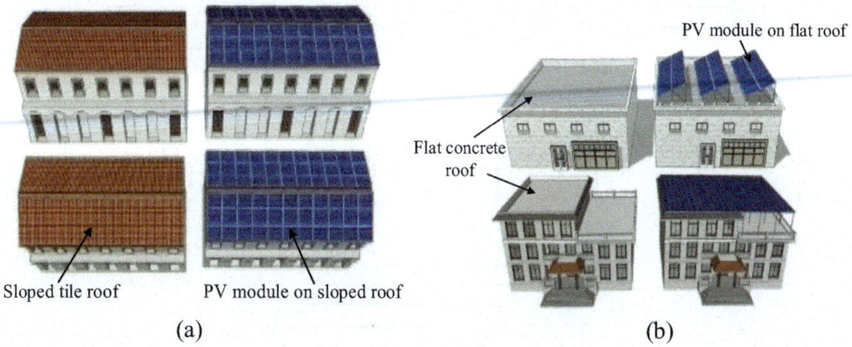

Fig. 3.30 Rooftop PV array. **a** Tile rooftop PV array. **b** Concrete rooftop PV array [8]

connected to the roof to install the PV modules. Therefore, the structure of the mounting system has a great impact on lightning-induced voltage. The typical rooftop PV arrays are shown in Fig. 3.30. They can be divided into two categories according to the roof shape. One is a sloped roof, such as the tile roof, and the other is a flat roof like the concrete roof. The mounting systems for PV modules are dependent on the roof, and the tile roof has no support leg. However, the mounting system for the PV array on flat roof has a number of legs, which is the same as that on flat ground.

The sloped roof structures, with different crossbars and legs, are presented in Fig. 3.31. The roof in Fig. 3.31a has 9 crossbars and 3 legs; the roof of Fig. 3.31b has 17 crossbars and 3 legs, and the one of Fig. 3.31c has 9 crossbars and 5 legs. In terms for the roof material, two common materials are considered, namely, tile roof and metal roof in Fig. 3.32. The roofs of houses are generally made of tile whereas the roofs of industrial buildings are almost made of metal. The flat rooftop has the two metal structures, as illustrated in Fig. 3.33. Since the rooftops have different structures, the distributions of induced current on the rooftops produced by the lightning EM field are not consistent, and hence the EM field near the rooftop PV array.

3.3.2 Modeling for Rooftop PV Array

The conductors of the rooftop PV system, such as roof structure and the mounting system, PV modules, cable, etc. are modeled based on their geometry information. The structural steels have a C-shaped cross section, as shown in Fig. 3.34. w_c, t_c and l_c represent the width, thickness and length of C-shaped steel. Their detail values are given in Table 3.14. To facilitate the analysis, the C-shaped cross section is represented by the cylindrical conductor. The equivalent radius r_c is defined by,

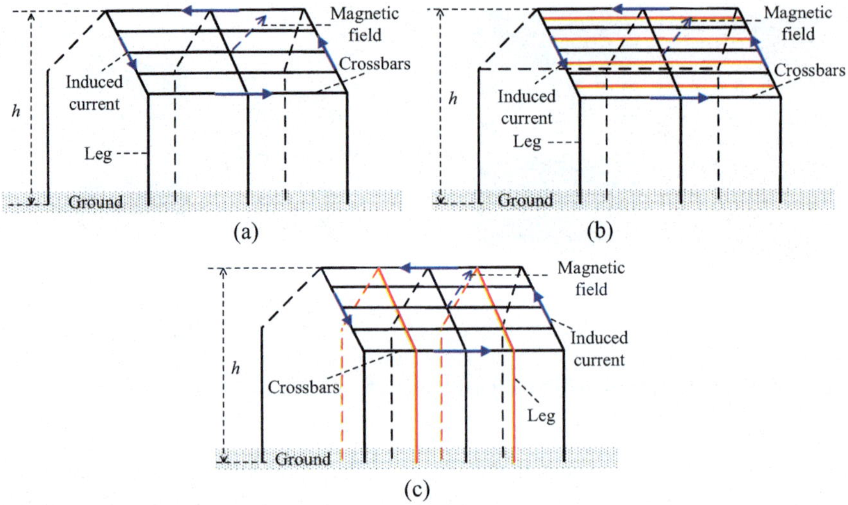

Fig. 3.31 Sloped rooftop PV array. **a** Case a. **b** Case b. **c** Case c [8]

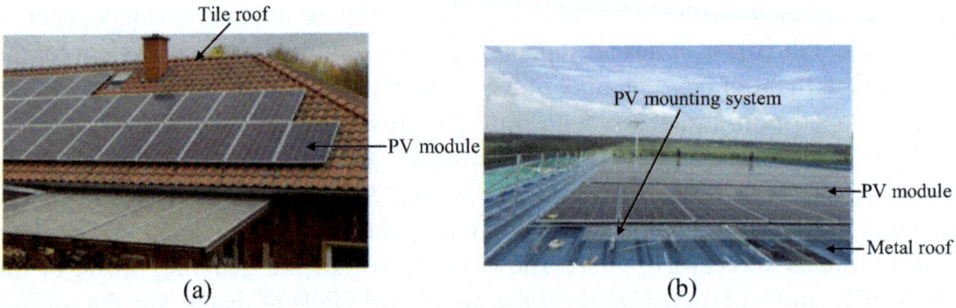

Fig. 3.32 PV array installed on rooftop with different materials. **a** Tile roof. **b** Metal roof [8]

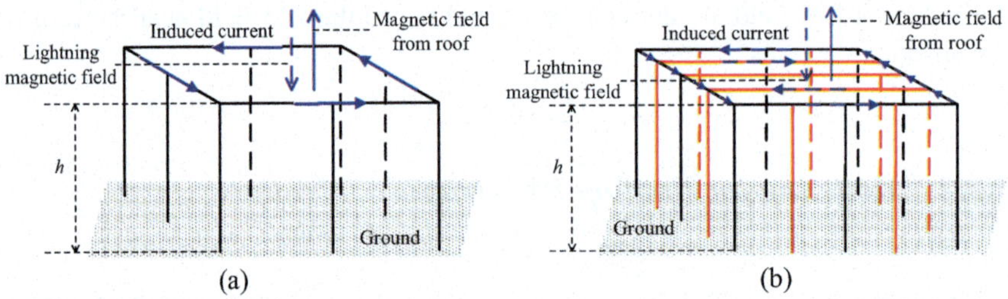

Fig. 3.33 Flat rooftop. **a** Case I. **b** Case II [8]

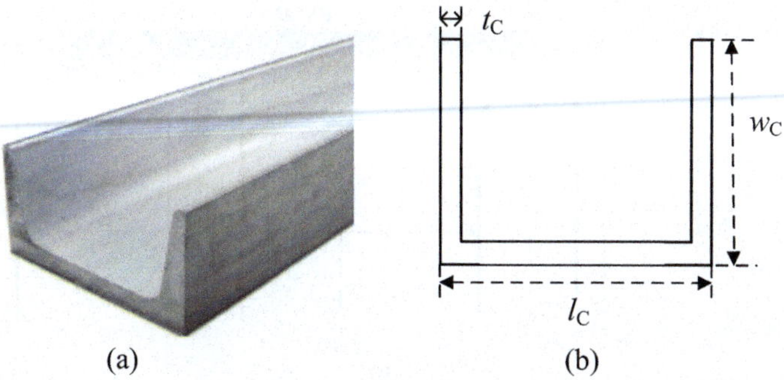

Fig. 3.34 C-shaped frame. **a** Photograph of metal frame. **b** Cross-sectional view

Table 3.14 Parameters of C-shaped structure

Quantity	t_c (mm)	w_c (mm)	l_c (mm)
PV Frame	3	30	30
PV mounting system	5	80	40

$$r_c = \sqrt{\frac{t_c(2w_c + l_c)}{\pi}} \tag{3.9}$$

Similarly, the roof structure is also equivalent to the cylindrical conductor with 5 cm in radius. The detailed sizes of sloped roof and flat roof are given in Fig. 3.35. The rooftop PV array consists of two rows of PV modules connected in series and has 26 PV modules. The end for the two rows of PV-module string are connected with an inverter, and the output voltage for the PV array is 491 V. The mounting system for PV modules is made of aluminum alloy and consists of six crossbars. The length and width of the mounting system for sloped rooftop and flat rooftop are the same, which are 10.4 and 2.55 m, respectively. The length and width of PV module are 1.32 and 0.57 m, respectively.

The details of PV module are presented in Fig. 3.36. The solar cells inside the PV module are connected in series by means of straight wires, which are regarded as an equivalent loop shown in Fig. 3.36. The DC cable and the loops of all PV modules are taken as the cylindrical conductor with 0.4 cm in radius.

The distance for the lightning directly strike to the PV array can be evaluated according to the rolling sphere method,

$$R_x = \sqrt{h_1(2h_r - h_1)} - \sqrt{h_x(2h_r - h_x)} \tag{3.10}$$

where h_r is the rolling sphere radius, and it is set to be 45 m; h_1 is the height of the rooftop PV array and h_x is set to be zero to represent the ground.

Consequently, the minimum distance for the indirect lightning strike to the rooftop PV array is approximately 26 m. Once the distance of lightning channel from the PV array is shorter than 26 m, the direct lightning may strike to the PV array. Taking

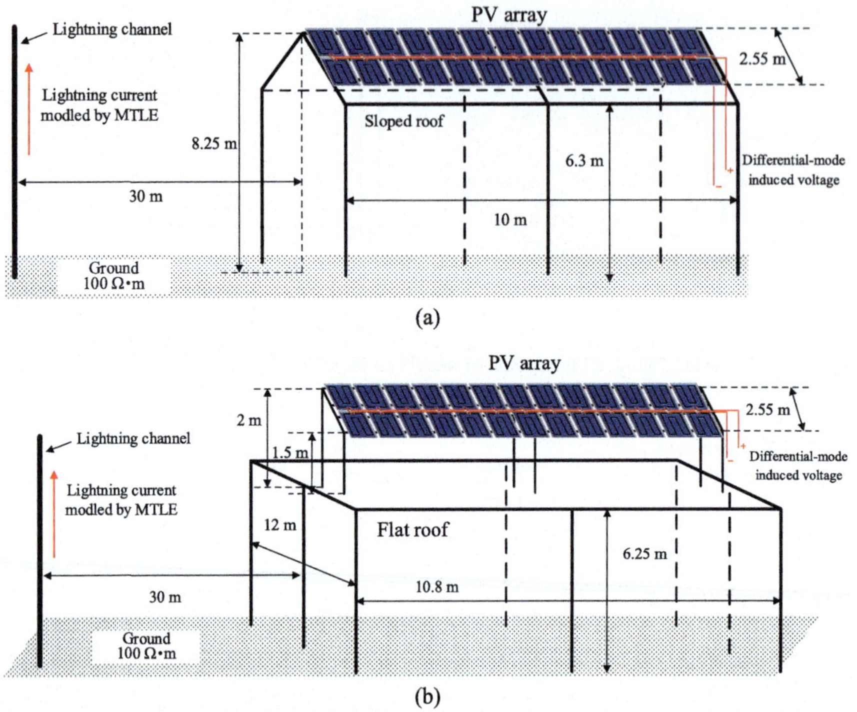

Fig. 3.35 Method of moments (MoM)-based model of rooftop PV array. **a** Sloped rooftop PV array. **b** Flat rooftop PV array [8]

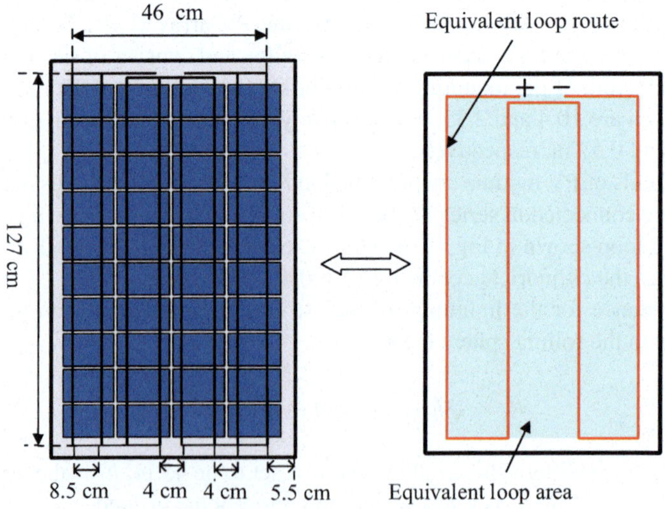

Fig. 3.36 Modeling for PV panel

into account the real-world scenario, the lightning strike at the ground on the left of PV array, and the distance is set to be 30 m. The ground is represented by the soil with 100 Ωm resistivity in this paper. The lightning-induced voltage of PV array on the lossy ground is evaluated by the accurate numerical solutions of Sommerfeld integrals.

Lightning EM transient is generally solved using the full wave models. The MoM is an efficient method especially for studying the EM transient of thin conductor structures, and the accuracy of method of moments (MoM) for calculating the induced voltage is relatively high, which can be used for modeling rooftop PV arrays. The Maxwell equations are solved by the MoM in the frequency domain. The metal surfaces with various shapes are divided into grids. Then, the equations of the electric field on the grids are calculated by the MoM to obtain the surface current and the scattered field produced by metal conductors.

Considering the EM couple among the PV array, PV mounting system and roof structure, the lightning transient is calculated by using the MoM-based model built in the software: FEKO [9]. The computation is carried out over the frequency range from 2.93 kHz to 1.5 MHz with the increment step of 2.93 kHz, and the number of the frequencies is 512. The time interval from 0 to 341 μs with the increment step of 1/3 μs, and 1024 points in IFFT are thus used to obtain the time-domain response. The mounting system, PV modules and DC cable are modeled as the combination of cylindrical segments, which satisfy the limitation in the ratio of conductor length to wavelength in the time range of interest.

Most of lightning is downward negatively charged, and the upward lightning is barely initiated except for tall objects over 100 m. Consequently, the first negative lightning stroke is considered, and the lightning waveform is 50 kA, 1/200 μs. In the MoM-based model, the lightning channel is modeled by the vertical conductor represented by cylindrical elements having distributed series resistance and inductance to realize the propagation speed of return-stroke current on it. The length of the lightning channel is set to be 1 km.

3.3.3 Lightning Electromagnetic Coupling of Roof

Since DC cables are connected to vulnerable electronics like inverter, the induced voltage between ± DC cables is one of the most prominent problems reported in the indirect lightning accidents. The differential-mode (DM) induced voltages for the PV array with and without roof shown in Fig. 3.35a are calculated, and the results are plotted in Fig. 3.37.

The results indicate that the induced voltage can reach up to 2–3 kV while neglecting the roof steels. It is larger than the maximum voltage of PV inverter (1 kV DC, IEC 62894) [10]. Considering surrounding conductors, such as the roof and PV mounting structure, the induced voltage is significantly reduced. The induced current on the roof steels generated by lightning EM field will generate an additional field,

Fig. 3.37 Lightning-induced voltage of sloped rooftop PV array [8]

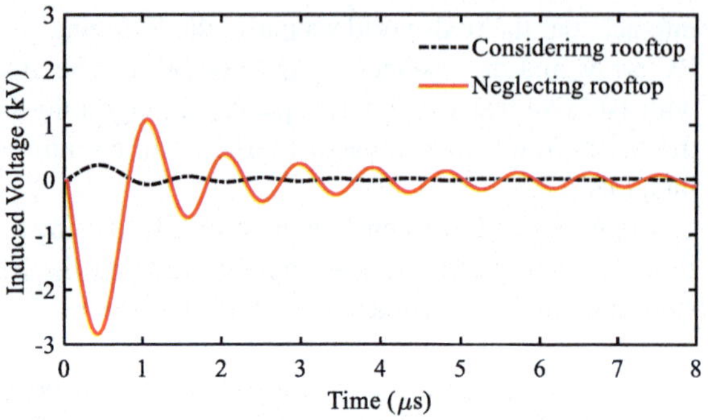

which weakens the lightning EM field. The voltage peak is thus not only reduced, but also the voltage polarity is opposite. In terms of the rooftop PV array, its loop area is relatively smaller and the distance away from the lightning channel is greater. Therefore, the nearby lightning EM field is relatively smaller whereas the additional EM field is larger. The lightning EM field can be fully compensated by the additional EM field. Since the direction of additional EM field is opposite to the direction of the lightning EM field, the polarity of the induced voltage on the roof is reversed, as observed in Fig. 3.37.

The induced voltage of rooftop PV array with three rooftop structures shown in Fig. 3.31 (Case a, Case b, Case c) are evaluated, as plotted in Fig. 3.38. The induced voltage of Case a is the smallest whereas the voltage peak of Case b is the greatest. The induced voltage of rooftop PV array rises as the number of crossbars increases. The voltage peak rises from 0.26 to 0.49 kV when the number of crossbars changes from 9 to 17. It can be concluded that the greater the number of legs for the rooftop structure, the higher the voltage peak.

The magnetic field generated by the induced current on the roof is greater than the lightning magnetic field due to the far distance from the lightning channel to the PV array. The lightning magnetic field is fully cancelled, and the magnetic field near the PV array is dependent on the induced current of rooftop. The more metal structures of the roof, the greater the induced current and hence the magnetic field

Fig. 3.38 Influence of roof structure on lightning-induced voltage [8]

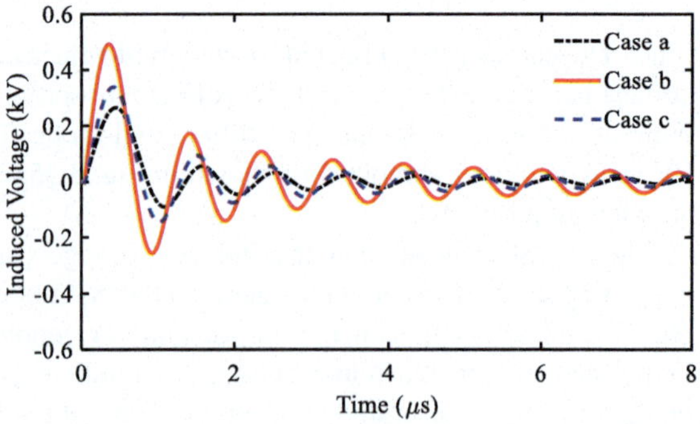

around the PV loop. The rooftop in Case a has fewer conductors, and its peak voltage is thus the smallest. Since the distance between the crossbars and PV array is shorter, the EM coupling among the crossbars and the PV loop is more remarkable than that of the legs. The induced voltage in Case b is thus larger than that of Case c. The lightning-induced DM voltage of rooftop PV array is almost determined by the induced current on the roof. In order to further study the effect of rooftop structure, the induced current on the three rooftop structures is computed, as follows:

Case a

The distribution of the induced current in Case a is presented in Fig. 3.39. It is clear that the current on the crossbar is larger than that on the leg. The impedance of the loop formed by legs and the ground is quite great due to the lossy soil with the resistivity of 100 Ωm, and the induced current on the loop is smaller. Moreover, the current of outside crossbar has higher peaks compared with that of inside crossbars. Since the outside crossbars consist of a larger loop, the more the lightning EM field passing through the loop and the greater the current induced on the outside crossbars, according to the Faraday's law.

Figure 3.40 indicates that the waveforms of the current on the crossbars and the legs are quite different. The waveform of the current on the crossbar is similar to the lightning-current waveform. It demonstrates that the current is determined by the magnetic flux passing through the crossbar loop. However, the waveform of the current on the leg is close to that of the induced voltage. The soil has a high resistivity and the loop of legs can be almost regarded as an open circuit. Therefore, the waveforms of the voltage and current are close to each other for the open circuit. The outside crossbars have the greatest influence on the induced voltage among the roof steels of the roof, and the PV modules installed on the outside crossbars are more vulnerable to the indirect lightning.

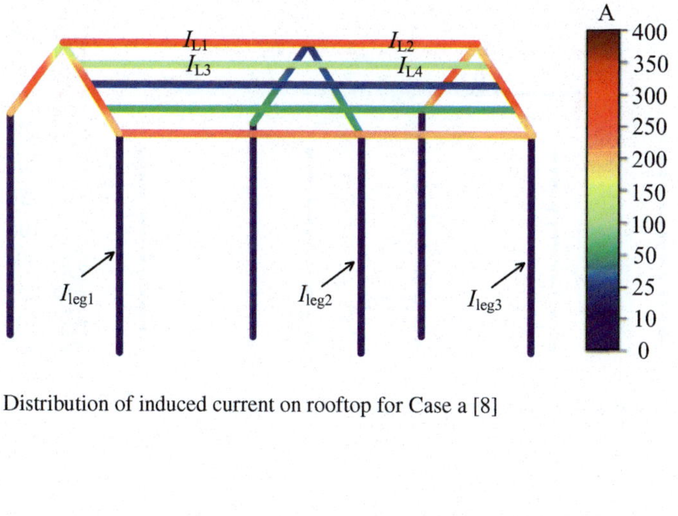

Fig. 3.39 Distribution of induced current on rooftop for Case a [8]

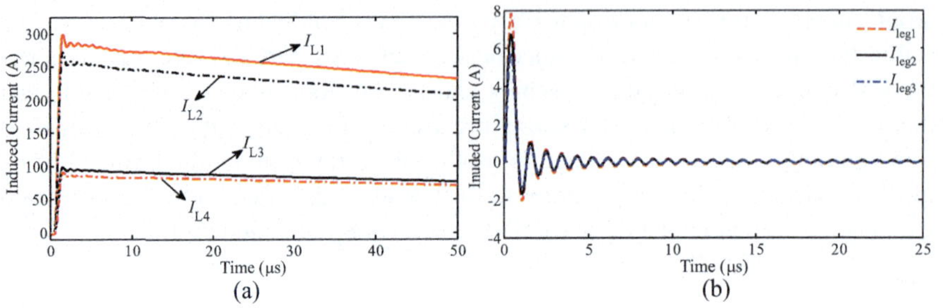

Fig. 3.40 Waveform of induced current on rooftop for Case a. **a** Induced current on crossbar. **b** Induced current on leg [8]

Case b

The induced current is presented in Figs. 3.41 and 3.42. The maximum current still appears on the crossbar whereas legs have less current. Similarly, the closer to the outside of the crossbar, the greater the induced current is. Since the lightning channel is at the left side of the roof, the larger current is induced on the crossbar and leg at the left side. In addition, the current on the leg of Case b is larger than that in Case a.

The current is mainly concentrated on the crossbars L_1 and L_2 in Case a, and the current in Case b is distributed in the conductors L_1 and L_3, L_2 and L_4 due to the shunting of multi-conductors. Since the number of crossbars is greater, the total current of L_1 and L_3 in Case b is slightly larger than that of L_1 in Case a. However, the current on the inside crossbars is significantly reduced in Case b. For instance, the current on L_5 and L_7, L_6 and L_8 is smaller than that on L_3 and L_4 of Case a. The result indicates that the PV modules are suggested to be installed on the middle for the roof with many crossbars in this case.

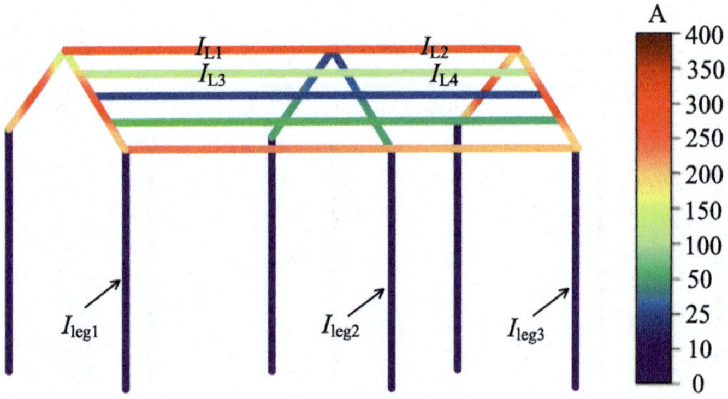

Fig. 3.41 Distribution of induced current on rooftop for Case b [8]

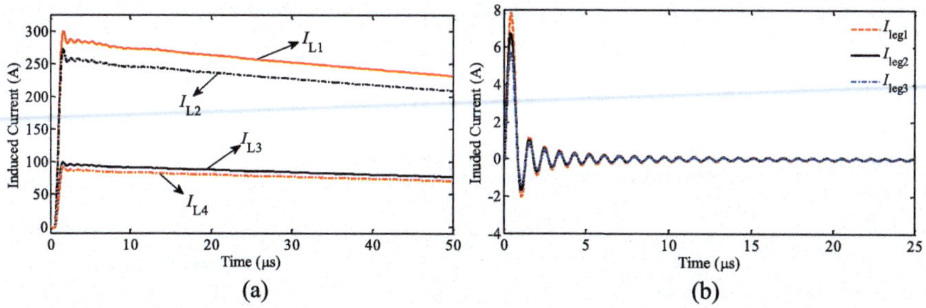

Fig. 3.42 Waveform of induced current on rooftop for Case b. **a** Induced current on crossbar. **b** Induced current on leg [8]

Case c

As shown in Figs. 3.43 and 3.44, the currents on L_1, L_2, L_3 and L_4 have relatively greater values whereas that on the middle crossbar is the smallest. The induced current on the crossbar is easily influenced by the leg of rooftop. Compared to the three legs, the main flowing route of the current on the rooftop is changed from the left side to the middle, which is the loop of i.e., L_2, L_3, L_9 and L_{10}. Consequently, in terms of the roof with many legs, the PV modules are recommended to be installed on both sides of the rooftop as possible in this case. Though the number of legs increases, the current flowing on the legs is close to that of Case a. The total current on the roof in Case c is still greater than that in Case a.

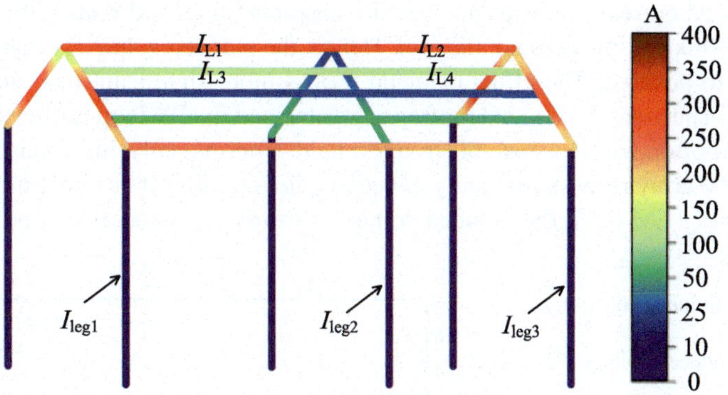

Fig. 3.43 Distribution of induced current on rooftop for Case c [8]

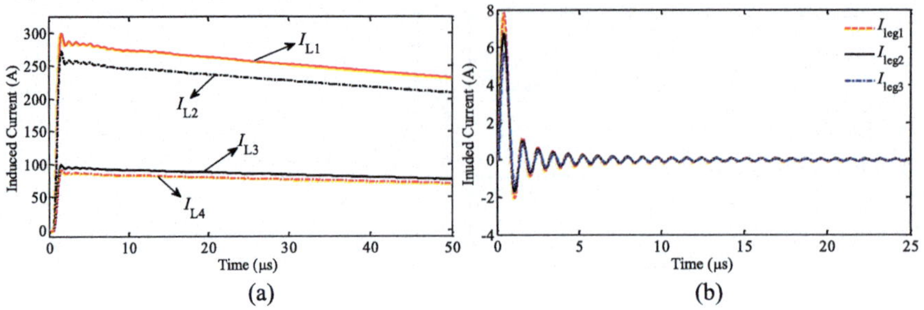

Fig. 3.44 Waveform of induced current on rooftop for Case c. **a** Induced current on crossbar. **b** Induced current on leg [8]

3.3.4 Influence of Rooftop Material, Height and Shape

The roof material has also a great impact on the lightning-induced current of the rooftop, and the induced voltage of PV array is inevitably affected. Considering the two roof materials, including the tile rooftop and the metal rooftop shown in Fig. 3.32, the induced voltage of the rooftop PV array shown in Fig. 3.35a are plotted in Fig. 3.45.

It is clear that the induced voltage on the metal rooftop is larger than that on the tile rooftop. The lightning-induced current on the metal rooftop is a key factor. It increases the induced voltage for positive polarity. However, the lightning EM field generates a negative voltage between the cables of PV array, as shown in Fig. 3.37. The induced current produces an opposite magnetic field, and reduce the lightning magnetic flux passing through the roof. Hence, the induced voltage is changed from negative to positive. The effect of metal roof is more significant than that of roof structure. The area of the metal rooftop is greater, and the PV loop is directly located over the rooftop within a zone of strongest field. Consequently, the voltage of metal rooftop PV array rises significantly. Since the tile roof almost has little influence on the lightning EM field, the induced voltage is nearly the same as that presented in

Fig. 3.45 Influence of roof material on lightning-induced voltage [8]

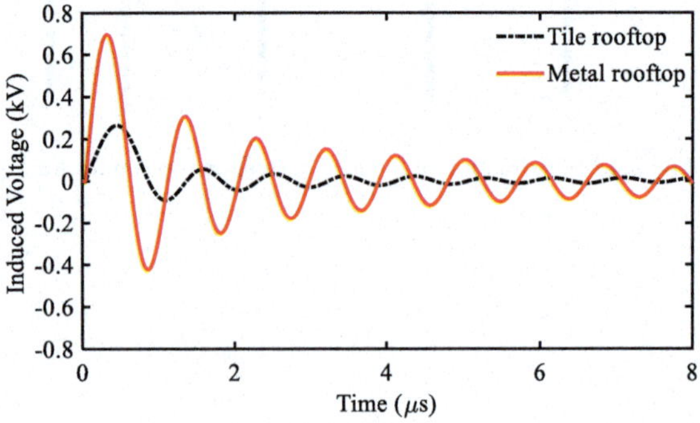

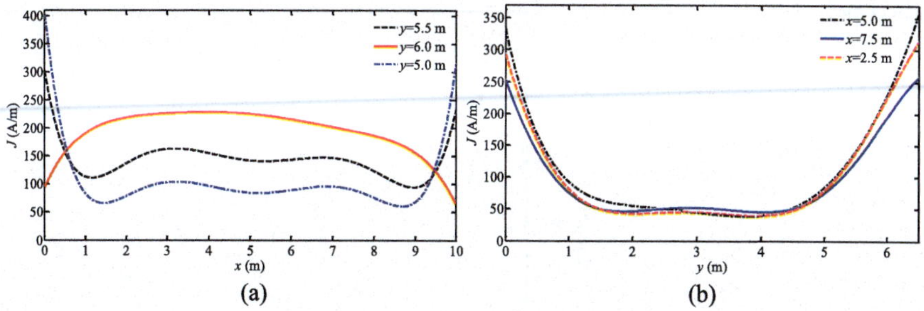

Fig. 3.46 Current density of metal rooftop. **a** Along x-direction. **b** Along y-direction [8]

Fig. 3.38. The current density induced on the surface of metal rooftop is presented in Fig. 3.46.

Figure 3.46a shows that the current density on both sides of the inside rooftop is the maximum whereas the greatest current density for the outside rooftop is in the middle. The induced current of outside rooftop is still larger than that of the inside rooftop, which is the same as the distribution of the current on the roof structure, as plotted in Fig. 2.46b. Therefore, the PV modules on the outside of the metal rooftop are more easily damaged by the indirect lightning.

The induced voltage of the rooftop PV array is closely related to the roof height. The two cases are considered, i.e., h is 8.25 and 5.25 m, as illustrated in Fig. 3.47. The DM induced voltage at DC cables is plotted in Fig. 3.48. Clearly, the lower the rooftop height, the smaller the induced voltage on PV array. As a higher roof has a greater impedance, the lightning induced current flowing on the conductor will produce a larger voltage, which result in the rise of the induced voltage at the output. Meanwhile, the greater the impedance of the roof structure is, the less induced current on the conductor is generated and hence the effect of the roof. Therefore, the induced voltage becomes larger when the roof height is higher.

The rooftop has many shapes, such as sloped rooftop and flat rooftop. Considering the two flat rooftops with different metal structures, such as Case I and Case II shown in Fig. 3.33. The PV array still consists of 26 PV modules in two rows, and the other

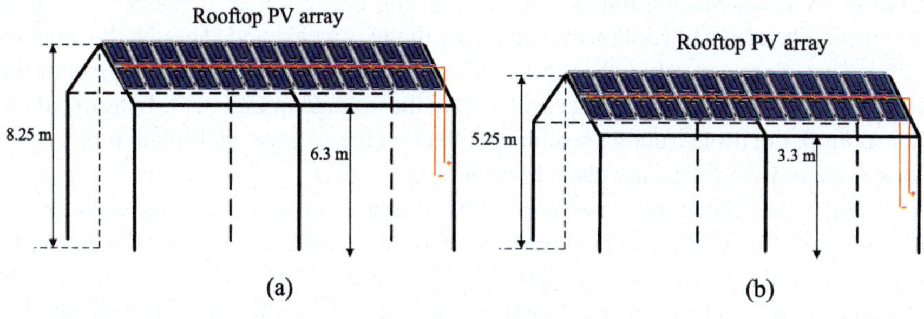

Fig. 3.47 Rooftop PV array with various height. **a** $h = 8.25$ mt. **b** $h = 5.25$ m

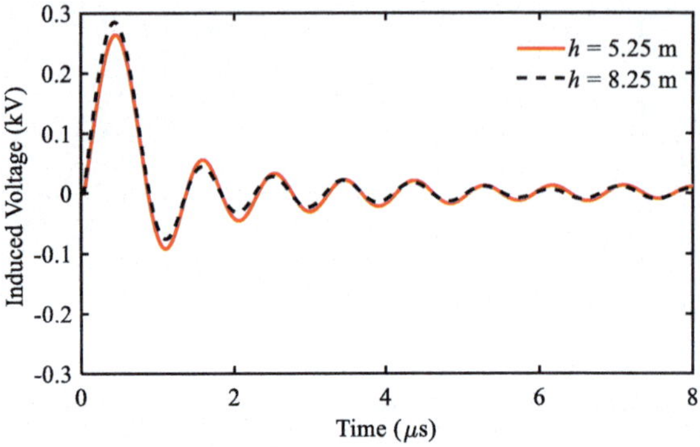

Fig. 3.48 Lightning-induced voltage under different roof height [8]

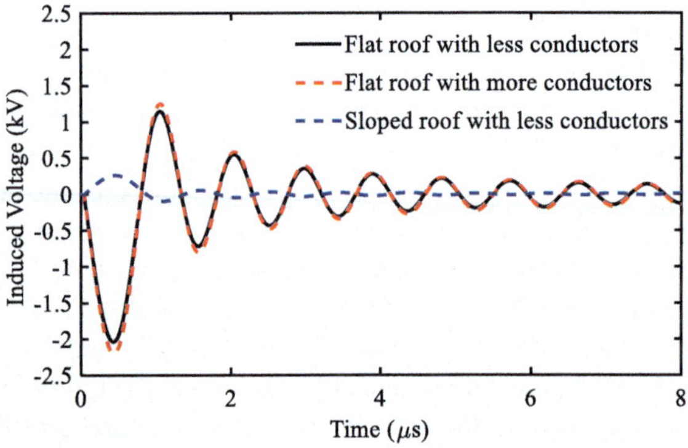

Fig. 3.49 Lightning-induced voltage of rooftop PV array with different rooftop shapes [8]

parameters are unchanged. The induced voltage is evaluated and compared with that of sloped rooftop PV array (Case a), as plotted in Fig. 3.49.

It is evident that the lightning induced voltage of PV array installed on flat rooftop is much larger than that on sloped rooftop. The voltage peak for the flat rooftop ranges from 2.04 to 2.24 kV whereas that on the sloped rooftop is approximately 0.26 kV. The crossbars of sloped rooftop are directly located in the bottom of the PV loop within a zone of greatest field reduction. However, the mounting structure of flat rooftop PV array has a number of legs, resulting in the longer distance to the roof structure. The effect of roof on the induced voltage is weakened. Though the induced voltage has a greater value, the 0.6–0.7 kV of voltage reduction due the effect of the flat rooftop still can be observed compared to the voltage in Fig. 3.37. In terms of the flat rooftop, the roof structure has minor effect on the voltage, as shown in Fig. 3.49, their voltage waveforms are close to each other.

Furthermore, the induced voltage under different flat rooftop heights is calculated, as plotted in Fig. 3.50. The peak voltage increases as the flat roof height rises. The induced voltage increases from 2.04 to 2.24 kV, the 9.8% of rise for the height changed from 6.25 to 10 m. The result indicates that the rooftop PV array is not only

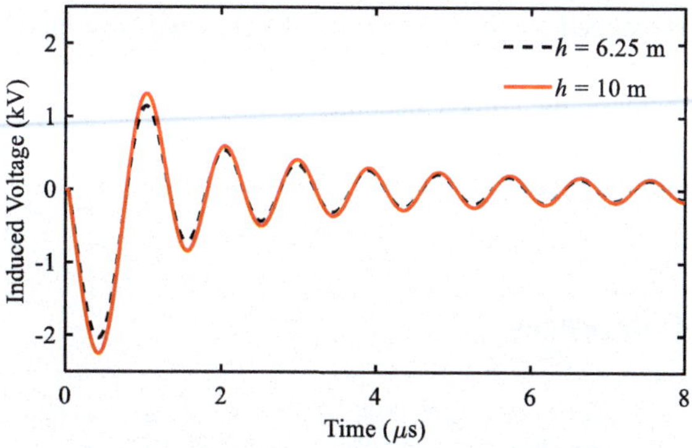

Fig. 3.50 Lightning-induced voltage of flat rooftop PV array [8]

more susceptible to the lightning strike due to the greater rooftop height, but also the lightning EM field would generate a severer surge on it.

3.4 Conclusion

The presence of encapsulated metal frame can produce an impulsive short-circuit current under lightning strike effect. An opposite magnetic field with respect to original lightning channel would be generated, and the induced voltage is significantly minimized. The mutual inductance between metal frame and internal loop is derived through numerical simulation; the reduction factor of frame strongly depends on the oblique angle and the lightning distance. The use of metal frame is preferred and in dispensable for regions with a great grounding flash density and at a high risk.

The lightning-induced voltage is strongly dependent on the ground conductivity due to the variation of the reflection coefficient at the interface. As the conductivity decreases, the peak value becomes higher and the waveform becomes a little wider. The characteristics of induced voltage of PV system installed on lake are completely different from that on flat ground as the water has a good conducting performance. The saltwater lake has a relatively lower lightning threat whereas the glacial lake has a great induced voltage. The voltage may change seasonally resulting from the seasonal variation of lake composition and hence the electric property. The mountain geography is divided into triangular, trapezoidal and V-shaped terrains in this article. Physically, the LEMF, along with the wave propagation path, leads to the distinction in the induced voltage on mountain. The voltage for PV system on V-shaped terrain is relatively higher than others. A steeper mountain generally has a lower LEMF, a longer wave propagation path and a greater energy loss. As a consequence, the induced voltage on PV will be fairly small.

The roof has a great effect on the lightning-induced voltage of PV array due to the induced current on the roof steels. The induced current generates an opposite EM field which weaken the original lightning EM field. Hence, the induced voltage is reduced and its polarity may be changed. The induced voltage is strongly affected by the roof structure and material. As the numbers of crossbar and supporting leg increase, the voltage becomes greater. Since the rooftop is larger and directly located over PV loop, the loop is in a zone of strongest field generated by the induced current, and the induced voltage on the metal rooftop is more severe.

References

1. Sun Q. Q, Zhong, X, et al.: Three-dimensional modeling on lightning induced overvoltage for photovoltaic arrays installed on mountain. J. Clean. Prod., vol. 288, 125084 (2021).
2. Yee, K.: Numerical solution of initial boundary value problems involving Maxwell's equations in isotropic media. IEEE Trans. Antennas Propagat 14(3), 302–307 (1966).
3. Mur, G.: Absorbing boundary conditions for the finite-difference approximation of the time-domain electromagnetic-field equations. IEEE Trans. Electromagn. Compat. 23(4), 377–382 (1981).
4. Liao, Z. P. Wong, H. L. Yang, B. et al.: A transmitting boundary for transient wave analysis. Sci. Sin. 17, 1063–1076 (1984).
5. Tian, M. H. Sheng, S. L. Wei, M. et al.: Calculation of the electromagnetic fields of lightning return stroke based on DU model. High. Voltage. Eng. 28(9), 34–36 (2002).
6. Zhong, X, Zheng, Z, Sun Q. Q, et al.: Investigation on lightning-induced voltage for
7. photovoltaic system considering electromagnetic coupling of rooftop structure. IEEE J. Photovolt., vol. 14, 503–511 (2024).
8. Photovoltaic inverters: Data sheet and name plate. IEC 62894 (2016).
9. Sun Q. Q, Zhong, X, et al.: Investigation on induced voltage of photovoltaic system on complex terrain. Electr. Power Syst. Res., vol. 201, 107549 (2021).
10. Altair FEKO 2019.3.2 User's Manual, EM Software and Systems. Stellenbosch, South Africa: S. A. (Pty), Ltd. (2019), http://www.feko.info.
11. Sheng, C. Yang, S. Shen, D. et al.: Safety and Effectiveness analysis of lightning protection for mountain photovoltaic power station. Sol. Energy. 04, 26–28 (2016).

Chapter 4
Novel Crossover Wiring of DC Cable for Photovoltaic Array Against Lightning Induced Overvoltage

4.1 Mechanism of Electromagnetic Coupling for Photovoltaic Mounting System

The lightning current will produce a strong EM radiation, and an impulse voltage will be induced in the DC cable around. The associated electronic device like the bypass diode and inverter, as illustrated in Fig. 4.1, may be damaged. The computational model for induced voltages is presented as follows.

4.1.1 EM Coupling of Lightning Channel

The lightning current is represented by the return-stroke model $i\ (z', t)$. The MTLE model is selected to represent the lightning current in this article [2],

$$i(z',t) = \begin{cases} P(z')\, i(0, t - z'/v) & t - z'/v \geq 0 \\ 0 & t - z'/v > 0 \end{cases} \tag{4.1}$$

where z' is the height of return current and t is the time. The lightning discharge is regarded as the current following up a vertical channel. The propagation speed of lightning current v typically ranges from c/3 to 2c/3. It is set to be $2c/3$ in this article, and c is the propagation speed of the EM wave in vacuum. The exponential decay P is dependent on the height z' with the factor $\lambda = 2000$,

$$P(z') = e^{-z'/\lambda} \tag{4.2}$$

The lightning channel-base current $i\ (0,\ t)$ is represented by Heidler function [3]. Given the current source, the vector potential $\boldsymbol{A}(r, z)$ generated by the lightning

Q. Sun et al., *Lightning Transients and Protection for Renewable Energy System*,
https://doi.org/10.1007/978-981-97-7267-4_4

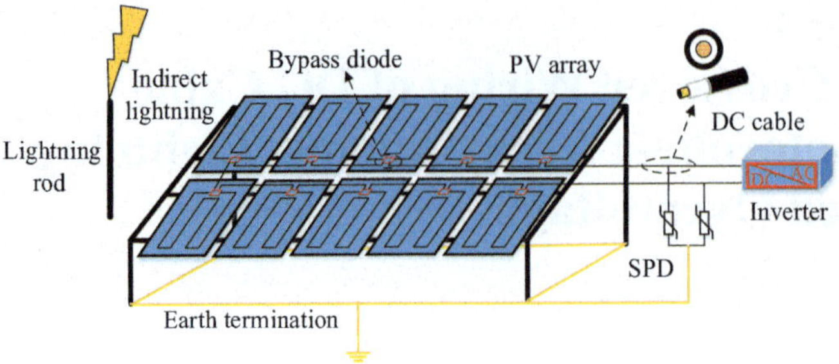

Fig. 4.1 Schematic diagram of a PV array [1]

current is obtained by:

$$A(r_s, t) = \frac{\mu_0}{4\pi} \int_{V'} \frac{J(r'_s, t - \frac{|r_s - r'_s|}{c})}{|r_s - r'_s|} dV' \tag{4.3}$$

where V', r_s and r'_s are given in Fig. 4.2a.

According to Maxwell's equations in isotropic medium, the magnetic field $\boldsymbol{H}_\phi$ of the lightning current integrated along the length h for the lightning channel by the dipole technique [4], as shown in Fig. 4.2b.

$$H_\phi(r, z) = \frac{\vec{e}_\phi}{4\pi} \int_{-h}^{h} \left[\frac{r}{R^3} i(z', t - R/c) + \frac{r}{cR^2} \frac{\partial i(z', t - R/c)}{\partial t} \right] dz' \tag{4.4}$$

According to Faraday's law, the differential-mode induced voltage U in the loop l of DC cable can be determined,

$$U = \int_l E \cdot \mathrm{d}l = -\int_s \frac{\partial B_\phi}{\partial t} \cdot \mathrm{d}S \tag{4.5}$$

The magnetic flux Φ is obtained by integrating the magnetic field over the surface S of the loop,

$$\Phi = \iint_S \mu_0 H_\phi(r, z)\, ds \tag{4.6}$$

For a PV array installed on flat ground, the surface of the loop for DC cable can be expressed by the normal vector (k_x, k_y, k_z) and a point (x_0, y_0, z_0) in Cartesian coordinates,

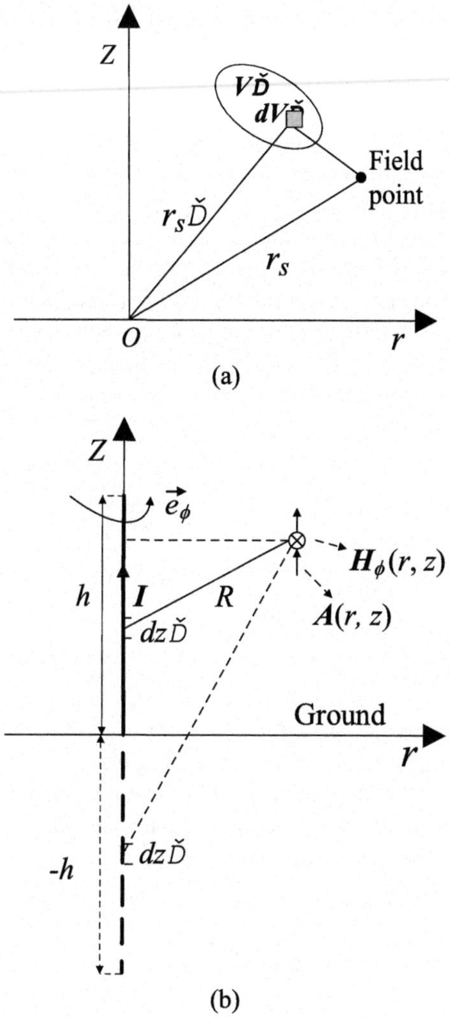

Fig. 4.2 EM-field model. **a** Coordinate system for current source and field point. **b** Lightning channel model in dipole technique

$$z(x, y) = \frac{z_0 + k_x(x_0 - x) + k_y(y_0 - y)}{k_z} \tag{4.7}$$

Substituting Eq. (4.7) into Eq. (4.6),

$$\Phi = \iint \frac{\mu_0}{\sqrt{x^2 + y^2}} \left[\frac{xk_y}{k_z\sqrt{x^2 + y^2}} - \frac{yk_x}{k_z\sqrt{x^2 + y^2}} \right] \times H_\phi\left(\sqrt{x^2 + y^2}, \frac{z_0 + k_x(x_0 - x) + k_y(y_0 - y)}{k_z} \right) dxdy \tag{4.8}$$

4.1.2 EM Coupling of Metal Bracket and Frame

The lightning EM field would generate induced currents on PV frames and the bracket. The currents will generate the scattered magnetic field, which has a remarkable influence on the spatial EM field. The total EM field is the sum of them. There are a large number of conductors in the bracket, and they are in different spatial position. Since the metal frames are connected to the bracket, the circuits of the bracket and frames have many branches. To facilitate the analysis, the induced currents in the bracket and PV frames are calculated individually.

In terms for the single loop, such as the frame, its equivalent circuit can be represented by RL model, as illustrated in Fig. 4.3. The induced current on the loop can be derived,

$$L_{\text{frame}}\frac{di_{\text{frame}}(t)}{dt}+R_{\text{frame}}\,i_{\text{frame}}(t)=U_{\text{frame}} \tag{4.9}$$

The frame of PV module is C-shaped in structure, as well as the bracket. The equivalent radius r of C-shaped conductor is defined as [4],

$$r=\sqrt{\frac{t_c(2w_c+l_c)}{\pi}} \tag{4.10}$$

where w_c, t_c and l_c represent the width, thickness and length of C-shaped structure, respectively, as illustrated in Fig. 4.3.

All loops in the PV bracket are represented by RL models. Taking the EM coupling between the loops into account, the matrix equation for the induced currents on PV brackets can be expressed as:

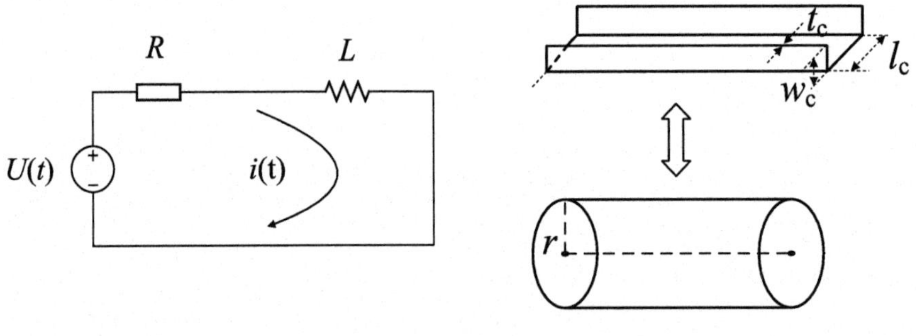

Fig. 4.3 Equivalent model of frame

Fig. 4.4 EM coupling model of PV bracket [1]

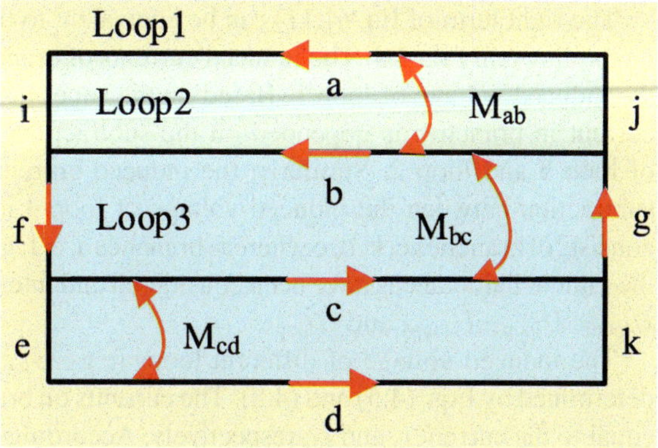

$$\mathbf{L}\cdot\begin{bmatrix}\frac{d\,i_a}{d\,t}\\ \frac{d\,i_b}{d\,t}\\ \frac{d\,i_c}{d\,t}\\ \frac{d\,i_d}{d\,t}\end{bmatrix}+\mathbf{R_B}\cdot\begin{bmatrix}i_a\\ i_b\\ i_c\\ i_d\end{bmatrix}=\begin{bmatrix}U_{\text{branch - a}}-M_{ab}\frac{d\,i_b(t)}{d\,t}\\ U_{\text{branch - b}}-M_{ab}\frac{d\,i_a(t)}{d\,t}-M_{bc}\frac{d\,i_c(t)}{d\,t}\\ U_{\text{branch - c}}-M_{cd}\frac{d\,i_d(t)}{d\,t}-M_{bc}\frac{d\,i_b(t)}{d\,t}\\ U_{\text{branch - d}}-M_{cd}\frac{d\,i_c(t)}{d\,t}\end{bmatrix}\tag{4.11}$$

where the mutual inductances (M_{ab}, M_{ac}, M_{bc} and M_{cd}) are used to represent the EM coupling of the loops for the bracket, as shown in Fig. 4.4.

L and RB are the inductance matrix and the resistance matrix of the equivalent circuit for PV bracket, respectively.

$$\mathbf{L}=\begin{bmatrix}L_a & 0 & 0 & 0\\ 0 & L_b & 0 & 0\\ 0 & 0 & L_c & 0\\ 0 & 0 & 0 & L_d\end{bmatrix}\tag{4.12}$$

And,

$$\mathbf{R_B}=\begin{bmatrix}R_a & 0 & 0 & 0\\ 0 & R_b & 0 & 0\\ 0 & 0 & R_c & 0\\ 0 & 0 & 0 & R_d\end{bmatrix}\tag{4.13}$$

where L_a, L_b, L_c and L_d represent the inductances of branches a, b, c and d, respectively. R_a, R_b, R_c and R_d represent the resistances of branches a, b, c and d, respectively. The detailed calculation of the resistances and inductances for PV frames and the bracket is presented in [6].

The right term of Eq. (4.11) can be equivalent to the induced voltages of various loops shown in Fig. 4.4. The induced currents of branches a and d are determined by the induced voltage of loop 1. Based on the superposition theorem [7], the induced current in branch b is depended on the subtraction between the induced voltages of loop 3 and loop 2. Similarly, the induced current of c can be obtained by the subtraction between the induced voltage of loop 4 and that of loop 5. The loop 4 consists of branches c, k, d, e whereas branches a, c, f, g, i, j form the loop 5. Therefore, the mutual inductances have been considered and integrated in the parameters U_{loop1}, U_{loop2}, U_{loop3}, U_{loop4} and U_{loop5}.

The induced voltage of different loops (e.g., U_{loop1}, U_{loop2}, U_{loop3}, etc.) can be determined by Eqs. (4.5) and (4.8). The currents on branches i, j and branches e, k are equal to the current i_a and i_d, respectively. According to the Kirchhoff's current law, the current i_f is the sum of i_a and i_b whereas the current i_g is the sum of i_c and i_d. The branches of the frames and bracket are regarded as the straight conductors. Then, the induced currents are used to solve the scattered magnetic field. The flowchart is presented in Fig. 4.5.

4.2 Wiring Scheme of DC Cable

4.2.1 Crossover Wirings and Traditional Wiring

PV modules, inverters and other electrical components in the system are connected by DC cables, which are usually installed beneath the bracket. The DC cables are designed with double-layer insulation, have high mechanical strength and great anti-corrosion performance. The configuration of the DC loop depends on the connection of PV panel, and three typical wiring schemes are given in Fig. 4.6.

The cable wiring has a great impact on the induced voltage caused by lightning, and the area of DC loop is a key factor. Since the loop area of Case a is the largest among the three wirings, the induced voltage is the highest. Currently, Case c is widely used in PV arrays due to the smaller loop area. The magnetic flux passing through the loop is reduced, and hence the induced voltage. However, the incidents about electronic devices connected in DC cable damaged by lightning are often reported. The width of DC loop in Case c is quite small, and it requires more cables to reduce the loop area [8]. Additionally, the loop areas contributed by PV modules will still lead to a large voltage.

The novel crossover connection of PV modules is proposed to further reduce the induced voltage, as shown in Fig. 4.7. The direction of the current is changed in crossover cables, resulting in the opposite electric potentials caused by lightning. In terms of Case d, the loop is divided into loop I and loop II, as illustrated in Fig. 4.8b. The EM field generates the electric potentials U_{I} and U_{II} in loop I and loop II, respectively. If the direction of the magnetic field in the two loops is consistent, U_{I} and U_{II} have the same direction. Due to the crossover, the polarity of U_{I} and U_{II} is

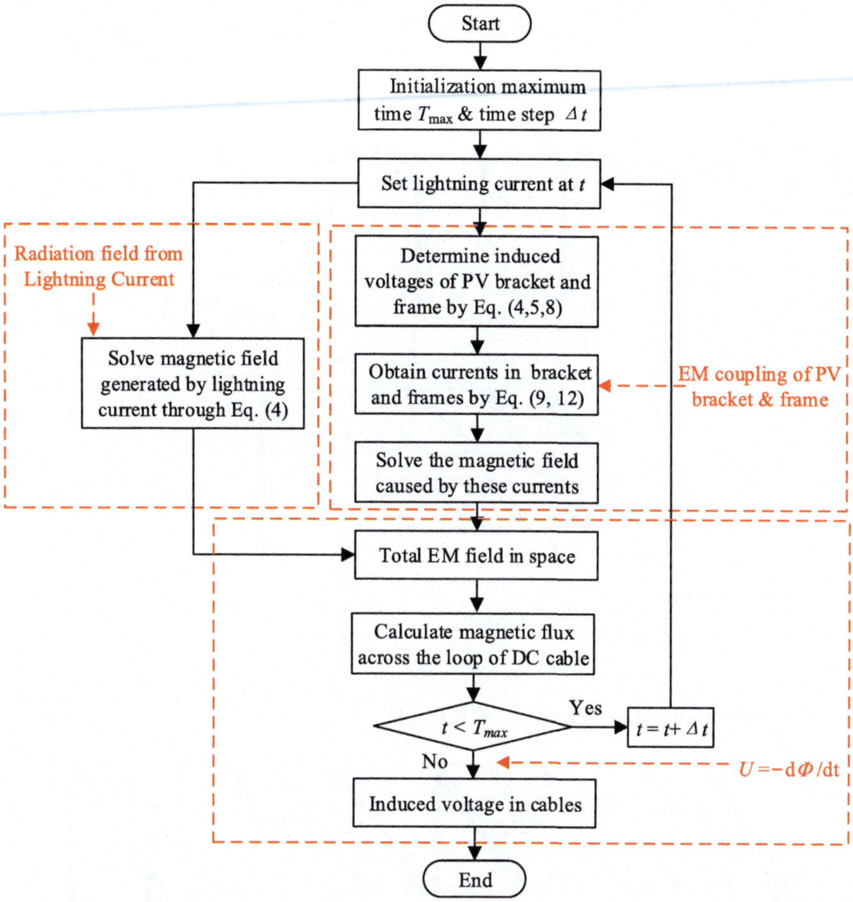

Fig. 4.5 Flowchart for computation induced voltage of DC cable [1]

opposite. The induced voltage of the whole loop is thus equal to $U_{\mathrm{I}} - U_{\mathrm{II}}$. Therefore, the resultant value of Case d is much lower than that of Case c. The whole loop can be regarded as loop I, loop II and loop III for Case e, as illustrated in Fig. 4.8c. Similarly, the electric potentials of these three loops are U_{I}, U_{II} and U_{III}. There are two crossovers, the polarity of U_{III} is different from that of U_{I} and U_{II}. Hence, the induced voltage of the whole loop is $U_{\mathrm{I}} - U_{\mathrm{III}} + U_{\mathrm{II}}$. Here, U_{I}, U_{II} and U_{III} can be calculated by Eqs. (4.9) and (4.12).

If the direction of the magnetic field passing through the loop is inconsistent, the induced voltage in the outputs of the cable with crossover wiring is different. Assume that half of the magnetic field across the loop is inward whereas the other half is outward, as shown in Fig. 4.8d–f. The total magnetic flux passing through the loop is mostly cancelled for Case c, and the induced voltage is lower. In contrast, the induced voltage of Case d is larger. Since the direction of the magnetic field

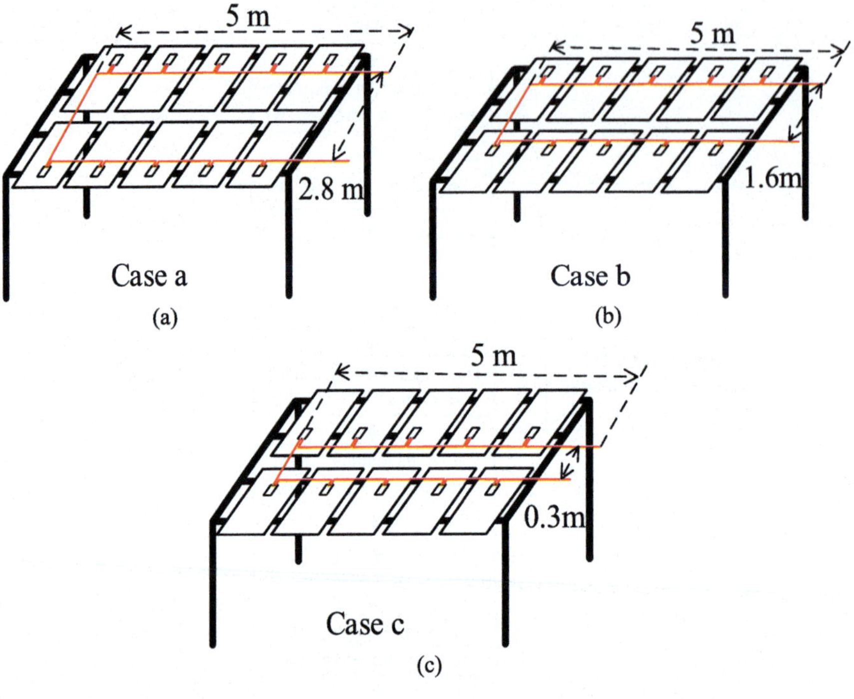

Fig. 4.6 Traditional wiring schemes of DC cable [1]

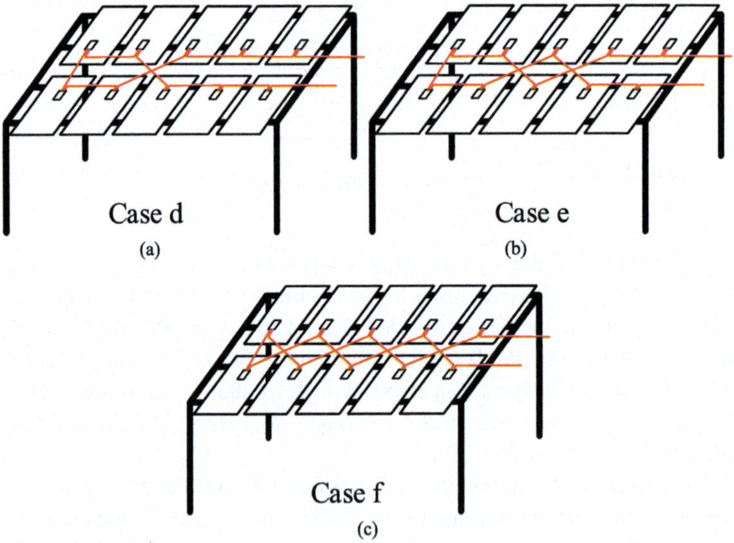

Fig. 4.7 Proposed crossover wiring schemes of DC cable [1]

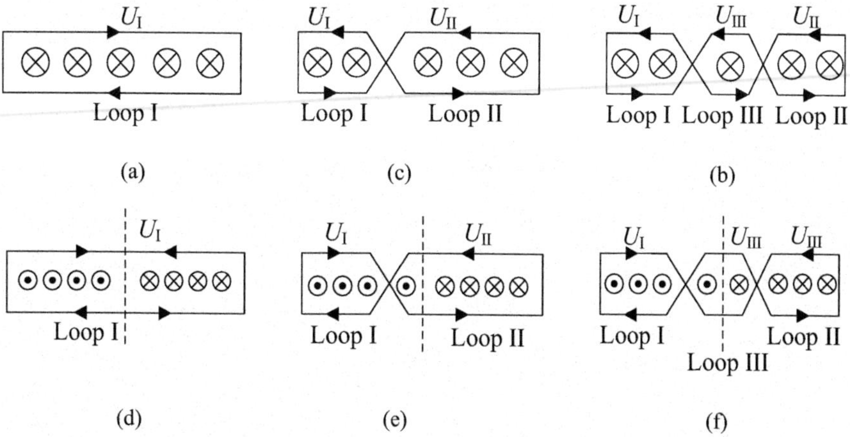

Fig. 4.8 Loop of DC cable. **a** Case c with single loop. **b** Case d with two loops. **c** Case e with three loops. **d** Case c with different magnetic field. **e** Case d with different magnetic field. **f** Case e with different magnetic field [1]

is opposite to that in Fig. 4.8b, the electric potential of loop I changes from U_{I} to $-U_{\text{I}}$. The induced voltage of Case II is thus $U_{\text{I}} + U_{\text{II}}$. In terms of Case e, the electric potential of loop I is equal to $-U_{\text{I}}$ whereas the electric potential in loop II is unchanged. The magnetic flux through loop III is small and can be approximately negligible. Hence, the induced voltage becomes $U_{\text{I}} - U_{\text{II}}$, which is relatively small.

Based on the above cable wirings, a novel crossover wiring is proposed, as shown in Case f of Fig. 4.7. The loop area is further reduced, and the induced voltages in Case f are reduced by several crossovers of cables. Moreover, the proposed cable wiring is symmetrical, which is beneficial to the compensation of magnetic flux.

4.2.2 Verification

The proposed model and the MoM-based model in our work are validated by the experimental results in [9]. The experimental setup is illustrated in Fig. 4.9. The impulse generator, including a simple half-wave rectifier, a capacitor bank and a triggered gap, is used to generate the current. It represents the lightning current, and is injected to the conductor, which is parallel to the PV panel with a distance of d. The steepness of the injected current is 0.5 kA/μs. The PV panel is darkened to avoid the light influence, and only the effect of the magnetic field caused by the impulse current is presented. The open-circuit voltage of the PV panel without frame is recorded by a digital oscilloscope.

The comparison for the calculation results of the proposed model and the experimental results is summarized in Table 4.1. The deviations between the measured

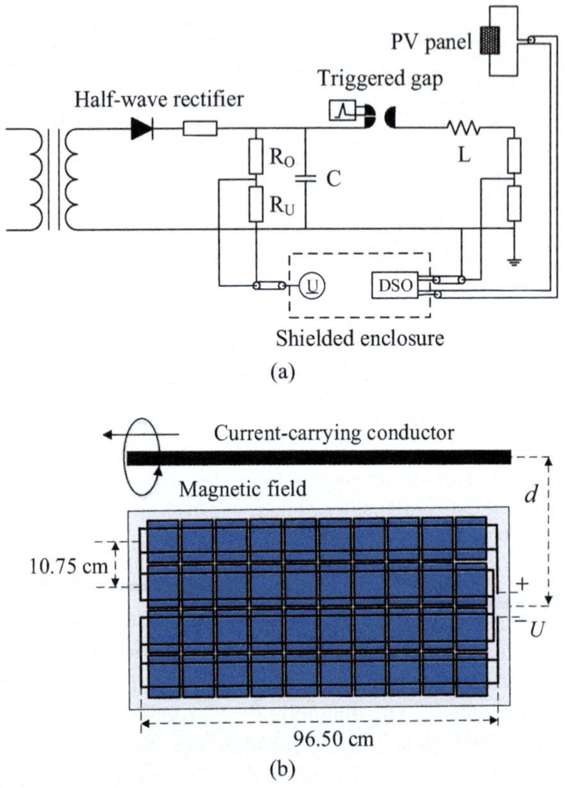

Fig. 4.9 Schematic diagram of experiment [9]. **a** Experimental setup. **b** Layout of conductor and PV panel

values and the calculated values are relatively small and reasonable, and their errors are all lower than 5%.

Table 4.1 Comparison for proposed model and experimental result (V)

d (m)	Measured [9]	Proposed model	MoM
1.0	14.8	14.5	14.8
0.9	17.6	17.0	17.4
0.8	21.3	20.2	20.7
0.7	26.0	24.5	25.0
0.6	31.9	30.2	30.9

4.2.3 *Effect of Crossover Wiring on Induced Voltage of Photovoltaic Array*

The lightning transients for the PV arrays with three typical wirings of cables (Case a, Case b and Case c in Fig. 4.6) are calculated, as shown in Figs. 4.10 and 4.11. The induced voltage of Case c is the lowest among these three cable wirings, either S_1 or S_5 is struck by lightning. Since the loop area formed is the smallest and hence the magnetic flux passing through the loop, Case c is set to be benchmark for comparison study. The waveforms of lightning current and its derivative in our work rises very slowly and seems have an initial time delay at the beginning. Hence, the voltage waveforms present the initial time delay.

The differential-mode induced voltage of the PV array is calculated by the proposed model and MoM, respectively. It is clear that the voltage peaks evaluated by the two methods are consistent with small deviation, as plotted in Fig. 4.13. The capacitances of PV bracket and frames to the ground, which are very small and

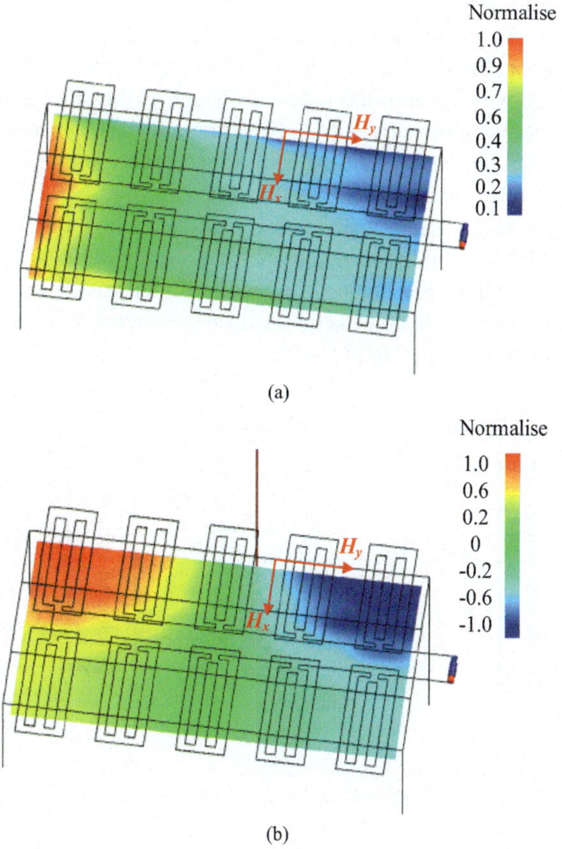

Fig. 4.10 Lightning transients for conventional wirings. **a** Magnetic field H_x near PV array when lightning strikes at S_1. **b** Magnetic field H_x near PV array when lightning strikes at S_5 [1]

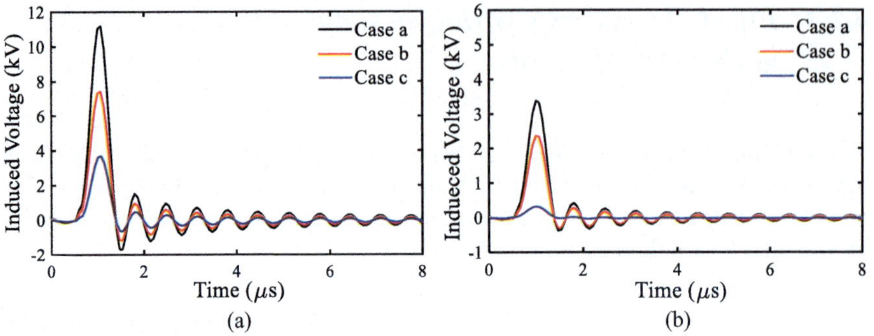

Fig. 4.11 Lightning induced voltage for conventional wirings. **a** Induced voltage when lightning strikes at S_1. **b** Induced voltage when lightning strikes at S_5 [1]

accounts for little, are neglected in the proposed model to facilitate the analysis. The RLC circuit often produces the oscillation. However, the induced currents on the bracket and frames, which are the RL circuits, have no oscillations on the tail of the waveform.

As shown in Fig. 4.12a, the peak voltage in Case d is remarkably minimized, about 70% reduction compared to that of Case c. The whole loop is divided into two loops in Case d, the induced voltage is the subtraction of the induced voltages in the two loops ($U_{\mathrm{I}} - U_{\mathrm{II}}$). Nevertheless, Case d can only reduce the induced voltage when the magnetic field across the loop are in the same direction, such as lightning striking at S_1 and S_2, as illustrated in Fig. 4.10a. For lightning striking at S_5, the magnetic field is in opposite direction, as shown in Fig. 4.10b. According to previous analysis, the induced voltage for Case d would change from $U_{\mathrm{I}} - U_{\mathrm{II}}$ to $U_{\mathrm{I}} + U_{\mathrm{II}}$. Therefore, the induced voltage is much greater than that of Case c, as plotted in Fig. 4.12b. To address this issue, a symmetrical crossover wiring is proposed in Case e, as illustrated in Fig. 4.7.

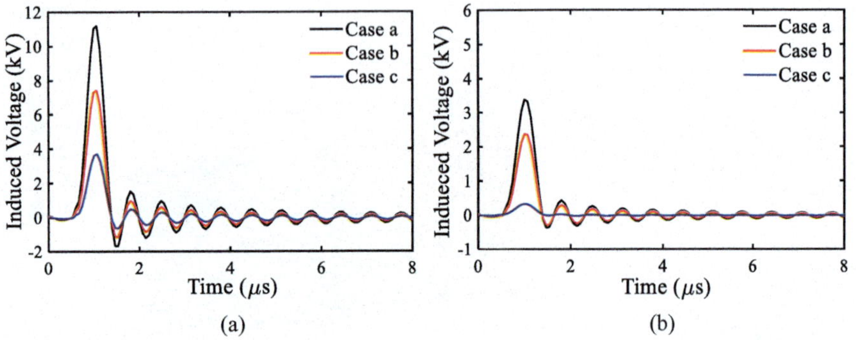

Fig. 4.12 Induced voltage for crossover wirings. **a** Lightning strikes at S_1. **b** Lightning strikes at S_5 [1]

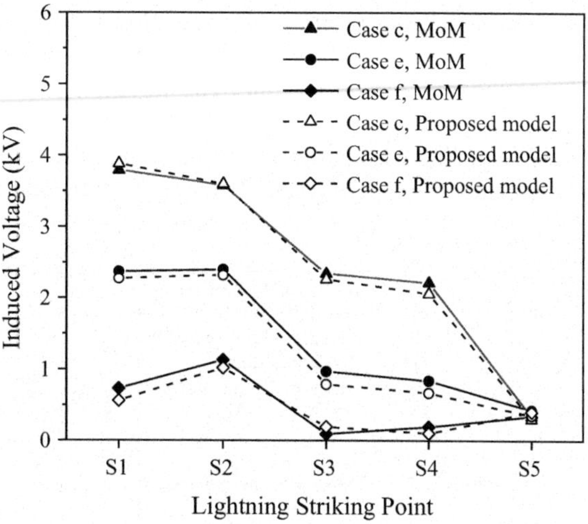

Fig. 4.13 Induced voltage between ± cables under different strike locations [1]

Though the cable wiring of Case e can reduce the induced voltage, the reduction is only about 37%, which is smaller than that of Case d, as shown in Fig. 4.12a. The induced voltage for Case e is equal to $U_{\mathrm{I}} - U_{\mathrm{III}} + U_{\mathrm{II}}$. When S_1 is struck by lightning, as illustrated in Fig. 4.8c, loop I is closer to the lightning channel, and the magnetic flux passing through loop I is greater. The area of loop III is smaller, and the flux can only cancel a part of the flux through loop I. Therefore, the drop of the induced voltage is not optimal for Case e. In contrast, the area of loop I is smaller than that of loop II in Case d. The magnetic fluxes of loop I and loop II are almost the same, leading to more offset compensation of the magnetic flux.

Based on Case e, an improved crossover wiring is proposed, as shown in Case f of Fig. 4.7. The induced voltages are calculated, and the peak voltages are presented in Fig. 4.13. The result shows that the wiring of Case f is the most effective in minimizing the induced voltage. If lightning strikes at S_1, the reduction of voltage for Case f can reach up to 80% compared with that of Case c, which is close to that of Case d. In terms of other lightning striking locations, its peak voltages are all smaller than that of Case c and e. Compared with Case e, the induced voltage is further decreased by 30%-40%. It is attributed to the following reasons: (i) the loop area of Case f is further decreased. (ii) The whole loop is divided into several subloops via crossovers. The polarity of the electric potential in each loop is opposite, and the voltages are partially offset. (iii) The cable wiring of Case f is symmetrical, which will reduce the magnetic flux passing through the DC loop.

Moreover, the external factors, including lightning current waveform and the distance between the lightning channel and PV array, are discussed. The first return stroke (100 kA, 10/350 μs) and the subsequent stroke (25 kA, 0.25/100 μs) are considered, and the induced voltages are plotted in Fig. 4.14. It demonstrates that the wiring scheme of Case f can reduce the induced voltage caused by the first return stroke, as well as the subsequent stroke. The subsequent stroke leads to a greater voltage, and its waveform is narrower. The voltage caused by the first return stroke is smaller than that of the subsequent stroke, however, the waveform becomes wider due to the long wavefront time. The variation of peak voltage versus the lightning distance is shown in Fig. 4.15. The peak voltage drops rapidly in the first several meters, and then decreases slowly. No matter the distance, the voltages of Case c are the greatest whereas those of Case f are kept the lowest.

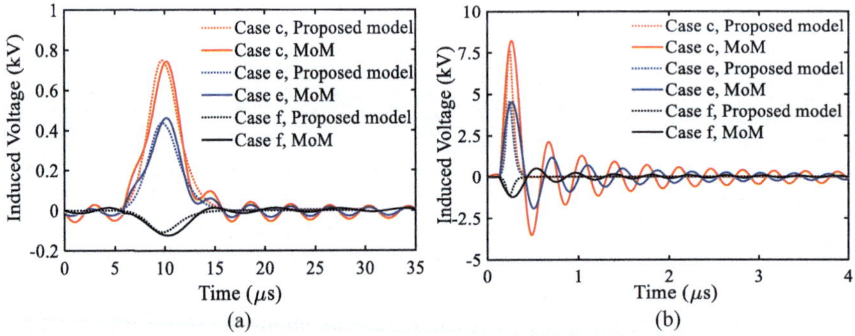

Fig. 4.14 Induced voltage under various lightning waveforms when S_1 is struck. **a** First return stroke. **b** Subsequent stroke [1]

Fig. 4.15 Induced voltage between ± DC cables versus distance between PV array and lightning channel [1]

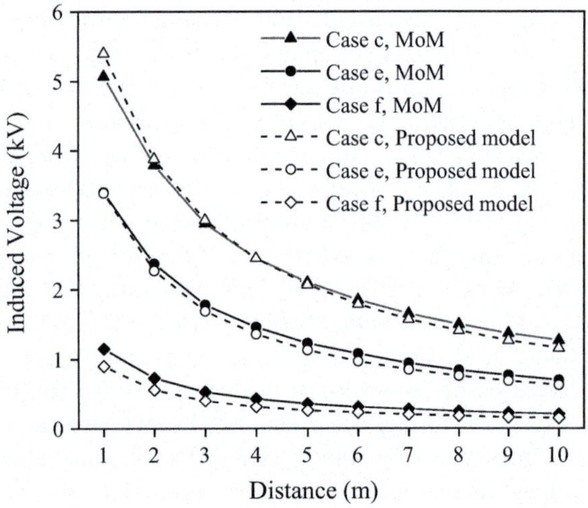

4.3 Evaluation of Proposed Crossover Wiring for DC Cable

The lightning induced voltages between ± DC cables under various lightning striking points are evaluated by using both the proposed model and the method of moments (MoM). The details of the MoM-based model for PV array are presented as follow:

4.3.1 *Simulation Study*

The overhead lines, towers and buildings are exposed to the direct lightning due to their tall structures, which has a great attracting distance in the case of lightning strike. However, the PV array is distinct from overhead lines, and the height of PV array is quite low and arranged very densely. The isolated lightning rods are often installed near PV arrays to intercept the lightning. The distance between the lightning rod and PV array ranges from 1 to 7.6 m [10, 11], depending on the height of the lightning rod. To avoid the shadow caused by the rod, its height is generally controlled. The lightning may strike to the lightning rods. Hence, the distance between the lightning rod and the PV array is very small, as shown in Fig. 4.16a.

The PV array under study consists of a bracket and ten PV modules connected in series by the DC cables. The oblique angle for PV panel is set to be 26.6°. The

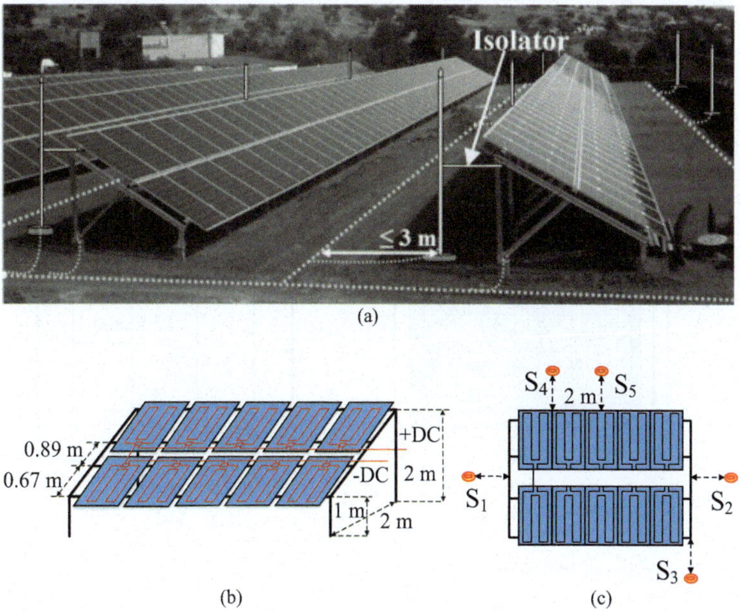

Fig. 4.16 PV array. **a** Example of isolated lightning rods for large-scale PV array applications [10]. **b** Configuration of PV array. **c** Lightning striking points

bracket structure has four grounding legs and six crossbars. Its length and width are 5 and 2.23 m, respectively, and the detailed parameters are given in Fig. 4.16b.

The five different lightning striking points (S_1, S_2, S_3, S_4, S_5) are investigated, as shown in Fig. 4.16c. The PV module is an assembly of four PV-cell strings, packaged with a metal frame. The PV-cell string is regarded as straight wires in series, as shown in Fig. 4.17. Its length and width are 1.34 and 0.6 m. The lightning protection zone for PV system is classified as III [12]. Lightning currents are random and complex. The actual stroke currents present multiple peaks and there are several strong concavities between these peaks [13]. To facilitate the analysis, the lightning current in this paper is represented by a single Heidler function according to IEC 62,305 [3]. The conclusions drawn in this paper can be also applied to other lightning current waveforms. The current peak is set to be 50 kA, and its waveform is 1 μs/ 200 μs.

To validate the proposed model in our work, the lightning transient of PV array is also calculated by using the MoM-based software FEKO [14]. The lightning channel is represented by cylindrical elements having distributed resistance and inductance, and the propagation velocity of the lightning current wave on it is included. The height of lightning channel is set to be 1 km. The mounting system, PV modules and DC cable are modelled as the combination of cylindrical segments, which satisfy the limitation in the ratio of conductor length to wave length in the time range of interest. If not mentioned, the other computations in this article are based on MoM.

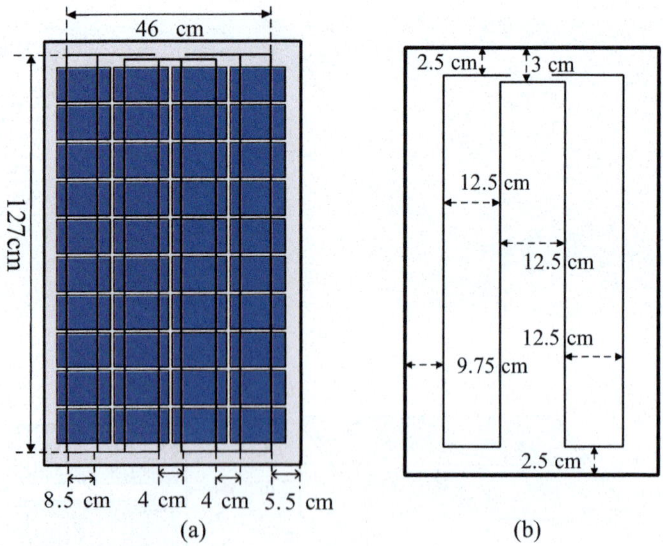

Fig. 4.17 Dimension of PV module

4.3.2 *Effect of PV Bracket and Frame*

The induced voltage in cable is affected by PV bracket and metal frame. The voltages of different wiring schemes without bracket and frame are presented in Figs. 4.18 and 4.17. The waveform of the induced voltage and the trend of peak values are similar to the results considering the bracket and frame. The induced voltage drops in Case d if lightning strikes at S_1. However, it increases when S_5 is struck. The cable wiring of Case f has the optimal performance, and the voltage can be reduced by approximately 75% compared to that of Case c.

The effect of PV bracket on the lightning induced voltage is seldom studied in previous literature. Table 4.2 shows that the bracket and metal frame can decrease the voltage, and the reduction is about 40–50%. The lightning EM field will generate the magnetic flux in the loop of bracket and frame. Meanwhile, the induced current of the loop would produce additional magnetic flux in opposite to the magnetic flux caused by lightning, according to the Faraday's law. Hence, the total magnetic flux in the loop of cable is reduced, and the induced voltage is smaller than that of PV string without the bracket and frame.

The impact of the bracket and frame is relatively minor for striking at S_5, as plotted in Figs. 4.12b and 4.19b. Since the structures of bracket and frame are also symmetrical, most of the magnetic flux across the loop of PV bracket and metal frame is cancelled. Thus, the induced current in the bracket and frame is minor. Frameless PV modules are increasingly popular in recent years. The leakage current is reduced, and its life span and efficiency are enhanced. However, they have a greater induced voltage and tend to aggregate the indirect lightning risk.

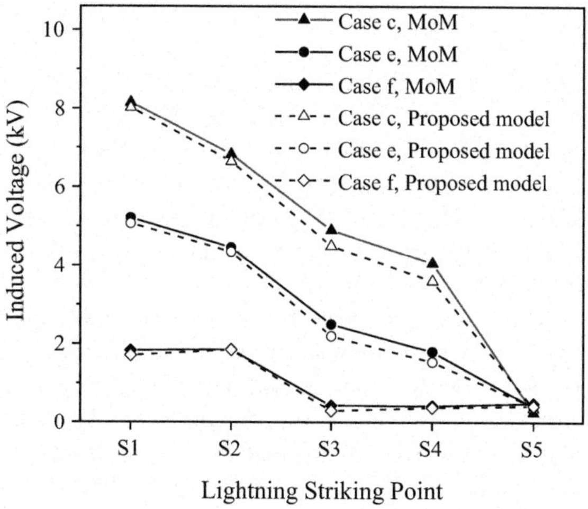

Fig. 4.18 Peak of Induced voltage between ± cables for frameless PV [1]

Table 4.2 Induced voltage in cable with different wirings (kV) [1]

Case	With bracket and frame		Without bracket and frame	
	MoM	Proposed model	MoM	Proposed model
c	2.21	2.09	4.05	3.90
d	2.56	2.43	3.64	3.48
e	0.84	0.70	1.80	1.56
f	0.19	0.12	0.41	0.39

Note Induced voltage is calculated when the lightning strikes at S_4

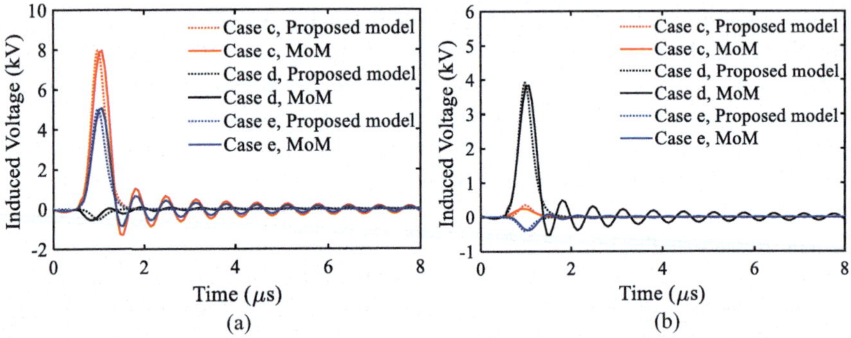

Fig. 4.19 Induced voltage of different wirings for frameless PV. **a** Lightning strikes at S_1. **b** Lightning strikes at S_5 [1]

4.3.3 Electromagnetic Effect of Mounting System

The structures of the mounting systems for PV arrays under various terrains are complex. There are many branches of bracket in different spatial position, and the detailed influence of the structure of bracket is unclear. The mounting systems can be classified as one-leg bracket, two-leg bracket and four-leg bracket. The mounting system with one-leg structure and two-leg structure are studied, as depicted in Fig. 4.20. Figure 4.21 shows that the induced voltage becomes lower as the number of grounding leg increases.

The legs are connected to the ground, and the induced current of the bracket is smaller when the number of leg is few. Therefore, the influence of bracket on the lightning EM field is weakened. The cable wiring of Case d tends to be affected by the structure of PV bracket. In terms of the bracket with two-leg structure, the induced voltage of Case d is greater than that of Case b when lightning strikes at S_5. However, the voltages in Case d are reduced for the bracket with one-leg structure. It is even lower than that of Case f if S_1 is struck. Nevertheless, the induced voltages of Case f are the smallest for both one-leg bracket and two-leg bracket in most cases. Moreover,

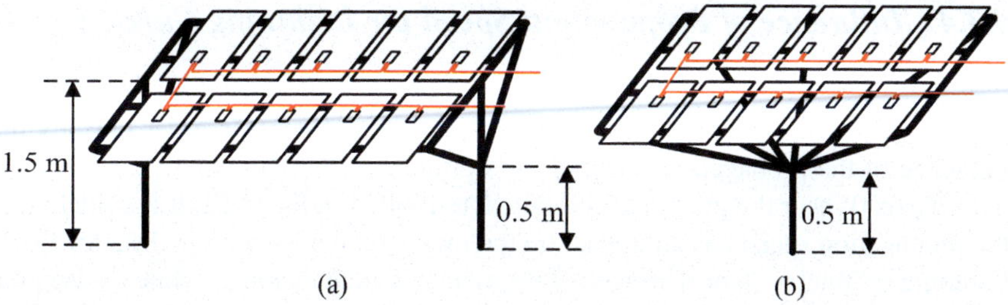

Fig. 4.20 Mounting system. **a** Two-leg structure. **b** One-leg structure

Fig. 4.21 Induced voltage of PV array with various mounting systems. **a** Two-leg structure. **b** One-leg structure [1]

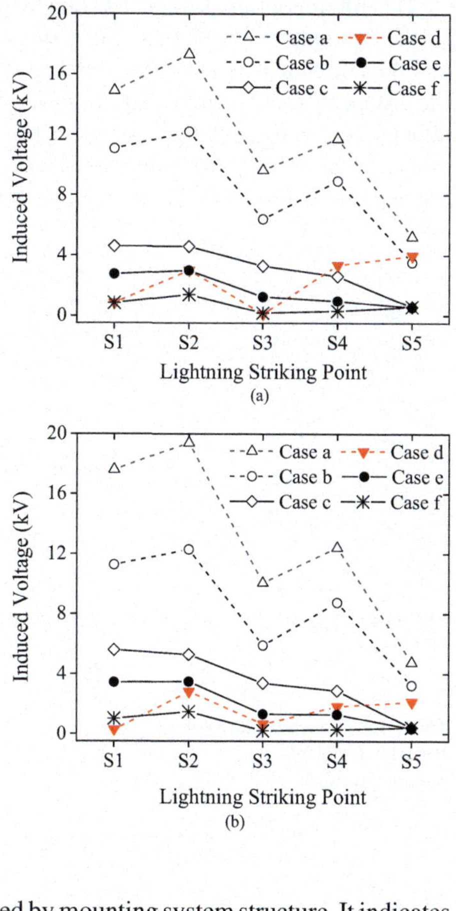

the cable wiring of Case f is hardly affected by mounting system structure. It indicates that the proposed crossover wiring scheme is suitable for various mounting systems.

4.3.4 Influence of Propagation Speed for Lightning Current

The propagation speed of lightning channel is not a constant. In order to study the influence of the propagation speed of the lightning current, the lightning magnetic field $\boldsymbol{B}_{\phi}$ $(r, 0)$ near the ground is computed, as plotted in Fig. 4.22. It is evident that the propagation speed has an impact on the magnetic field when the distance to the lightning channel is 50 m. The peak of the magnetic field increases while its rise time decreases, due to the greater speed v which increases the high frequency component of the EM field. Nevertheless, for the area near the lightning channel, such as $r =$ 2 m, the effect of the speed v on the magnetic field is less.

The differential-mode induced voltage of DC cable in the PV array is determined by the lightning magnetic field. Since the peak of the magnetic field becomes larger and the rise time is reduced, the induced voltage will rise as speed v increases. For the distance r is 50 m, the voltage has a remarkable reduction, which can reach up to about 56% compared than the previous results, as shown in Table 4.3. However, the voltage is hardly affected by the speed v when the distance r is 2 m, and the variation range of the induced voltage is less than 1%.

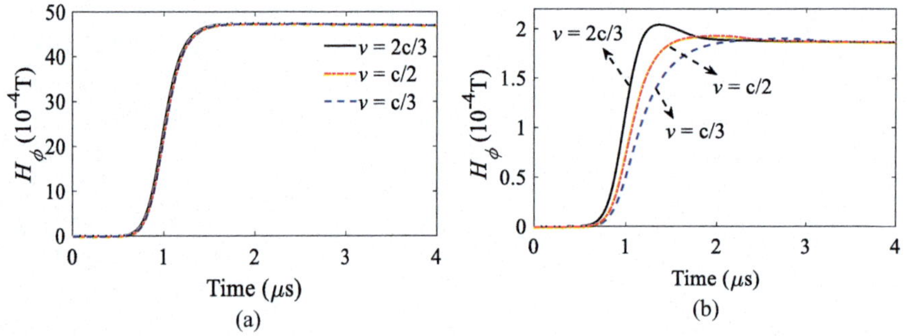

Fig. 4.22 Magnetic field near ground at different speeds of lightning current. **a** $r = 2$ m; **b** $r =$ 50 m [1]

Table 4.3 Induced voltage of PV array for different propagation speeds [1]

r (m)	v (m/s)	Voltage (kV)
2	2c/3	3.90
	c/2	3.87
	c/3	3.85
50	2c/3	0.49
	c/2	0.42
	c/3	0.32

4.3.5 Effect of Soil and Grounding Resistance of Bracket

The lightning-induced voltages on PV arrays are calculated by the accurate numerical solutions of Sommerfeld integrals to analyze the effect of soil. As shown in Fig. 4.23**a**, the induced voltage on the soil with the finite conductivity is larger than that on a perfectly-conducting ground plane, which is consist with the conclusion in [15]. The PV array is distinct from the distribution line. The induced voltage of the PV array is determined by the lightning magnetic field. Since the magnetic field is affected by the soil, the magnetic field on the soil with finite conductivity is greater.

Considering the realistic soil, the conductivity is set to be 0.001 S/m and the relative permittivity of the soil is 10 in the MoM simulations. The induced voltages of PV arrays with various cable wirings is calculated for lightning striking at S_1, as plotted in Fig. 4.23b. It can be revealed that the soil shows the same impact under different wiring schemes. The conclusions regarding the wiring scheme of DC cables are still valid. Additionally, the induced voltages of the PV array with the different grounding resistances of each leg for the PV bracket are evaluated, as listed in Table 4.4.

Table 4.4 indicates that the induced voltage of the PV array rises as the grounding resistance of the bracket increases. Principally, the currents induced on the frames can reduce the induced voltage of PV array according to Lenz's law. Since a high grounding resistance of the bracket decreases the currents induced on PV frames, the effect of the frames on the induced voltage is weakened and the voltage becomes

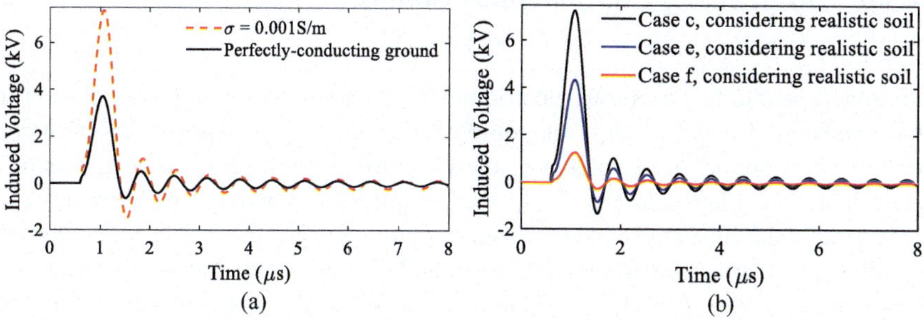

Fig. 4.23 Induced voltage of PV array on realistic soil. **a** Differences between the perfectly-conducting ground and the soil with the finite. **b** Induced voltages for various cable wirings [1]

Table 4.4 Induced voltage for different grounding resistances of bracket [1]

Grounding Resistance (Ω)	1.5	3	10	30	50	100
Case c (kV)	4.74	4.79	5.11	5.52	5.63	5.71
Case f (kV)	0.84	0.86	0.95	1.01	1.02	1.02

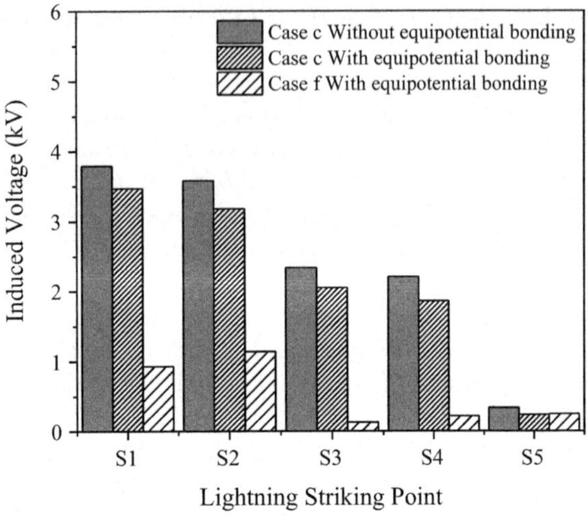

Fig. 4.24 Effect of equipotential bonding on induced voltage of PV array

greater. The induced voltages are close to each other for the low grounding resistances. The grounding resistance plays a key role in the case of directing lightning strike. However, in the case of indirect lightning strike, the influence of a low grounding resistance on the induced voltage is minor.

4.3.6 Influence of Equipotential Bonding

In order to avoid the hazardous electric potential caused by the lightning current in PV bracket and metal frames, the equipotential bonding is required for the metal frames [12]. The induced voltage of the PV array with the equipotential bonding is evaluated, as plotted in Fig. 4.24. The equipotential bonding of PV frames can reduce the induced voltage, and the drop is approximately 8–15%. The equipotential bonding of PV array restricts the EM field, and the hazardous electric potential thus drops in the cables. Clearly, the equipotential bonding is not contradictory to the proposed crossover wiring. The voltages of Case f with the equipotential bonding are still at a lower level. Furthermore, the proposed wiring has a more significant drop of lightning induced voltage than that of the equipotential bonding.

4.3.7 Crossover Wiring for Multiple Photovoltaic Arrays

In a large-scale PV plant, several modules are installed above the mounting system, forming a unit, and a number of PV units are connected to an inverter. The crossover

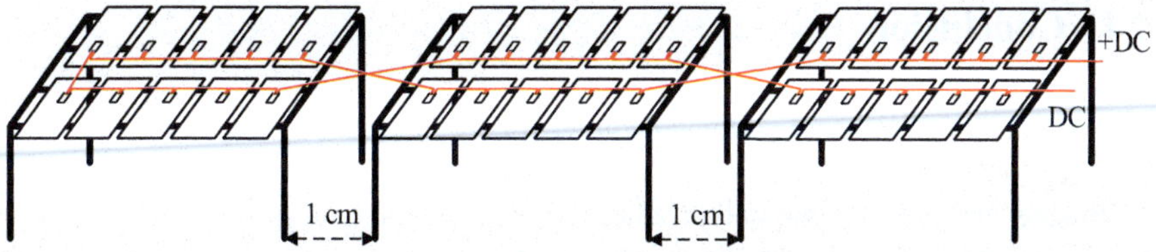

Fig. 4.25 Multiple PV arrays connected in series with crossover wiring

wiring scheme is not only used to connect PV modules, but also suitable for the cable connection between PV units. Three units are cross-connected, as shown in Fig. 4.25, the induced voltages between the DC outputs are compared with that of the conventional connection.

The induced voltage of PV units with conventional cable wiring reaches up to 7.0 kV, as plotted in Fig. 4.26. The voltage is decreased by more than 50% and it drops to 3.1 kV when the PV units are connected by crossover cables. However, the peak voltage still threatens the inverters. To further reduce the induced voltage, PV modules are connected with the proposed crossover wiring as well. The peak voltage is 0.71 kV, which is similar to the induced voltage for single PV unit. The result indicates that the crossover wiring can effectively minimize the lightning induced voltage. The voltage does not rise rapidly versus the number of PV units. Consequently, the proposed crossover wiring scheme is very suitable for large-scale PV plants.

Fig. 4.26 Induced voltage of PV array with equipotential bonding [1]

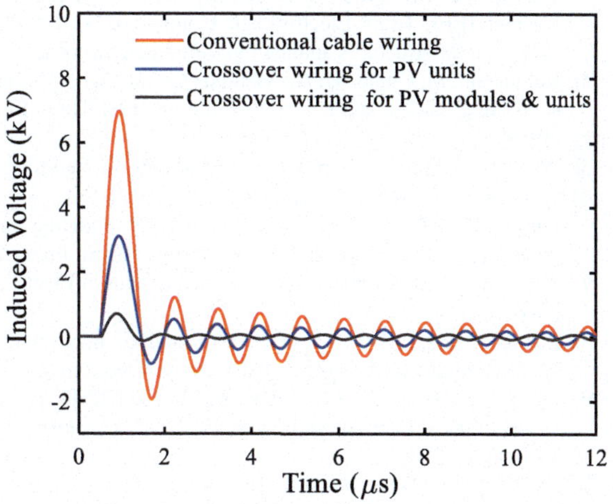

4.4 Conclusion

The mechanism of lightning induced voltage for PV system is elucidated, and the EM coupling between the bracket and the cable is studied. Since the induced current in the bracket and metal frame will produce a magnetic field in opposite to the lightning magnetic field, the induced voltage is reduced.

A novel crossover wiring scheme of cable for PV systems is proposed, and the induced voltages between ± DC cables with various wirings are compared. The results demonstrate that the proposed wiring can significantly reduce the induced voltage. It is easy to implement and suitable for various conditions without any additional protective devices.

Moreover, the symmetrical loop of cable is beneficial to the offset of lightning magnetic flux. The magnetic flux is influenced by the oblique angle of PV array, and the induced voltage rises as the PV oblique angle increases. The induced voltage can also be minimized by the equipotential bonding of PV frames.

References

1. Sun Q. Q, Zhong, X, et al.: A novel crossover wiring of DC cable for photovoltaic system against lightning-induced overvoltage. IEEE Trans. Electromagn. Compat. 65(2), 475–485 (2023).
2. Ramarao, G. Chandrasekaran, K.: Lightning Channel-Base-Current Estimation Using Engineering Return-Stroke Models from Measured Magnetic Field Based on Deconvolution Method. IEEE Trans. Electromagn. Compat. 62(3), 798–806 (2020).
3. Protection of structures against lightning: Part I general principles. IEC 62305–1 (2010).
4. Rubinstein, M. Uman, M. A.: Methods for calculating the electromagnetic fields from a known source distribution: application to lightning. IEEE Trans. Electromagn. Compat. 31(2), 183–189 (1989).
5. Wang, Y. Zhang, X. and Tao, S.: Modeling of lightning transients in photovoltaic bracket systems. IEEE Access. 7, 12262–12271 (2019).
6. Sun, Q. Zhong, X. et al.: Three-dimensional modeling on lightning induced overvoltage for photovoltaic arrays installed on mountain. J. Clean Prod. 288, 125084 (2021).
7. Alexander, C. K. Sadiku, M. N.O. Fundamentals of Electric Circuits. New York: McGraw-Hill, 2010.
8. Zhang, Y. Chen, H. Du, Y. et al.: Considerations of photovoltaic system structure design for effective lightning protection. IEEE Trans. Electromagn. Compat. 62, 1333–1341 (2020).
9. Stern, H. Karner, H. C. "Lightning induced EMC phenomena in photovoltaic modules," Proceedings of International Symposium on Electromagnetic Compatibility, Dallas, TX, USA, 442–446 (1993).
10. Zhang, Y. Li, B. Du, Y. et al.: Effective grounding of the photovoltaic power plant protected by lightning rods. IEEE Trans. Electromagn Compat. 63(4), 1128–1136 (2021).
11. Charalambous, C.A. Kokkinos, N.D. and Christofides, N.: External lightning protection and grounding in large-scale photovoltaic applications. IEEE Trans. Electromagn. Compat. 56, 427–434 (2014).
12. Ahmad, N. I. Ab-Kadir, M. Z. A. Izadi, M. et al.: Lightning protection on photovoltaic systems: a review on current and recommended practices. Renew. Sust. Energ. Rev. 82, 1611–1619 (2018).
13. De Conti, A. Visacro, S.: Analytical Representation of Single- and Double-Peaked Lightning Current Waveforms. IEEE Trans. Electromagn. Compa. 49(2), 448–451 (2007).

14. Altair FEKO 2019.3.2 User's Manual, EM Software and Systems. Stellenbosch, South Africa: S. A. (Pty), Ltd., (2019), http://www.feko.info.
15. Hossain, A. Ahmed, R.: Analysis of indirect lightning phenomena on solar power system. J. Electr. Eng. 21, 127–133 (2014).

Chapter 5
Novel Solar-Cell String Wiring of Photovoltaic Module for Reducing Lightning Induced Overvoltage

5.1 Solar-Cell String Wiring of Photovoltaic Module

5.1.1 Crossover Connection of Solar-Cell String

One solar cell only can provide a very low voltage in full sunlight [1]. To obtain a larger output voltage, tens of solar cells encapsulated inside a PV module are connected in series. The structure of solar-cell string is presented in Fig. 5.1, the metal grid of a cell is connected to the back of another cell by using thin wire. The metal grid is the negative polarity of cell, and its positive polarity is the back of cell.

In terms of common wiring for solar-cell string, the positive string and negative string are arranged side by side and connected in series, as illustrated in Fig. 5.2a. Based on the conventional wiring, a new solar-cell string wiring is proposed to reduce the lightning induced voltage of PV panel shown in Fig. 5.2b. In the proposed wiring, the third and fourth solar-cell strings are connected by crossover wires. The induced voltage polarity of the middle string is reversed by crossover wiring among these strings. To keep the output voltage of PV panel unchanged, the solar-cell strings need to be connected in series. Therefore, the solar-cell strings with the same polarity are arranged in parallel for the proposed wiring.

The new solar-cell string wiring is easy to implement and cost-effective for the manufacturer. On the one hand, the solar-cell structure is unchanged and is still suitable for the process of manufacturers. On the other hand, the proposed wiring scheme also connects the solar cells in series, and the output power of PV panel is the same as that of the conventional wiring. Since the solar-cell strings are in crossover series, a two-layer wiring is presented in Fig. 5.1. The additional ethylene vinyl acetate (EVA) film is used in the middle of the two-layer wiring for the insulation. As illustrated in Fig. 5.2, a PV panel with the proposed solar-cell string wiring is encapsulated with aluminum frame, the cover glass, EVA film, solar-cell string, EVA film and backsheet.

Q. Sun et al., *Lightning Transients and Protection for Renewable Energy System*,
https://doi.org/10.1007/978-981-97-7267-4_5

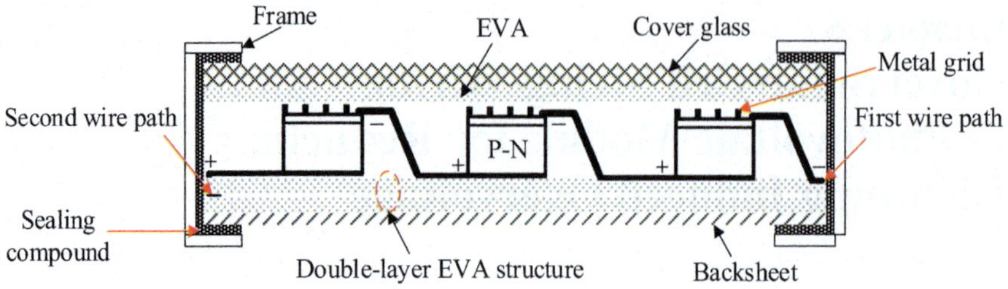

Fig. 5.1 Schematic diagram of solar-cell string

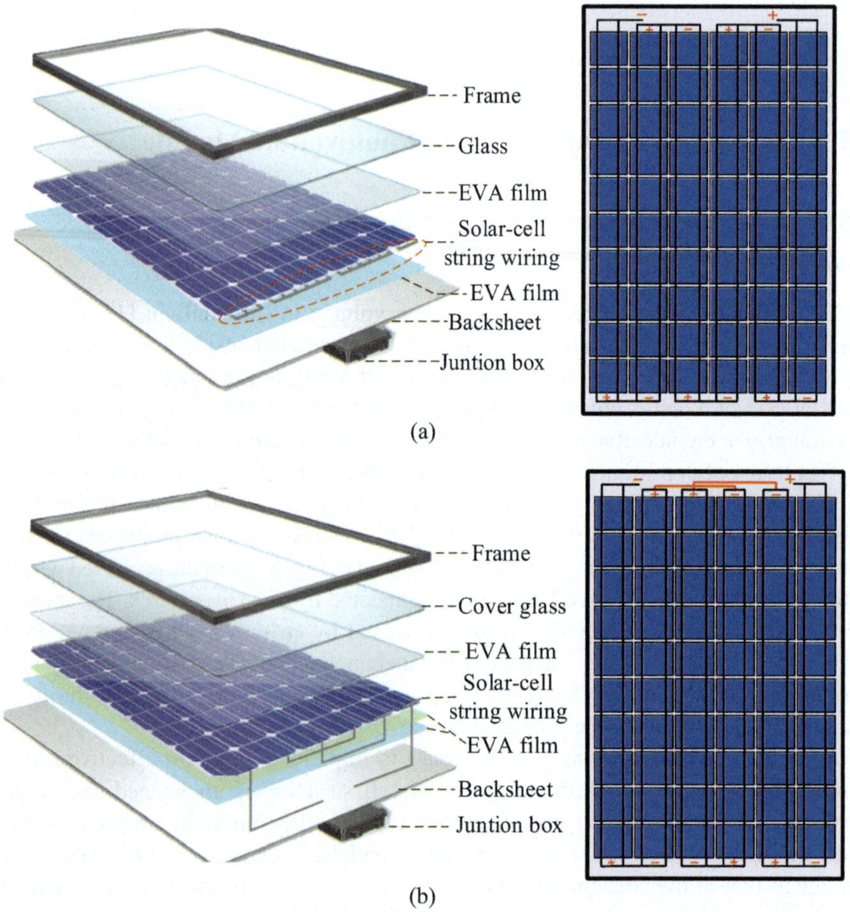

Fig. 5.2 PV panel. **a** Common wiring. **b** Proposed wiring [2]

5.1.2 Mechanism of Induced Voltage Reduction

The solar cell is a P–N junction, which is often represented by a diode in operation. The equivalent circuit of the solar-cell string inside a PV module is given in Fig. 5.3. The polarities of the lightning induced voltages ($\boldsymbol{U_1}$, $\boldsymbol{U_2}$, $\boldsymbol{U_3}$) for the three solar-cell strings depend on the lightning EM field. Due to the small loop area of PV module, the direction of the lightning EM field passing through the PV module is almost constant. Therefore, the polarities of U_1, U_2, U_3 are the same, as illustrated in Fig. 5.3.

As shown in Fig. 5.3a, the induced voltage $\boldsymbol{U}$ between the outputs of the PV panel with common solar-cell string wiring is the sum of the voltage of the three strings,

$$\boldsymbol{U}=\boldsymbol{U}_1+\boldsymbol{U}_2+\boldsymbol{U}_3 \tag{5.1}$$

Since the solar-cell strings proposed in Fig. 5.3b are connected by crossover wires, the induced voltage $\boldsymbol{U_2}$ of the solar string is reversed in the equivalent circuit of solar-cell string. Hence, the induced voltage $\boldsymbol{U}$ between the outputs of the PV panel with proposed solar-cell string wiring is the subtraction of the induced voltages for the three strings,

$$\boldsymbol{U}=\boldsymbol{U}_1-\boldsymbol{U}_2+\boldsymbol{U}_3 \tag{5.2}$$

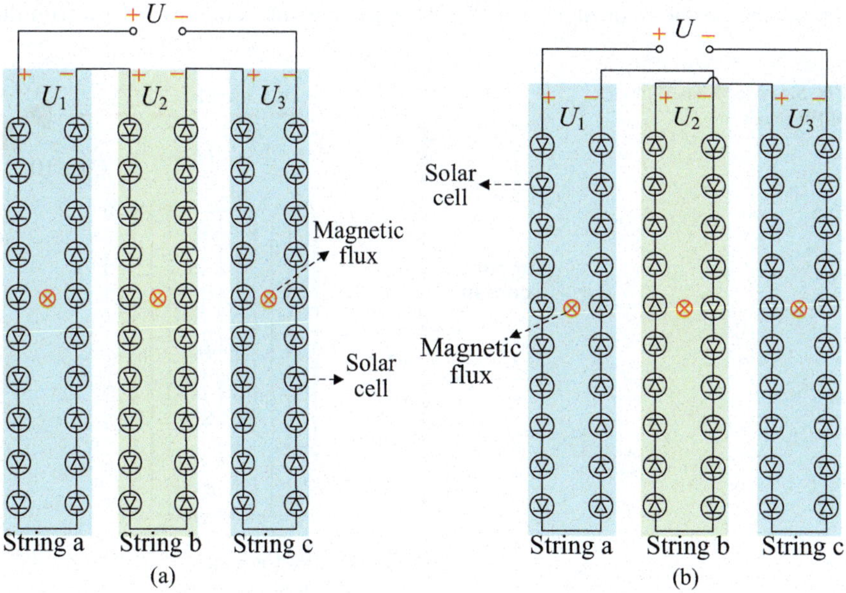

Fig. 5.3 Equivalent circuit of solar-cell string. **a** Common wiring. **b** Proposed wiring [2]

The loop inside a PV module is mainly formed by the solar-cell strings. For the conventional wiring scheme, the lightning induced voltages of all solar-cell strings are added to the out-put of PV module. However, the DM voltage of the PV module with proposed wiring scheme can be reduced by the reversed voltage of the middle solar-cell string.

5.1.3 *Effect of Novel Solar-Cell Wiring for Ideal PV Panel*

In order to validate the effectiveness of the proposed solar-cell string wiring, single PV module shown in Fig. 5.4 is studied. It consists of six rows of solar cells with 60 cells connected in series. The length and width of PV module are 0.98 and 0.66 m, respectively. To facilitate the computation, the diodes are supposed closed when cells are in operation. The solar-cell strings are thus represented by the wires with a radius of 0.1 cm in this paper. The frame of PV module is a C-shaped structure, and its thickness, width and length are 0.3, 1 and 2 cm.

The method of moments (MoM) is used to calculate the induced voltage of the PV panel. The MoM-based model is built in the software FEKO [3], as illustrated in Fig. 5.4. The impulse current is set to be 2 m in length, and the distance to the PV panel is 2 m. The current waveform is 1/200 μs and its amplitude is set to be 50 kA to represent the impulse EM field. The computation is carried out over the frequency range from 2.93 kHz to 1.5 MHz with the increment step of 2.93 kHz, and the number of the frequencies is 512. The time interval from 0 to 341 μs with the

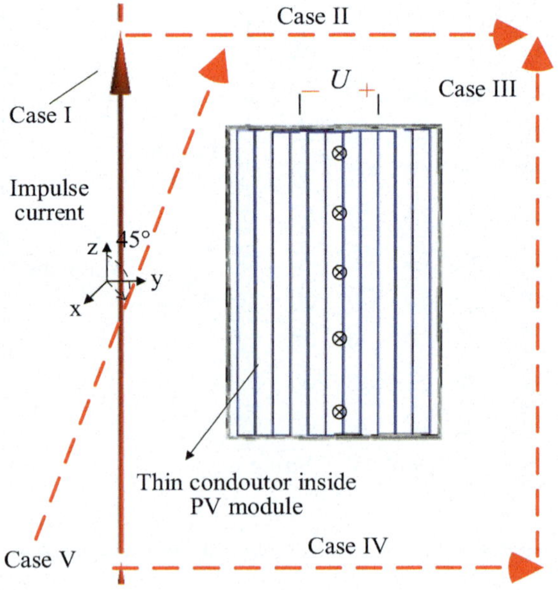

Fig. 5.4 MoM-based model of PV panel

increment step of 1/3 μs, and 1024 points in inverse fast Fourier transform are thus used to obtain the time-domain response. More details about the calculation have been given in [4].

The induced voltages of two PV modules with and without metal frame are calculated, respectively, as plotted in Fig. 5.5. Compared with conventional solar-cell string wiring, the proposed wiring can significantly reduce the DM voltage of PV module. The reduction is more than 50% for both PV modules with and without frame. Moreover, the induced voltages of PV module at various positions of the impulse current in Fig. 5.4 are evaluated. Table 5.1 shows that the DM voltages of the PV panel with the proposed solar-cell string wiring are all in a relatively safe range. Though the impulse current arrangement of Case II is symmetrical to that of Case IV, the solar-cell string of a PV panel is not fully symmetrical. The top of the solar-cell string is slightly different with its bottom. Therefore, the peak voltage in Case II is not the same with that in Case IV for the proposed wiring. Nevertheless, the two values are very close to each other.

In the ideal case, the impulse current is parallel to the PV panel, the magnetic flux passing through the solar-cell string is the largest and hence the induced voltage U_2 of the second cell string. In addition, only a PV panel is considered and the

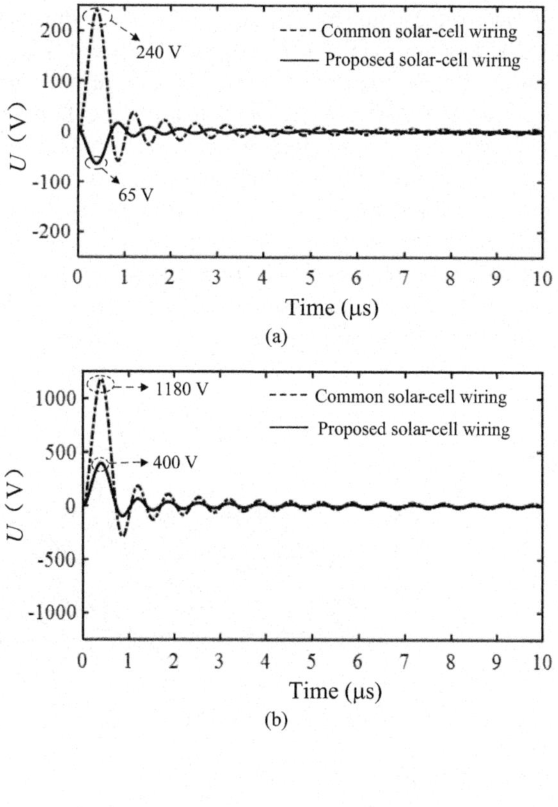

Fig. 5.5 Induced voltage of PV module. **a** With frame. **b** Without frame [2]

Table 5.1 Induced voltage peak of PV module (V) [2]

Relative position	Conventional solar-cell wiring	Proposed solar-cell wiring
Case I	240	65
Case II	218	70
Case III	240	65
Case IV	218	72
Case V	167	48

other conductors of PV system are neglected. Therefore, the difference between the induced voltages of conventional wiring and proposed wiring is the greatest.

5.2 Effect of Solar-Cell Wiring on Photovoltaic Array

Taking a PV array on flat ground into account, as shown in Fig. 5.6, it consists of ten PV panels named P1, P2… P10. The details of PV module are given in Sect. 5.1.3, and the equivalent loop of solar-cell string is presented in Fig. 5.3. Similarly, the solar-cell string is represented by the conductor with a radius of 0.1 cm. The maximum voltage of DC loop for the PV array U_{DCmax} is 880 V. The length and width of the mounting system for PV array are 3.28 and 1.94 m, respectively. To facilitate analysis, metal frames and PV bracket are modelled by the cylindrical conductors.

Lightning rods are often installed near PV bracket. To avoid the shadow, the rod of PV array cannot be too high and its height is set to be 3 m. The distance between the rod and PV array is 2 m. Suppose lightning directly strikes at lightning rod and the channel is vertical to ground with the length of 1 km. The MTLE model is used to model the return current. The propagation speed of lightning current on the channel is set to be 2×10^8 m/s and the decay constant is 2000 m. The peak of the lightning current is 50 kA, and its waveform is set to be 1/200 μs to model the first return

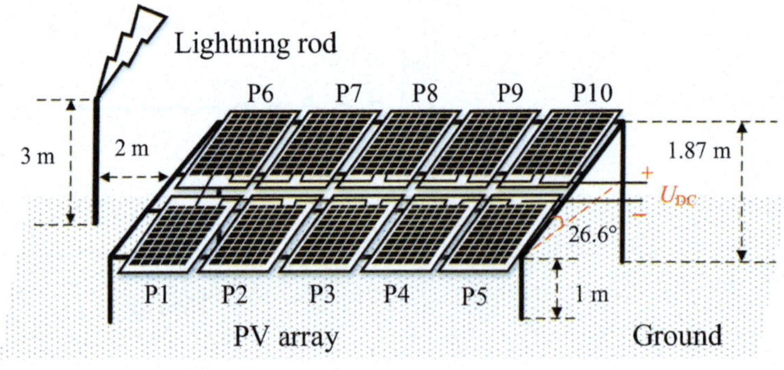

Fig. 5.6 PV array on flat ground [2]

stroke. Lightning currents are random and complex. To facilitate the analysis, the lightning current is represented by the Heidler function [5]. The proposed wiring is also suitable for other lightning current waveforms.

5.2.1 Perfectly-Conducting Ground

The PV array is mirrored to consider a perfectly-conducting ground, and the induced voltage between the outputs of PV array is plotted in Fig. 5.7. Compared to conventional wiring scheme, the voltage drop of proposed wiring is approximately 1.10 kV. The DC loop of a PV array is made up of the equivalent circuit inside PV module and the cables used to connect PV modules. The proposed wiring will reduce the lightning induced voltages between the outputs of PV modules, and hence the induced voltage of PV array.

The DM induced voltage of individual PV panel is calculated, as illustrated in Fig. 5.8a. When the traditional solar-cell string wiring is replaced by the proposed wiring, the voltages of most PV panels drop by around 30%-40%. A few PV panels have 53% voltage reduction (P4) and 4% voltage reduction (P5). The induced voltage of rare PV panel increases. For instance, the voltage of P1, changes from 0.18 to 0.27 kV. If ignoring the PV bracket, the voltage of P1 reduces from 0.79 to 0.48 kV. The lightning withstand voltage of a PV panel is determined by the BPD. Regarding our case, it is about 0.13 kV [6]. The result shows that the proposed wiring can protect 30% BPDs of PV panels from indirect lightning strikes. The proposed wiring reduces the number of the SPDs used in PV systems and the lightning damage to solar cells.

As plotted in Fig. 5.8b, the proposed wiring can also decrease the induced voltages of frameless PV panels. It is clear that most of the PV panels without metal frame have larger induced voltages. The lightning EM field will generate the induced current on the metal frame. According to Lenz's law, the direction of the induced current would generate an EM field to weaken the original EM field caused by lightning current, and hence the induced voltage of PV panel.

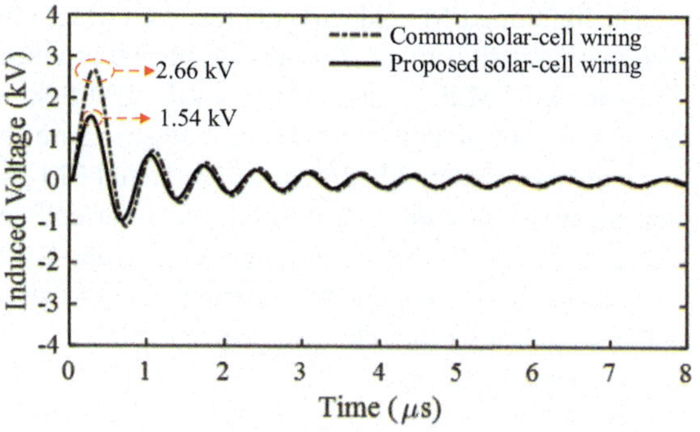

Fig. 5.7 Induced voltage of PV array on perfectly-conducting ground [2]

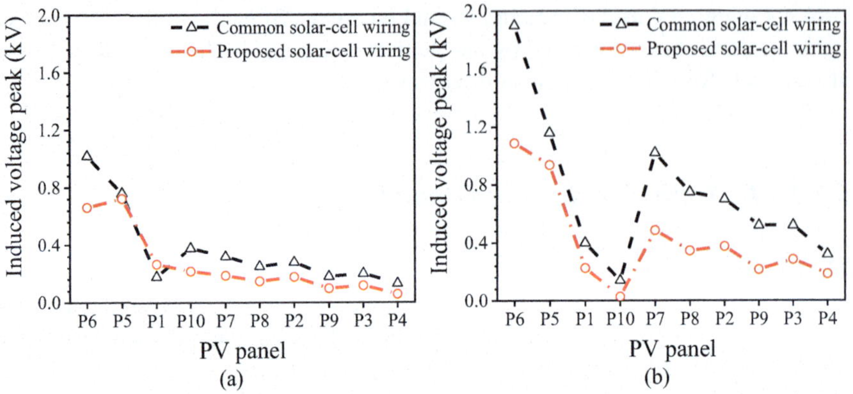

Fig. 5.8 Induced voltage of PV panel on perfectly-conducting ground. **a** PV panels with frame. **b** PV panels without frame [2]

Moreover, the induced voltage of PV panel is not only dependent on the lightning EM field, but also influenced by the metal frames, the brackets as well as the supporting legs. The PV array consists of many metal conductors, and the circuits of the bracket and frames have a number of branches, and the lightning induced current on the bracket is irregular. Therefore, though the width of PV array is just about 3–4 m, the induced voltages for PV panels at various positions of the bracket are quite different.

As illustrated in Fig. 5.9, the equivalent loop of a PV panel includes not only the circuit of solar-cell strings but also the additional wires connected to the outputs inside the junction box. Additionally, there is a double-layer wiring in the PV panel with proposed wiring. The upper layer wiring and down layer wiring also form another loop shown in Fig. 5.9. The induced voltages of these loops cannot be cancelled by the proposed wiring. Moreover, Table 5.2 indicates that the proposed wiring has none effect on the induced voltages of the first and third column solar-cell strings (U_1 and U_3). But for the second column solar-cell string, its voltage peak is decreased (U_2). Since the reverse voltage U_2 is minimized by the proposed wiring, according to Eq. (5.2), the voltage reduction of a single PV panel is not greater than 50%.

The lightning EM field will generate the surface current on metal structures, due to its remarkable skin effect. The surface current produces an additional EM field, called the scattered EM field, and influences the EM field in space. Since the positions of the wires connected the second solar-cell string are changed by the crossover wiring, more reverse scattered EM field and reflected EM wave are generated by the wires. The magnetic flux passing through the solar-cell string is reduced and hence the induced voltage of the second solar-cell string U_2. Compared to the conventional solar-cell string wiring, the metal structures of other solar-cell strings are unchanged in the proposed wiring. Hence, the peak voltages of U_1 and U_3 are the same for the two wirings.

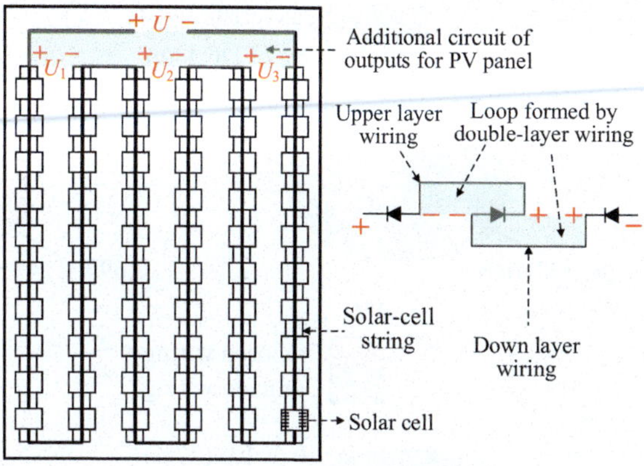

Fig. 5.9 Loop of PV panel [2]

Table 5.2 Induced voltage of solar-cell string [2]

PV panel	Common wiring (kV)			Proposed wiring (kV)		
	U_1	U_2	U_3	U_1	U_2	U_3
P1	− 0.39	0.12	0.16	− 0.39	0.06	0.16
P2	0.10	0.10	0.07	0.10	0.05	0.07
P3	0.11	0.07	− 0.02	0.11	0.03	− 0.03
P4	0.13	0.06	− 0.06	0.13	0.03	− 0.06
P5	0.18	− 0.01	− 0.93	0.18	− 0.02	− 0.93
P6	0.66	0.23	0.14	0.66	0.13	0.14
P7	0.17	0.11	0.04	0.17	0.05	0.04
P8	0.14	0.09	0.02	0.14	0.04	0.02
P9	0.12	0.06	− 0.02	0.12	0.03	− 0.02
P10	0.15	0.09	0.14	0.15	0.07	0.14

Note They are maximum values and represent vector voltages

The lightning EM field will generate the surface current on metal structures, due to its remarkable skin effect. The surface current produces an additional EM field, called the scattered EM field, and influences the EM field in space. Since the positions of the wires connected the second solar-cell string are changed by the crossover wiring, more reverse scattered EM field and reflected EM wave are generated by the wires. The magnetic flux passing through the solar-cell string is reduced and hence the induced voltage of the second solar-cell string U_2. Compared to the conventional solar-cell string wiring, the metal structures of other solar-cell strings are unchanged in the proposed wiring. Hence, the peak voltages of U_1 and U_3 are the same for the two wirings.

Nevertheless, the solar-cell wiring proposed in this paper can still significantly reduce the induced overvoltage of most PV panels. Figure 5.10c indicates that the total voltage of ten PV panels decreases from 1.82 to 0.69 kV, comparing the conventional

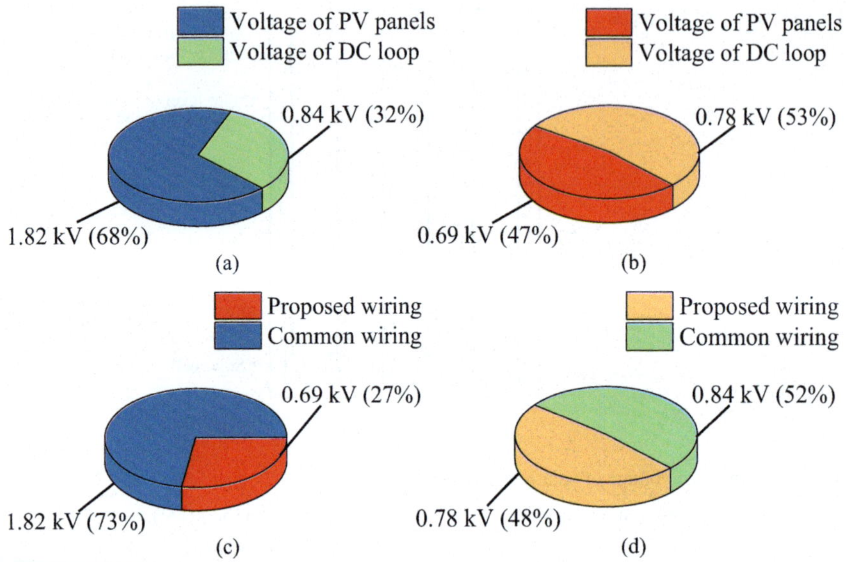

Fig. 5.10 PV array on perfectly-conducting ground. **a** Composition of induced voltage of PV array with common wiring. **b** Composition of induced voltage of PV array with proposed wiring. **c** Comparison of total voltages of PV panels with different wirings. **d** Comparison of total voltage on DC cable with different wirings [2]

wiring and proposed wiring. The induced voltage of a PV array can be divided into the sum of the voltages for PV panels and the induced voltage on DC cables. As far as the common wiring is concerned, the total voltage of PV panels, accounts for 68% of the induced voltage for PV array, is much larger than the total voltage on DC cables, as shown in Fig. 5.10a. But for the proposed wiring, the contribution for the voltages of all PV panels declines significantly, and accounts for only 47% shown in Fig. 5.10b, which is lower than that of the voltage on cables. However, the proposed wiring has little influence on the induced voltage on DC cables, as illustrated in Fig. 5.10d.

5.2.2 Soil with High Resistivity

The grounding system of PV array is individually arranged by bare conductor, which is used as a natural grounding system [7]. The induced voltages of PV array on the lossy ground with the resistivity of 100 and 1000 Ωm is presented in Fig. 5.11, respectively. The soil resistivity has a great influence on the lightning induced voltage of PV array. As the soil resistivity increases, the voltage peak rises significantly. Though the PV array is on the ground with high soil resistivity, the proposed solar-cell string wiring can still reduce the induced voltage of PV array.

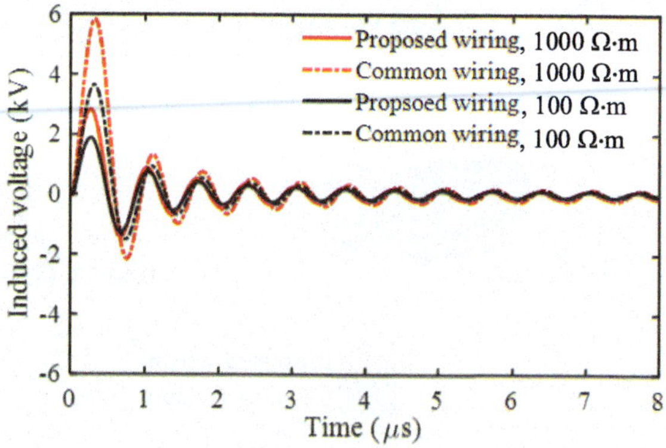

Fig. 5.11 Induced voltage of PV array on lossy ground [2]

As shown in Fig. 5.12, the induced voltage of most PV panels on lossy ground can be reduced by the proposed wiring, and the voltage reduction is in the range from 38 to 50%. A few PV panels have about 14% voltage drop, such as P1 and P6. Since the induced current on the supporting legs of the PV bracket is greater, the effect of the current is remarkable and the voltages of P5 and P10 are relatively larger.

Nevertheless, the total of output voltages for all PV panels decreases significantly compared to that of the conventional wiring, as illustrated in Fig. 5.13. As the soil resistivity increases, the total voltage of PV panels with proposed wiring is almost constant whereas that of traditional wiring will rise significantly. The induced voltage of PV array with common wiring is also determined by the sum of voltages for all PV panels, which accounts for 66%. In contrast, the total induced voltage on DC cables plays a key role in the voltage of PV array with proposed wiring while the total voltage of PV panels accounts for only 35%, as shown in Fig. 5.13b.

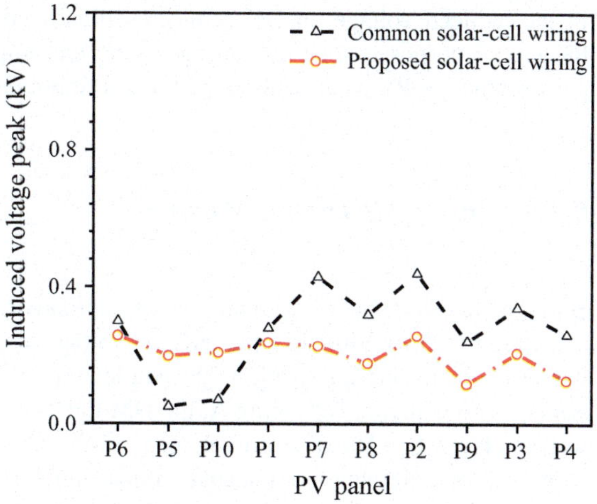

Fig. 5.12 Induced voltage of PV panel on soil with resistivity of 100 Ωm [2]

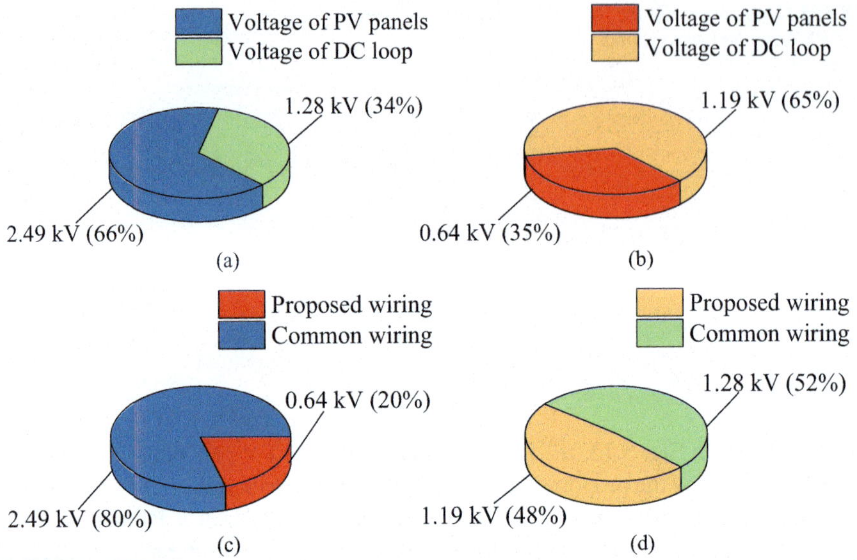

Fig. 5.13 PV array on soil with resistivity of 100 Ωm. **a** Composition of induced voltage of PV array with common wiring. **b** Composition of induced voltage of PV array with proposed wiring. **c** Comparison of total voltages of PV panels with different wirings. **d** Comparison of total voltage on DC cable with different wirings [2]

Comparing Fig. 5.10c with Fig. 5.13c, it is evident that the performance of proposed wiring on lossy soil is better than that on perfectly-conducting soil. In terms of the proposed wiring, the voltage rise of PV array is generated by high soil resistivity and the difference of the voltage on cables is the major contribution. Since the proposed wiring changes the circuit of PV module, there is no remarkable effect in the induced voltage on DC cables, as illustrated in Fig. 5.13d. Consequently, the use of proposed solar-cell string wiring can improve the lightning protection performance of PV panel without additional lightning protection measures.

5.2.3 *Other Alternative Wirings*

Based on the same principle, a number of alternative wiring schemes are studied, as shown in Fig. 5.14. Similarly, the induced voltage of one solar-cell string is opposite to that of other solar-cell strings in the whole loop of a PV panel, due to the crossover wires. For the alternative wiring A, the DM induced voltage $\boldsymbol{U}$ of PV panel is equal to $\boldsymbol{U}_1 - \boldsymbol{U}_2 + \boldsymbol{U}_3$. The induced voltage $\boldsymbol{U}$ of the PV panel with alternative wiring B is the voltage subtraction of the solar-cell strings I, II and III:

$$\boldsymbol{U}=\boldsymbol{U}_{\mathrm{I}} - \boldsymbol{U}_{\mathrm{II}}+\boldsymbol{U}_{\mathrm{III}} \tag{5.3}$$

To study the effect of the two wiring schemes, the induced voltages of PV panels within the PV array, which is presented in Section. III. C, are calculated and compared with that of the proposed solar-cell wiring. As plotted in Fig. 5.15, both wiring schemes A and B can decrease the induced voltages of these PV panels in the middle of the PV array. The voltage reduction of the wiring B is close to that of the proposed wiring, and the wiring A has a slightly larger voltage drop. However, the PV bracket has a greater influence on the wiring A. It is evident that the wiring A leads to greater voltages induced on the PV panels near the supporting legs of PV mounting system.

5.3 Rooftop Photovoltaic Array with Proposed Solar-Cell String Wiring

The lightning protection for the rooftop PV array is only by means of the traditional measures for building. As illustrated in Fig. 5.16, the lightning protection tape is installed on four sides of the roof. The down conductors are installed at the four corners of the tape. The rooftop PV system, as shown in Fig. 5.16, consists of four PV arrays, named array A, array B, array C and array D. The detailed parameters are given in Section III. The lightning rod is installed on the lightning protection tape at left side of PV array A, and it is 3 m in length. Lightning directly strikes at the lightning rod, and the lightning current is the same as that in Section III.

The rooftop PV array includes a number of metal structures, such as solar-cell string, metal frame, lightning protection tape and building, etc. Considering the EM coupling of these metal structures, the MoM-based models of the building and the rooftop PV array shown in Fig. 5.16 are built in FEKO [3]. The steel structure of building is represented by the combination of cylindrical segments based on its geometry, which satisfy the limitation in the ratio of conductor length to the wavelength in the frequency range of interest.

5.3.1 Assessment of Proposed Crossover Solar-Cell String Wiring

The induced voltage of rooftop PV array with the conventional wiring and the proposed wiring is calculated and compared, as shown in Figs. 5.17 and 5.18.

Clearly, the induced voltages of most rooftop PV modules with proposed wiring are lower. For the four PV arrays, 29 PV panels have smaller voltages and the voltages of 4 PV panels are similar to those of conventional wiring. Only seven PV panels with proposed wiring have relatively larger voltages, and they are mainly around the supporting legs of the PV bracket, such as P5 & P10. Since the induced current on the supporting legs is larger, the EM field generated by the induced current has a great influence on the induced voltages of PV panels, especially for the area with

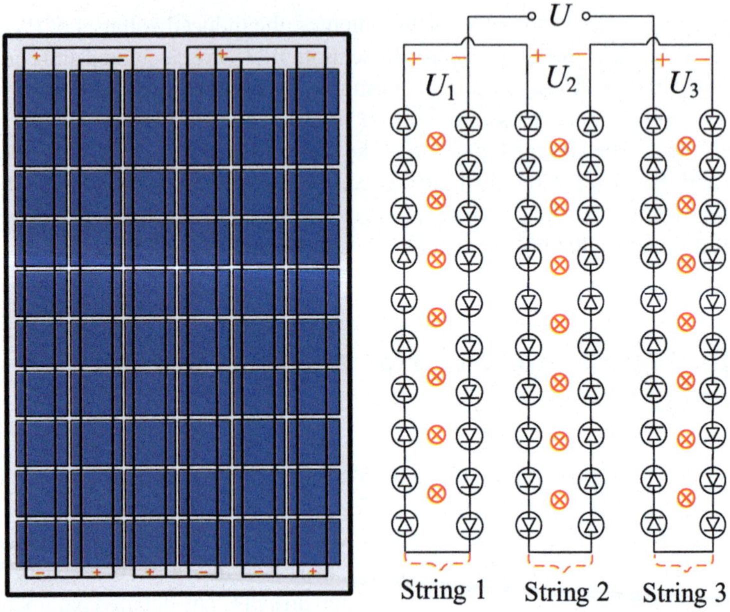

(a)

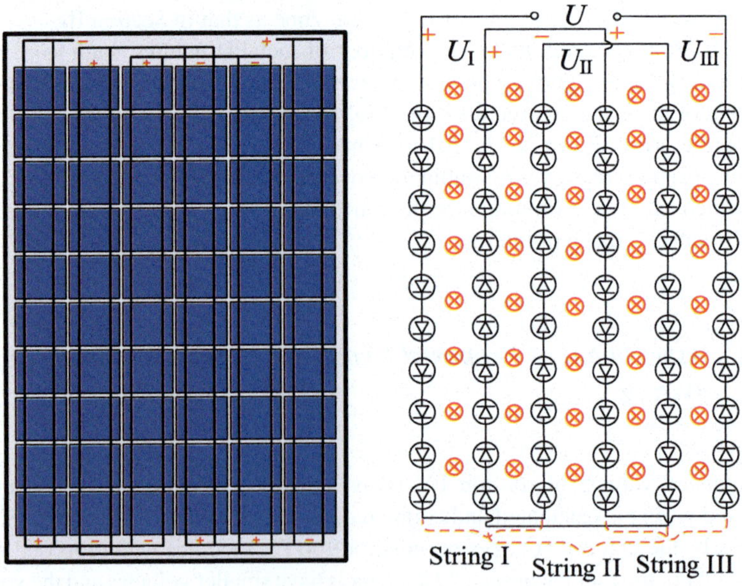

(b)

Fig. 5.14 Other alternative solar-cell string wirings. **a** Wiring A. **b** Wiring B [2]

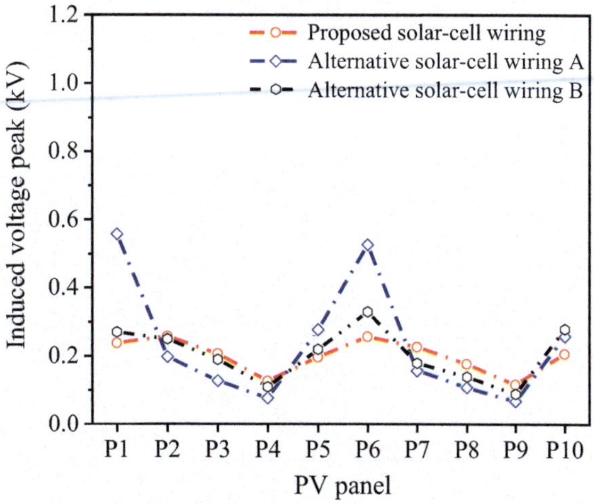

Fig. 5.15 Induced voltage of PV panel with various solar-cell wirings [2]

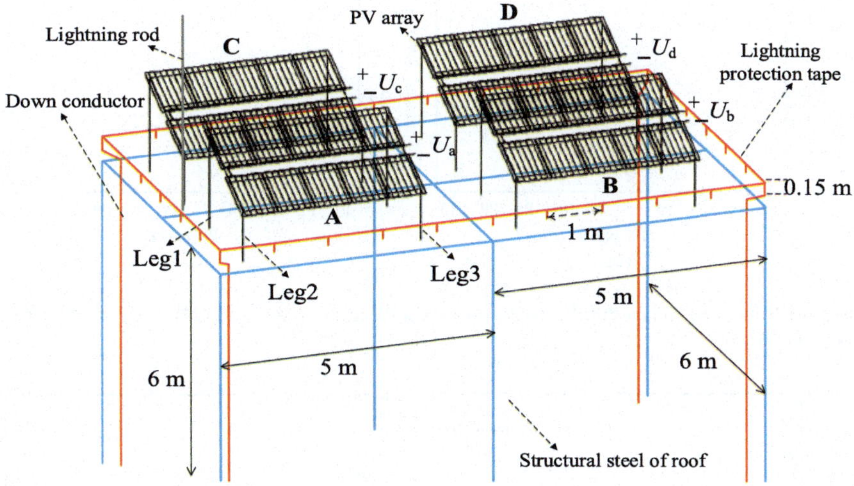

Fig. 5.16 Rooftop PV array

weaker lightning EM field. The farther distance to the rod, the less the magnetic flux through the solar-cell string and hence the smaller the effect of the proposed wiring. Nevertheless, the novel wiring in this paper can still reduce the induced voltage of 70% rooftop PV modules.

The lightning protection tape has a great impact on the induced voltage. Since the lightning rod is connected to the tape, the lightning current will flow on the tape and changes the magnetic flux passing through rooftop PV array. Though the distance between the rod and PV array A is similar to that between the rod and PV array C,

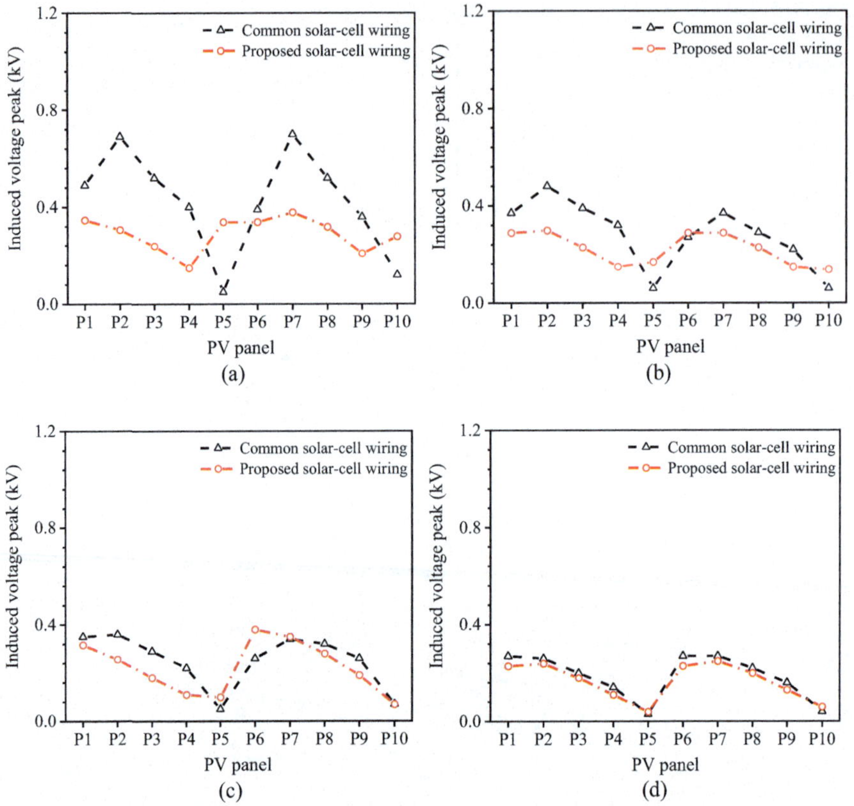

Fig. 5.17 Induced voltages of PV panels on roof: **a** PV array A. **b** PV array B. **c** PV array C. **d** PV array D [2]

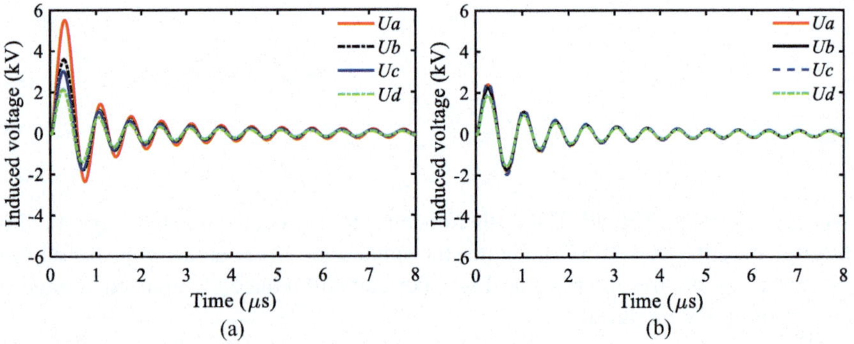

Fig. 5.18 Induced voltage of rooftop PV array. **a** Common wiring. **b** Proposed wiring [2]

the induced voltages of most PV panels for PV array A are larger than those on PV array C, as shown in Fig. 5.17a, c.

Though the voltage reduction for all rooftop PV modules cannot be performed by the proposed wiring, the induced voltages between the outputs of four rooftop PV arrays are decreased. Figure 5.18 shows that the voltage of PV array A is declined from 5.57 to 2.42 kV and that of array B is reduced from 3.60 to 2.24 kV. The voltage reduction of array C and array D are 18% and 13%, respectively. Therefore, the proposed wiring is not only suitable for the PV array installed on flat ground but also the rooftop PV array.

5.3.2 Influence of Lightning Protection Tape and Down Conductors

The influence of the lightning protection tape and the down conductors on the induced voltage of rooftop PV array is studied. There are three cases as follows, Case a: four down conductors are connected to the corners of the tape; Case b: only a down conductor is connected to the base of lightning rod; Case c: down conductors are installed at four corners of the tape and the rod is connected to additional down conductor.

Figure 5.19 shows that the rooftop PV array with this novel wiring has a smaller induced voltage in the three cases. Since the down conductor connected to the lightning protection tape provides a low impedance path for lightning current, the induced voltages of the PV arrays using the two wirings in Case c are the lowest among three cases. The down conductors and the lightning protection tape will influence the distribution of the EM field in space. It is possible that the magnetic flux passing through each solar-cell string has different directions. For some PV panels in Case b, the magnetic flux passing through the second solar-cell string would be opposite to that through the other solar-cell strings, and the induced voltage of the second solar-cell string is reversed. In contrast, the direction of the magnetic flux through the second solar-cell string is almost the same as that through the other solar-cell strings for most PV panels in Case a. Due to the crossover wiring, the voltage peak of the PV array in Case b is greater that in Case c for the proposed wiring.

The metal on the roof is often connected to the lightning protection tape. Taking the supporting leg of PV bracket connected to the tape into account, the induced voltage of PV array is listed in Table 5.3. Connecting the supporting leg to the lightning protection tape can decrease the induced voltage of rooftop PV array. The leg of PV bracket connected to the tape is similar to the equalizing ring. It can make high voltage caused by lightning be evenly distributed around the bracket, and the voltage of PV array is thus reduced. Moreover, the proposed wiring scheme can still reduce the induced voltage of rooftop PV array when the supporting leg is connected to the lightning protection tape. By using the proposed wiring and connecting leg near the lightning rod to the tape, the induced voltage will become the lowest.

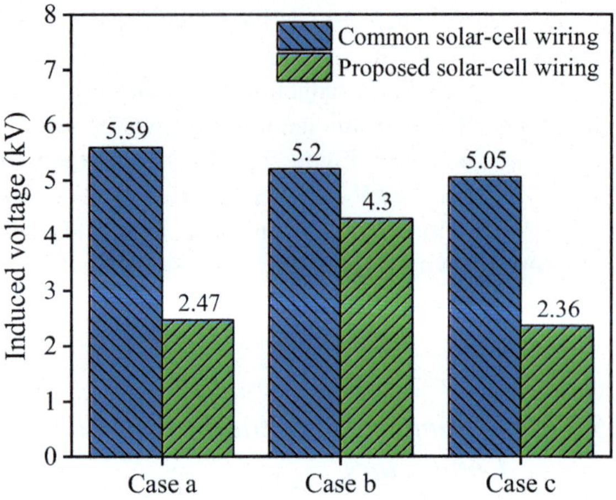

Fig. 5.19 Induced voltage of PV array A under three cases [2]

Table 5.3 Induced voltage of PV array (kV) [2]

Connected supporting leg	Common solar-cell string wiring	Proposed solar-cell string wiring
None	5.59	2.47
Leg1	4.26	1.00
Leg2	5.28	2.01
Leg3	4.74	1.60

As plotted in Fig. 5.20, the proposed wiring scheme can decrease the induced voltages of these PV panels in the middle of the bracket. When leg3 is connected to the lightning protection tape, the lightning current will flow from leg3 into the PV bracket. For the PV panels close to the supporting legs of the bracket, the proposed wiring cannot effectively reduce their voltage due to the lightning current on the bracket. Since the lightning current around leg3 becomes larger, the nearby EM field is greater and the panel P5 has a relatively high voltage. Consequently, except for the PV panels near the supporting legs, the proposed wiring is recommended for other PV panels to protect all PV panels against lightning as possible.

Additionally, the lightning-induced voltage of rooftop PV arrays on lossy ground is calculated to anylyze the influence of soil, as listed in Table 5.4. The voltage peak on a lossy soil is larger than that on the perfectly-conducting ground, which is consistent with the conclusion for the PV array on flat ground. However, the proposed solar-cell string wiring can still reduce the induced voltage of rooftop PV array on lossy ground.

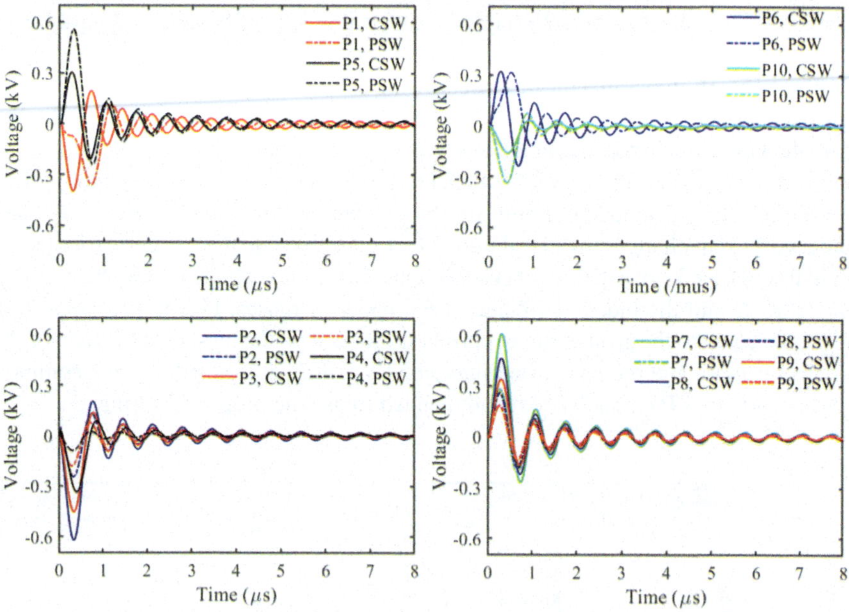

Fig. 5.20 Induced voltage of PV panel when leg3 is connected to the lightning protection tape. *Note* Common Solar-cell Wiring (CSW). Proposed Solar-cell Wiring (PSW) [2]

Table 5.4 Induced voltage of rooftop PV array considering lossy soil [2]

Soil resistivity	Common wiring (kV)	Proposed wiring (kV)
0	5.59	2.47
100 Ωm	6.87	3.03

5.4 Evaluation of Proposed Solar-Cell String Wiring of Photovoltaic Module

Both for the PV array on flat ground and the rooftop PV array, the PV panels near the supporting legs of the bracket have greater induced voltages using the proposed solar-cell string wiring. The PV mounting system is the main reason based on the above analysis. To elucidate the mechanism, the model and parameter of calculation are presented in Sect. 5.2.3, and the induced voltages of ten frameless PV panels shown in Fig. 5.6 are evaluated.

The proposed wiring can greatly reduce the induced voltage of the frameless PV panel, as shown in Fig. 5.21a. The reduction of voltage is much larger than that of the frame PV panel. Since the metal frame influences the lightning EM field, the magnetic flux passing through each solar-cell string declines. The reverse voltage U_2 is reduced by the frame and the voltage drop becomes smaller. Nevertheless, the frameless PV panels P5 and P10 near the supporting legs of bracket still have larger voltage. The frame has no influence on the proposed wiring.

Figure 5.21b shows that the proposed wiring can reduce the induced voltages of all PV panels when the PV bracket is not considered. Moreover, the closer to the lightning rod, the larger the induced voltage of PV panel. The proposed wiring is suitable for the EM field passing through a PV panel has the same direction. The lightning EM field produces the induced current on the bracket, as illustrated in Fig. 5.22. The induced current changes the distribution of the EM field and the direction for the EM field through each solar-cell string is different. Hence, the voltage U_2 maybe negative in Eq. (5.2) and the induced voltage of PV panel increases. Due to the supporting legs connected to the ground, more current is induced on the legs and the effect of the bracket is more significant. In summary, the PV panels at legs still use the common wiring and other PV panels are recommended to use the proposed wiring.

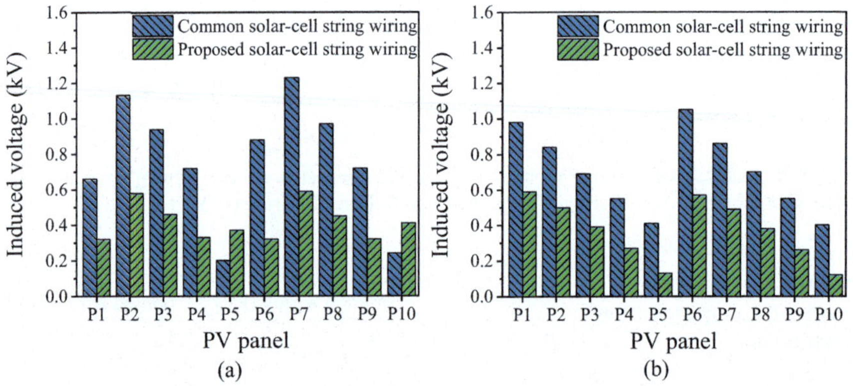

Fig. 5.21 Induced voltages for PV panels. **a** Frameless PV panel considering bracket. **b** Frame PV panel without considering bracket [2]

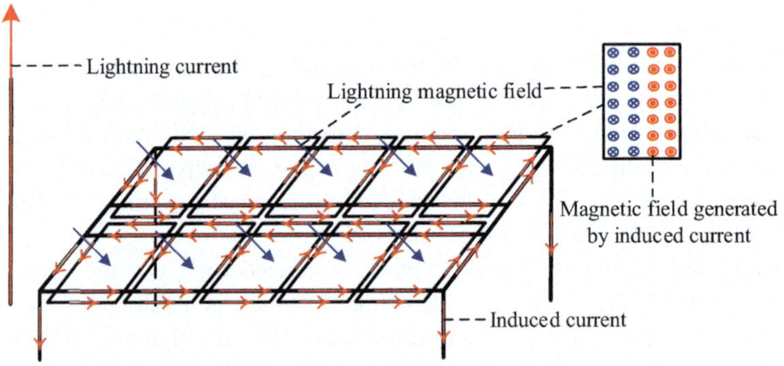

Fig. 5.22 Induced current on PV bracket and frame [2]

In order to mitigate the effect of bracket, the PV backsheet cored with the aluminum (Al) foil is proposed. In the proposed wiring with the mitigation procedure, the backsheet of PV panel at the supporting leg is cored with Al foil, and the proposed wiring is still used for all PV panels, as illustrated in Fig. 5.23.

As presented in Fig. 5.24, the backsheet cored with the Al foil can significantly minimize the induced voltages of PV panels at legs (like P1, P6, P5 and P10). The EM field will generate the eddy current on the Al foil. The direction of the eddy current is so that its EM field weakens the original EM field, and hence the effect of the bracket. The more details and validation of this mitigation method are given in Sect. 6. Moreover, the induced voltage of other PV panels can also be minimized by the proposed solar-cell wiring with the mitigation procedure.

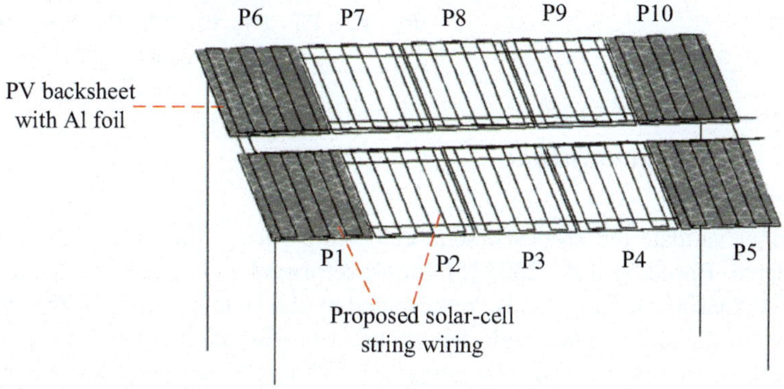

Fig. 5.23 Mitigation procedure [2]

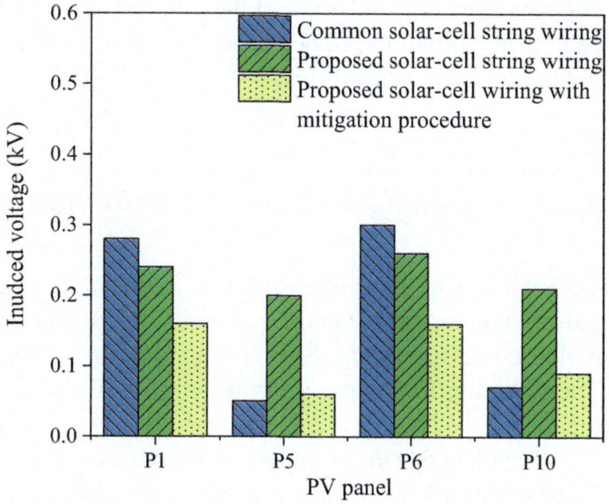

Fig. 5.24 Implementation of mitigation procedure [2]

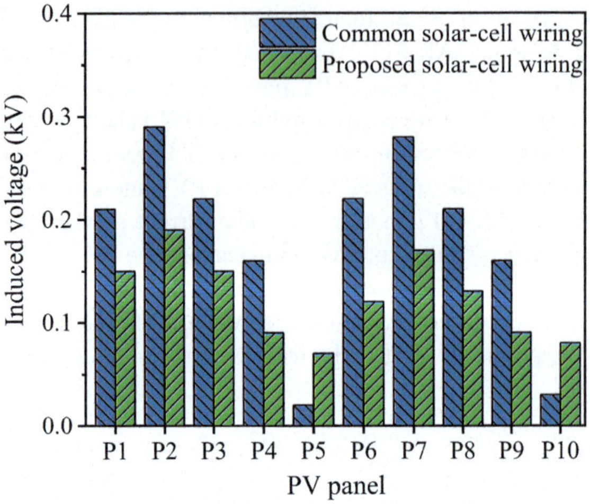

Fig. 5.25 Induced voltages of PV panels generated by subsequent stroke [2]

5.4.1 Subsequent Stroke

To further validate the proposed solar-cell string wiring, the subsequent stroke is considered. Based on IEC 62305 [5], the waveform of subsequent stroke current is 25 kA, 0.25/100 μs, and it is also represented by the Heidler function. The induced voltages of PV array on the flat lossy ground (100 Ωm) are presented at Fig. 5.25.

The result demonstrates that the proposed wiring scheme can decline the induced voltages of most PV panels caused by the subsequent stroke. The voltage reduction of each PV panel in case of the subsequent stroke is relatively close to that of the first lightning stroke. Similarly, the PV panels at the supporting legs have higher voltages due to the influence of the induced current on the bracket. Therefore, it can be concluded that the performance of the proposed wiring is also applied to other lightning current waveforms.

5.4.2 Lightning Striking at Nearby Taller Building

Taking into account another possible scenario, the lightning striking at the lightning rod on another taller building around the rooftop PV array. The size of the taller building is given in Fig. 5.26. The parameters of rooftop PV array are unchanged. The lightning EM transient on the rooftop PV array A is calculated, as shown in Fig. 5.27.

It is evident that the proposed solar-cell string wiring is also superior to the conventional wiring when the lightning strikes at nearby taller building. The induced voltage of the outputs for the rooftop PV array is reduced by using the proposed wiring. As shown in Fig. 5.27a, the voltage reduction reaches up to 42% compared with the

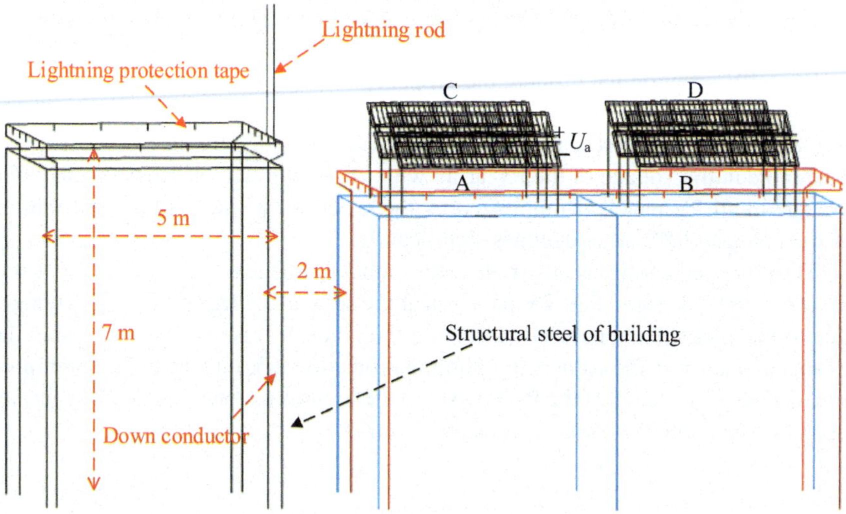

Fig. 5.26 Lightning strikes at nearby taller building in case of rooftop PV array

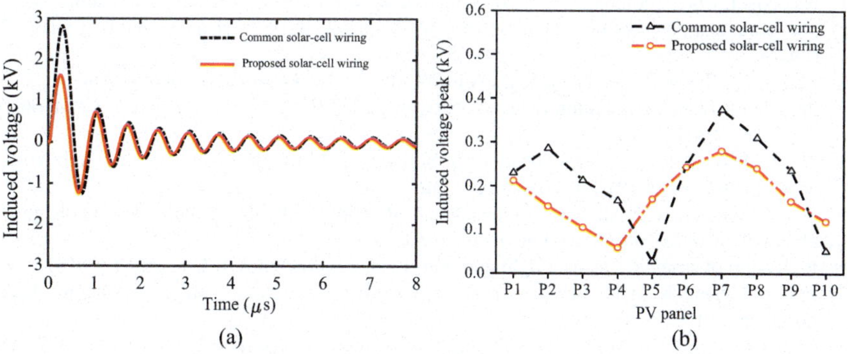

Fig. 5.27 Lightning strikes at nearby building. **a** PV array. **b** PV panel [2]

conventional wiring. In addition, the proposed wiring can still decrease the induced voltages of most PV panels, as plotted in Fig. 5.27b.

5.5 Conclusion

A novel solar-cell string wiring is proposed in this paper. The lightning induced voltage of a solar-cell string is re-versed by this crossover wiring, and the lightning induced voltage of PV panel is thus reduced. The results demonstrate that this new wiring can decrease the DM voltage of most PV panels without additional lightning protection devices.

For a PV array on flat ground, the proposed wiring can also reduce the induced voltage of the outputs for PV array. The drop of the voltage becomes more remarkable as the soil resistivity rises. The sum of the induced voltage for all PV panels plays a key role in the voltage of the PV array with traditional wiring. However, the total voltage of all PV panels is minimized by the proposed wiring, and its contribution to the voltage of PV array declines significantly.

The proposed wiring can also reduce the induced voltages of about 70% PV panels in a rooftop PV system. Few PV panels near the supporting legs have larger voltages due to the effect of the induced current on the bracket. It is necessary to install the down conductors at the corners of lightning protection tape and the base of lightning rod. Connecting the leg of the PV bracket to the lightning protection tape can further decrease the induced voltage of rooftop PV array.

References

1. Naxakis, I. Christodoulou, C. Perraki, V. et al.: Degradation effects on single crystalline silicon photovoltaic modules subjected to high impulse-voltages. IET Sci. Meas. Technol. 11, 563–570 (2017).
2. Zhong, X, Zheng, Z, Sun Q. Q, et al.: A novel solar-cell string wiring of photovoltaic
3. module for reducing lightning-induced overvoltage. IEEE Trans. Electromagn. Compat. 66(1), 204–215 (2024).
4. Altair FEKO 2019.3.2 User's Manual, EM Software and Systems. Stellenbosch, South Africa: S. A. (Pty), Ltd., (2019), http://www.feko.info.
5. Sun, Q. Zhong, X. et al.: Three-dimensional modeling on lightning induced overvoltage for photovoltaic arrays installed on mountain. J. Clean Prod. 288, 125084 (2021).
6. Protection of structures against lightning: Part I general principles. IEC 62305–1 (2010).
7. Zhong, T. Kang, K. Li, H.: Breakdown Accident analysis of large area photovoltaic bypass diode. Sol. Energy. 12, 53–60 (2018).
8. Charalambous, C.A. Kokkinos, N.D. and Christofides, N.: External lightning protection and grounding in large-scale photovoltaic applications. IEEE Trans. Electromagn. Compat. 56, 427–434 (2014).

Chapter 6
Effect of Eddy Current for Photovoltaic Module Generated by Lightning Electromagnetic Field

6.1 Mechanism of Eddy Current in Photovoltaic Module

A PV module is encapsulated with metal frame, cover glass and waterproof backsheet to prevent the oxidation of solar cells. The lightning EM field will produce an impulse voltage on the PV module, and can damage the bypass diodes in the junction box. Meanwhile, the lightning EM field would also generate the eddy currents, which are concentrated on the aluminum foil in backsheet and the back of the Al-BSF solar cell, as shown in Fig. 6.1. The computational model of the eddy current inside a PV module is presented as follows.

6.1.1 Eddy Current Generated by Lightning Electromagnetic Field

The electric field $\boldsymbol{E}$ and the magnetic field $\boldsymbol{H}$ in isotropic medium generated by a current source $\boldsymbol{J}$ (r) can be determined as,

$$\boldsymbol{E}(r) = -j\omega\mu \int_V g(r, r')\boldsymbol{J}(r')dr' - \nabla\frac{j}{\omega\varepsilon} \int_V g(r, r')\nabla \cdot \boldsymbol{J}dr' \tag{6.1}$$

$$\boldsymbol{H}(r) = \int_V \nabla \times g(r, r')\boldsymbol{J}(r')dr' \tag{6.2}$$

where ε and μ are the permittivity and permeability of the medium, respectively. ω represents the angular frequency of the current source $\boldsymbol{J}$. g (r, r') is the Green's function and is defined as,

Q. Sun et al., *Lightning Transients and Protection for Renewable Energy System*,
https://doi.org/10.1007/978-981-97-7267-4_6

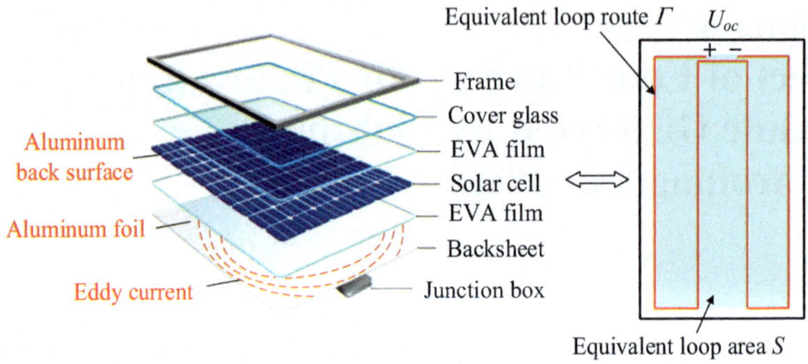

Fig. 6.1 Schematic diagram of a PV module [1]

$$g(r, r') = \frac{e^{-j\omega\sqrt{\mu\varepsilon}|r-r'|}}{4\pi|r - r'|} \tag{6.3}$$

Here, $|r - r'|$ is the distance between the source point r and field point r'. The eddy currents caused by lightning EM field have the skin effect due to the high frequency of lightning EM field. The eddy currents are mainly concentrated on the metal surfaces. Hence, the volume integral in Eq. (6.1) can be replaced by the surface integral, and the scattered field $\boldsymbol{E}^{sca}$ $(\boldsymbol{r})$ produced by eddy currents can be expressed as,

$$\boldsymbol{E}^{sca}(r) = -j\omega\mu \int\limits_S g(r, r')\boldsymbol{J}(r')ds' - \nabla\frac{j}{\omega\varepsilon}\int\limits_S g(r, r')\nabla \cdot \boldsymbol{J}ds' \tag{6.4}$$

The boundary condition of electric field on metal surface is,

$$\boldsymbol{n} \times [\boldsymbol{E}^{sca}(r) + \boldsymbol{E}^{inc}(r)] = 0 \tag{6.5}$$

where $\boldsymbol{n}$ is the unit normal vector of the metal surface, and $\boldsymbol{E}^{inc}$ (r) is the electric field generated by the lightning current.

The metal surface of a PV module includes the frame, the back of aluminum-back surface field (Al-BSF) solar cell, the aluminum foil embedded in backsheet. The eddy current on the copper wire can be almost negligible due to the small surface area. Taking into account the boundary condition, the integral equation of electric field for the eddy currents on PV panels can be derived as,

$$\begin{aligned}\boldsymbol{n} \times \boldsymbol{E}^{inc}(r) &= \boldsymbol{n} \times j\omega\mu \int\limits_S g(r, r')[\boldsymbol{J}_1(r') + \boldsymbol{J}_2(r') + \boldsymbol{J}_3(r')]dr' \\ &+ \vec{n} \times \frac{j}{\omega\varepsilon}\nabla \int\limits_S g(r, r')\nabla \cdot [\boldsymbol{J}_1(r') + \boldsymbol{J}_2(r') + \boldsymbol{J}_3(r')]dr'\end{aligned} \tag{6.6}$$

where the surface S consists of the frame surface S_1, the surface of foil S_2 and the Al-BSF surface S_3. $\boldsymbol{J}_1, \boldsymbol{J}_2$ and $\boldsymbol{J}_3$ are the eddy current densities of these surfaces.

The surfaces S_1, S_2 and S_3 are discretized into triangular units, and their numbers are N_1, N_2 and N_3, respectively. The eddy current of surfaces is determined by,

$$\begin{cases} \boldsymbol{J}_1 = \sum_{n=1}^{N_1} \alpha_n \boldsymbol{f}_n(r') \\ \boldsymbol{J}_2 = \sum_{n=N_1+1}^{N_1+N_2} \alpha_n \boldsymbol{f}_n(r') \\ \boldsymbol{J}_3 = \sum_{n=N_1+N_2+1}^{N_1+N_2+N_3} \alpha_n \boldsymbol{f}_n(r') \end{cases} \tag{6.7}$$

$\boldsymbol{f}_n$ (r) is the Rao-Wilton-Glisson (RWG) function [2],

$$\boldsymbol{f}_n(r) = \begin{cases} \frac{l_n}{2A_n^+} \quad \rho_n^+ \; r \in T_n^+ \\ \frac{l_n}{2A_n^-} \quad \rho_n^- \; r \in T_n^- \\ 0 \quad else \end{cases} \tag{6.8}$$

where A_n^+ and A_n^- are the areas of two triangles T_n^+ and T_n^-, as shown in Fig. 6.2. $\boldsymbol{\rho}_n^+$ and $\boldsymbol{\rho}_n^-$ are the two unit vectors corresponding to the vertices of the two triangles. l_n is the length of the common side for two triangles.

Equation (6.6) can be solved by using MoM, and it can be extended to a set of equations by the Galerkin method [3].

$$Z \cdot \alpha = V_m \tag{6.9}$$

Fig. 6.2 Rao-Wilton-Glisson (RWG) function [2]

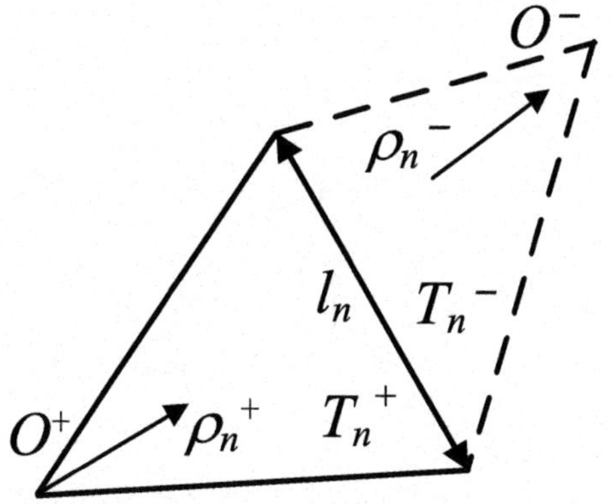

The matrix equation is composed of N ($N = N_1 + N_2 + N_3$) independent equations, and the impedance matrix Z can be obtained by,

$$Z_{mn} = j\omega\mu \int_{S_m} \boldsymbol{f}_m(r) \cdot \sum_{n=1}^{N} \int_{S_n} g(r, r')\boldsymbol{f}_n(r')dr' + \frac{j}{\omega\varepsilon} \int_{S_m} \boldsymbol{f}_m(r) \cdot \sum_{n=1}^{N} \nabla \int_{S_n} g(r, r')\nabla \cdot \boldsymbol{f}_n(r')dr' \tag{6.10}$$

The matrix $V_{\rm m}$ is defined by,

$$V_m = \int_{S_m} \boldsymbol{f}_m(r) \cdot \boldsymbol{E}^{inc}(r)dr \tag{6.11}$$

6.1.2 Lightning Electromagnetic Field in Frequency Domain

The MTLE model is used to represent the lightning return-stroke current $i\,(z', t)$ [4],

$$i(z', t) = i(0, t - z'/v)\, u(t - z'/v)e^{-z'/\lambda} \tag{6.12}$$

where z' is the height of lightning current. v is the propagation speed of the return-stroke current, which is set to be 2c/3 (c is light speed) [5]. The decay factor λ is taken as 2000 m [6], and $u\,(t)$ is the unit step function.

The lightning current needs to be periodically extended to satisfy the requirement of the Fourier series. Given the calculation time T, the lightning current is decomposed by [7]:

$$i(t) = \frac{a_0}{2} + \sum_{k=1}^{n} b_k \sin(k\omega t) \tag{6.13}$$

The factors in Eq. (6.13) can be determined by,

$$\begin{cases} a_0 = 0 \\ b_k = \dfrac{2}{T} \displaystyle\int_0^T i(t) \sin\left(\dfrac{k\pi t}{T}\right) dt \end{cases} k = 1, 2, 3...., n \tag{6.14}$$

Hence, $i\,(t)$ after Fourier series can be expressed as,

$$i(t) = \sum_{k=1}^{n} I_k \sin(k\omega t) = I_1 \sin \omega t + \cdots + I_n \sin n\omega t \tag{6.15}$$

The lightning channel is 'mirrored' to consider the perfectly conducting ground, as shown in Fig. 6.3. For the current I_k with the angular frequency $k\omega$, the electric field and magnetic field generated by I_k can be obtained by [8],

$$\begin{aligned} E_{rk}(r,z) &= \frac{3Z_0 I_k r}{4\pi} \int_{-h}^{h} \frac{z - z'}{\left[r^2 + (z - z')^2\right]} dz' \\ &= \frac{3Z_0 I_k r}{8\pi} \left[\frac{1}{r^2 + (h - z)^2} - \frac{1}{r^2 + (h + z)^2} \right] \end{aligned} \tag{6.16}$$

$$\begin{aligned} E_{zk}(r,z) &= \frac{Z_0 I_k}{4\pi} \int_{-h}^{h} \frac{2(z - z')^2 - r^2}{\left[r^2 + (z - z')^2\right]} dz' \\ &= \frac{Z_0 I_k}{4\pi} \left(2 + \frac{1}{2r^2}\right) \left[\frac{h - z}{r^2 + (h - z)^2} + \frac{h + z}{r^2 + (h + z)^2} \right] \end{aligned} \tag{6.17}$$

$$\begin{aligned} H_{\phi k}(r,z) &= \frac{I_k r}{4\pi} \int_{-h}^{h} \frac{1}{\left[r^2 + (z - z')^2\right]^{3/2}} dz' \\ &= \frac{I_k}{4\pi r} \left[\frac{h + z}{r^2 + (h + z)^2} + \frac{h - z}{r^2 + (h - z)^2} \right] \end{aligned} \tag{6.18}$$

where Z_0 is the wave impedance in vacuum and h represents the length of the lightning channel. The distance between the PV array and the lightning channel is shorter, and the effect of the propagation delay on the near field is minor. Hence, Eqs. (6.16)–(6.18) neglect the propagation delay.

Substituting E_{rk} and E_{zk} into Eq. (6.11) as the incident field $\boldsymbol{E}^{\text{inc}}$ to obtain the matrix V_{m}, and the eddy currents $\boldsymbol{J}_1$, $\boldsymbol{J}_2$, and $\boldsymbol{J}_3$ for the angular frequency $k\omega$ can be determined.

6.1.3 Induced Voltage of Photovoltaic Module Considering Effect of Eddy Current

The lightning discharge will produce a strong impulse EM field and the DM voltage could be induced on a loop. For a single PV module, a number of solar cells are connected in series, as illustrated in Fig. 6.1, forming an equivalent loop. Based on Faraday's Laws, the DM voltage U_{oc} is,

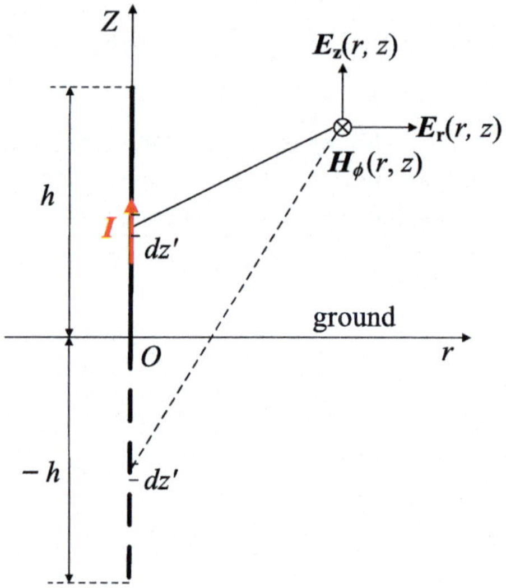

Fig. 6.3 Lightning EM-field in cylindrical coordinates [1]

$$U_{oc} = \int_{\Gamma} \boldsymbol{E} \cdot dl = -\int_{S} \frac{\partial \boldsymbol{B}}{\partial t} \cdot dt = -\mu \int_{S} \frac{\partial \boldsymbol{H}}{\partial t} \cdot dt \tag{6.19}$$

Considering the eddy currents of PV modules, the magnetic field in the loop is no longer the magnetic field $\boldsymbol{H}_\phi$ caused by lightning current, but the sum of the scattered magnetic field $\boldsymbol{H}^{sca}$ generated by the eddy current and $\boldsymbol{H}_\phi$.

$$\boldsymbol{H} = \boldsymbol{H}_\phi + \boldsymbol{H}^{sca} \tag{6.20}$$

$\boldsymbol{H}^{sca}$ includes $\boldsymbol{H}_1{}^{sca}$ caused by the induced current of frame, $\boldsymbol{H}_2{}^{sca}$ and $\boldsymbol{H}_3{}^{sca}$ generated by the eddy currents on aluminum foil and Al-BSF surface, respectively.

$$\boldsymbol{H}^{sca} = \boldsymbol{H}_1^{sca} + \boldsymbol{H}_2^{sca} + \boldsymbol{H}_3^{sca} \tag{6.21}$$

In order to obtain $\boldsymbol{H}_1{}^{sca}$, $\boldsymbol{H}_2{}^{sca}$ and $\boldsymbol{H}_3{}^{sca}$, it needs to integrate the eddy current density along the surface,

$$\begin{cases} \boldsymbol{H}_1^{sca}(r) = \int_S \nabla \times g(r, r')\boldsymbol{J}_1(r')ds' \\ \boldsymbol{H}_2^{sca}(r) = \int_S \nabla \times g(r, r')\boldsymbol{J}_2(r')ds' \\ \boldsymbol{H}_3^{sca}(r) = \int_S \nabla \times g(r, r')\boldsymbol{J}_3(r')ds' \end{cases} \tag{6.22}$$

The inverse Fourier transform (IFFT) is used to obtain $\boldsymbol{H}^{sca}$ in time domain. According to the Fourier series for lightning current in Eq. (6.15), the time-domain response of the lightning magnetic field $\boldsymbol{H}_{\phi}$ can be obtained directly by the sum of the magnetic field generated by the currents $I_1 \ldots I_k$ without IFFT,

$$\boldsymbol{H}_{\phi} = \sum_{k=1}^{n} H_{\phi k} \sin k\omega t \tag{6.23}$$

6.2 Eddy Current Characteristics of Photovoltaic Module

The eddy current generated by the lightning EM field will occur on the surfaces of the aluminum foil and Al-BSF solar cells inside a PV module. The direction of these currents is such that their EM fields weaken the lightning EM field, and hence the DM induced voltage of the PV module. To study the effect of the eddy current, two different PV modules with and without frame are discussed as follow.

6.2.1 Eddy Current on Aluminum Back Surface Field (Al-BSF) of Solar Cell

Many silicon solar cells are based on P-type silicon wafers, which is doped with boron. The phosphorus diffusion helps form P–N junction, and the P-type solar cells usually use the full-area Al-BSF [9], as shown in Fig. 6.4a. In the Al-BSF, the front side is the phosphorus diffusion, and the back side is a full-area Al back contact. The lightning EM field will produce the eddy current on the surfaces of full-area Al back for Al-BSF solar cells.

The metal frame has a C-shaped cross section. w_c, t_c and l_c represent the width, thickness and length of C-shaped frame. Their detail values are set to be 3, 30 and 30 mm, respectively. To facilitate the analysis, the C-shaped cross section of the frame is represented by the cylindrical conductor. The equivalent radius r_c is defined in Sect. 6.4.

In order to make the eddy currents on the Al-BSF solar cells more remarkable for analysis its fundamental influence, the conductor is set to be parallel to the PV module, as illustrated in Fig. 6.4b. The distance between the conductor and the PV module is 0.7 m. Since the conductor is paralleled to the PV module, the magnetic flux passing through the PV module is the largest. To represent the lightning EM field, an impulse current, with the waveform of 50 kA, 1/200 μs, is injected at the left end of the conductor. The impulse current flows from the left end to the absorbing boundary condition at the right end of the conductor. The PV panel consists of 40

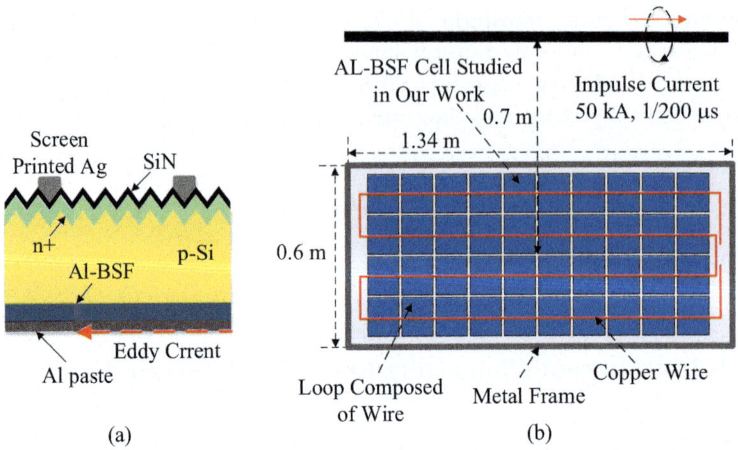

Fig. 6.4 PV panel with Al-BSF solar cells. **a** Al-BSF solar cell. **b** Calculation model [1]

AL-BSF cells connected in series, and its length and width are 0.98 and 0.445 m, respectively.

In terms of the PV panel without frame, the eddy currents on the Al backs of solar cells are given in Fig. 6.5a. The eddy current density becomes larger when the distance is closer to the current source, and its direction is counter clockwise. The direction of the impulse current is from left to right, and the direction of the EM field is inward. According to Lenz's law, the eddy current will produce a magnetic field that weakens the original EM field. Therefore, the direction of the eddy current is counter clockwise. The eddy current density is related to the strength of lightning EM field. The stronger the EM field strength is, the greater the density of the eddy current on Al backs.

The lightning EM field will also generate induced currents in the metal frame. For the PV panel with frame, Fig. 6.5b shows the eddy current in the Al backs of solar cell. The eddy current is different from that of the frameless PV panel. The EM field causing the eddy current includes the lightning EM field and the EM field generated by the induced currents on the frame. Compared to the frameless PV panel, the eddy current density near the frame is remarkably reduced. The lightning EM field is weakened by the EM field produced by the induced current on the frame. Hence, the EM field strength near the frame is smaller. Furthermore, the eddy currents are larger at the top and bottom corners in the right side of the PV panel. The eddy current at the right top and bottom is determined by the induced current on the frame, and its direction is thus opposite to that of the eddy current without frame. The lightning EM field at the right top and bottom is weaker due to the greater distance away from the lightning current. Therefore, the total EM field is larger and hence the eddy current.

To further analyse the eddy current distribution of Al-BSF solar cells, the eddy current densities along x-direction and y-direction are calculated for the AL-BSF cell (See the red circle in Fig. 6.5), as plotted in Fig. 6.6. The area with the highest eddy

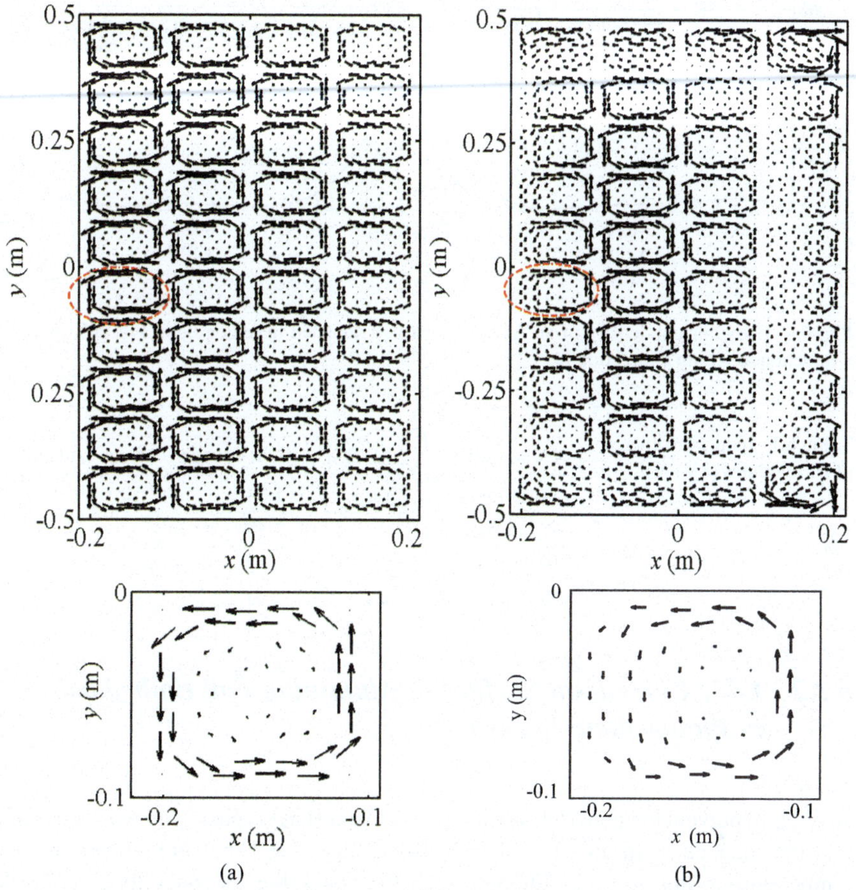

Fig. 6.5 Eddy current on Al-BSF solar cells. **a** Frameless PV panel. **b** Frame PV panel [1]

current density is the sides of the Al back when the PV panel is frameless. The eddy current density in the centre of the Al back is the smallest. Considering the effect of frame, the eddy current density decreases. The eddy current along *x*-direction is asymmetrically distributed, and its density is lower near the frame and increases away from the frame.

The eddy current on the Al backs can weaken the lightning EM field, and the induced DM voltages of PV panels are minimized in some extent, as shown in Table 6.1. Moreover, the PV frame will reduce the eddy current, and the effect of the eddy current on the induced voltage becomes weaker. The frameless PV module with AL-BSF solar cells can reduce the voltage by 13.5%. However, the voltage of the frame PV module is minimized by only 10.1%.

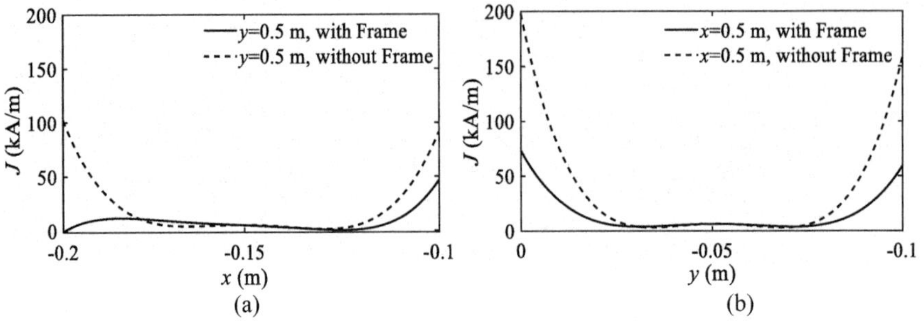

Fig. 6.6 Eddy current density of an Al-BSF solar cell depicted in red circle in Fig. 6.5. **a** Along *x*-direction. **b** Along *y*-direction [1]

Table 6.1 Induced voltages of PV modules (kV) [1]

PV Module	AL-BSF	DM voltage	Drop rate (%)
With frame	Yes	1.24	10.1
With frame	No	1.38	
Without frame	Yes	4.62	13.5
Without frame	No	5.34	

6.2.2 *Eddy Current on Surface of Aluminum Foil Embedded in Photovoltaic Backsheet*

In order to prevent the penetration of water, the backsheet is packaged with the frame and the solar cells. In addition, the backsheet can reflect the sunlight and reduce temperature to improve the efficiency. The PET backsheet is generally used in PV panels. The PET backsheet has excellent mechanical strength, good transparency and high gloss due to the mirror effect after being coated with Al foil. The PV backsheet with Al foil is illustrated in Fig. 6.7, and the lightning EM field also produces the eddy current on the foil surface.

Taking into consideration the PV module with PET backsheet that is coated with aluminum, as shown in Fig. 6.8a. The area of the foil is $0.97 \times 0.425\ m^2$ and its thickness is 0.1 mm, which is 7.5 mm away from solar cells. The amplitude of impulse current is still 50 kA, and the waveform is 1/200 μs. Figure 6.8b shows that the distribution of the eddy current on the aluminum foil. The direction is counter clockwise to produce an EM field to compensate the lightning EM field.

Similarly, the eddy current is mainly concentrated on the outside of the aluminum foil, as plotted in Fig. 6.9. The density of the eddy current on the center of the foil is weaker. The PV frame also reduces the eddy current density, and the eddy current near the frame is thus smaller. For the PV module with frame, the eddy current is approximately symmetrically distributed about *x* or *y* axis. According to

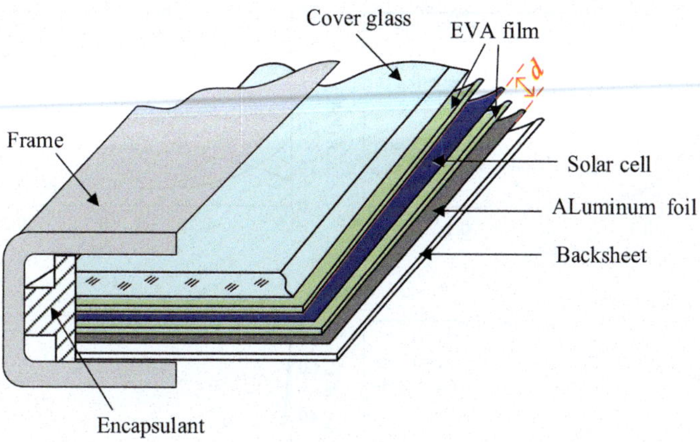

Fig. 6.7 PV backsheet with aluminum foil [1]

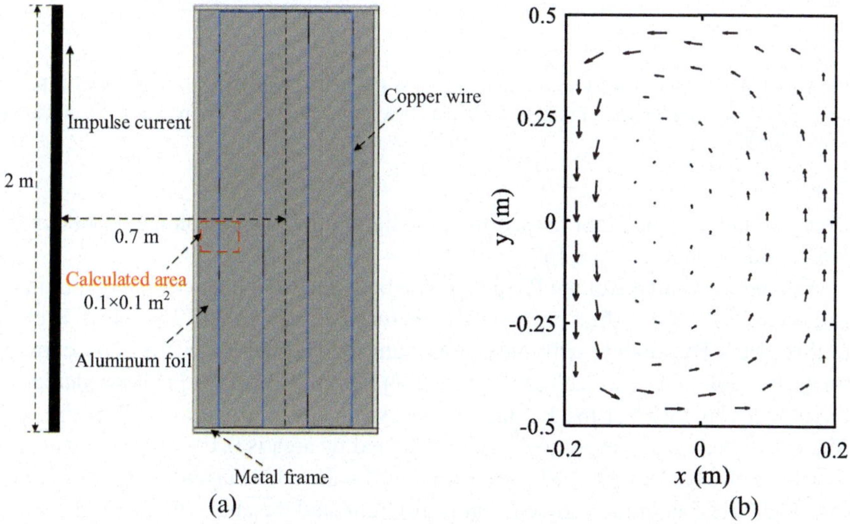

Fig. 6.8 PV module with aluminum foil. **a** Calculation model. **b** Eddy current on aluminum foil [1]

the distribution of eddy current, it is concluded that the outside of the aluminum foil is the main path of the eddy current caused by lightning EM field.

The eddy current on the aluminum foil has a greater impact on the induced DM voltage of the PV panel. As plotted in Fig. 6.10, the voltage drops from 1.50 to 0.13 kV for the frame PV module, and the reduction can almost reach up 90%. In terms of the frameless PV module, the voltage decreases from 5.82 to 0.69 kV, which is reduced by 88%. The area of the aluminum foil is larger, and the internal loop of PV module is directly located over the aluminum foil within a zone of strongest field reduction.

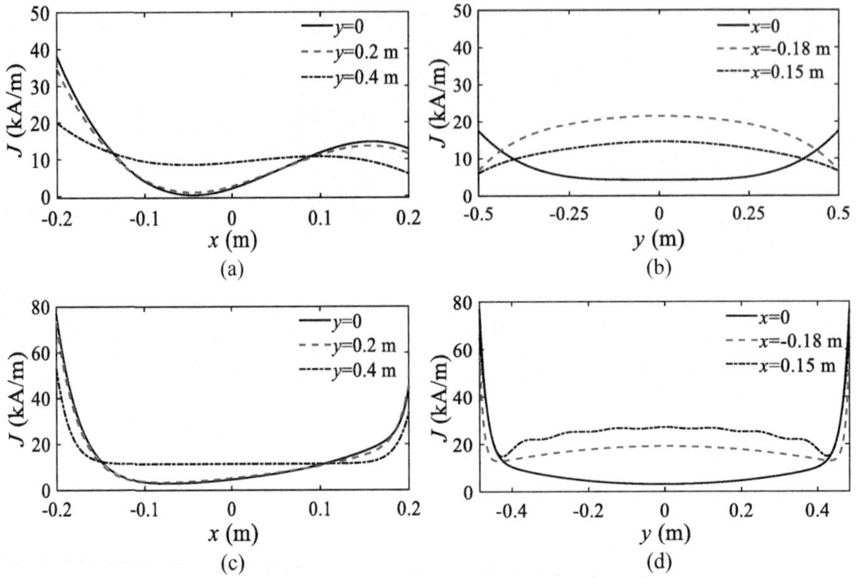

Fig. 6.9 Eddy current density of aluminum foil. **a** Frame PV panel along *x*-direction. **b** Frame PV panel along *y*-direction. **c** Frameless PV panel along *x*-direction. **d** Frameless PV panel along *y*-direction [1]

Consequently, a significant reduction of the lightning induced voltage occurs on the PV module with aluminum foil.

The induced voltage of the PV panel with both AL-BSF solar cells and aluminum foil is computed for further analysis. The induced voltage is 0.17 kV, which is close to that of the PV module with only aluminum foil. Furthermore, the eddy currents on the back of an Al-BSF cell are presented in Fig. 6.11. It is obtained by calculating the model composed of the PV module shown in Fig. 6.4 and the foil in Fig. 6.8. The calculated area is presented in Fig. 6.8, and its area is $0.1 \times 0.1\ m^2$. The eddy current on the back of AL-BSF cell is affected by the eddy current on the aluminum foil. The direction of the eddy current is different, and the area with the higher eddy current density changes from the outside of solar cell to the negative direction of *x* axis.

Fig. 6.10 Induced DM voltage of PV module. **a** PV module with frame. **b** PV module without frame [1]

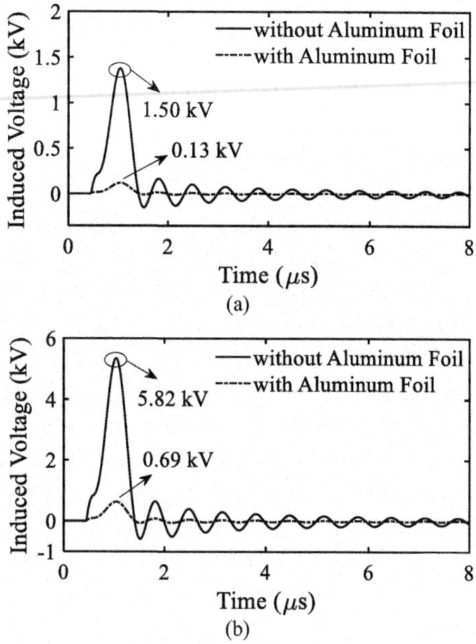

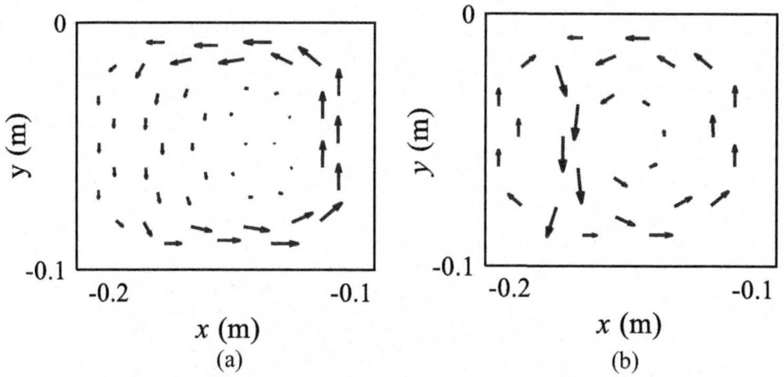

Fig. 6.11 Eddy current of an Al-BSF solar cell. **a** Without aluminum foil. **b** With aluminum foil [1]

6.3 Effect of Backsheet Cored with Aluminum Foil on Lightning Induced Overvoltage of Photovoltaic Array

A small PV array, including ten PV panels with PET backsheet that is coated with aluminum, is studied in this section. The induced DM voltages of PV array are calculated, and the eddy current on the aluminum foil at different lightning strike points is investigated.

As illustrated in Fig. 6.12, the length and width of the PV array are 5 and 2.23 m, respectively. The voltage of PV panel is 36 V, and the DM voltage U_{DC} of the DC output is 360 V. The mounting system for PV array consists of four support legs and six crossbars. Considering the EM couple between the PV array and the mounting system, the lightning transient of the PV array is computed by using the MoM-based software FEKO. The computation is carried out over the frequency range from 2.93 kHz to 1.5 MHz with the increment step of 2.93 kHz, and the number of the frequencies is 512. The time interval from 0 to 341 μs with the increment step of 1/3 μs, and 1024 points in IFFT are thus used to obtain the time-domain response. The mounting system, PV modules and DC cables are modeled as the combination of cylindrical segments, which satisfy the limitation in the ratio of conductor length to wavelength in the time range of interest. The thickness of aluminum foil is 0.1 mm, which is 5 mm away from the solar cells.

The lightning rod is quite likely to be struck by the lightning. In order to reduce the shadow generated as possible, the length of the lightning rod is generally short and located 2 m away from the PV array. The propagation of the lightning current on the rod is fast and the influence of the multiple reflections is small, due to the short length of rod. Hence, the lightning rod is regarded as the base of the lightning channel.

The lightning channel is represented by cylindrical elements having distributed series resistance and inductance to realize a 2c/3 propagation speed of the return-stroke current on it, and its length is set to be 1 km. Lightning currents are random and complex. The first stroke currents present multiple peaks and its strong concavity at the wavefront [10]. For a qualitative study, a single Heidler function is used to represent the first negative stroke current, and its waveform is 50 kA, 1/200 μs.

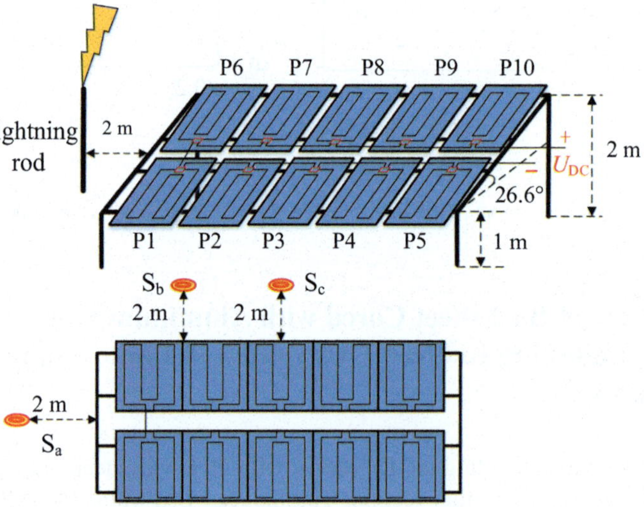

Fig. 6.12 Dimension of PV array

The conclusions drawn in this paper can be also applied to other lightning current waveforms. The three lightning strike points S_a, S_b and S_c are considered, as shown in Fig. 6.12.

6.3.1 Induced Voltages of Each Photovoltaic Module with Aluminum Foil

The induced DM voltages of PV modules, as well as the induced voltage between ± DC cables at the output of the PV array, are calculated for lightning striking at S_a. The results indicate that the aluminum foils of PV modules can greatly minimize the induced voltages. The induced voltages of the PV modules without aluminum foils are the range of 4–582 V, and they will drop to 7–90 V for the PV modules with the aluminum foils, as plotted in Fig. 6.13. Although the eddy current is produced on the aluminum foil, the eddy current loss is so small that hardly lead to overheating due to the short time of lightning current.

Since the induced voltage of each PV module is decreased, the induced voltage between ± DC cables at the output is reduced by 42.7%, dropping from 4.89 to 2.80 kV. It is concluded that the use of aluminum foil can effectively lower the damage of lightning EM field to PV systems, both for PV modules and the inverter.

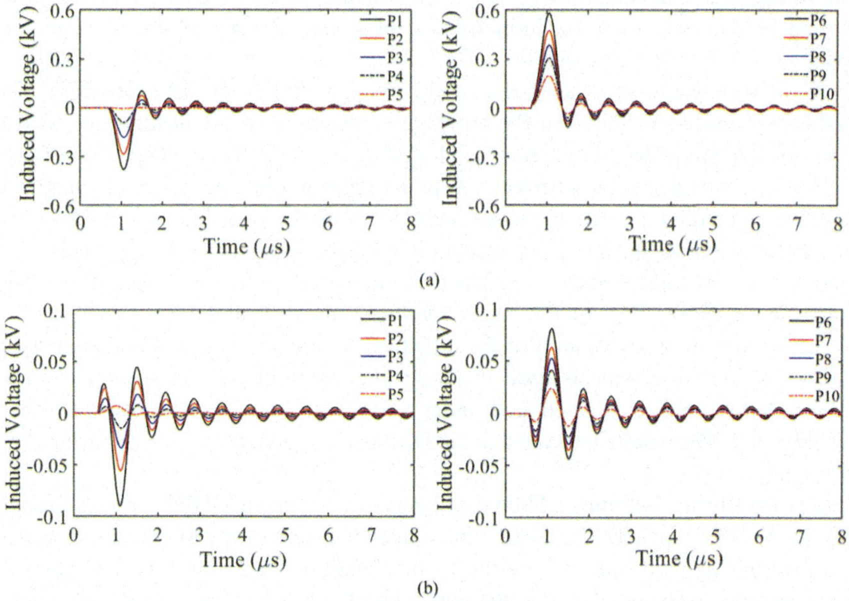

Fig. 6.13 Induced voltages of PV panels in a PV array. **a** Without aluminum foils. **b** With aluminum foils [1]

Fig. 6.14 Component for induced voltage of PV array [1]

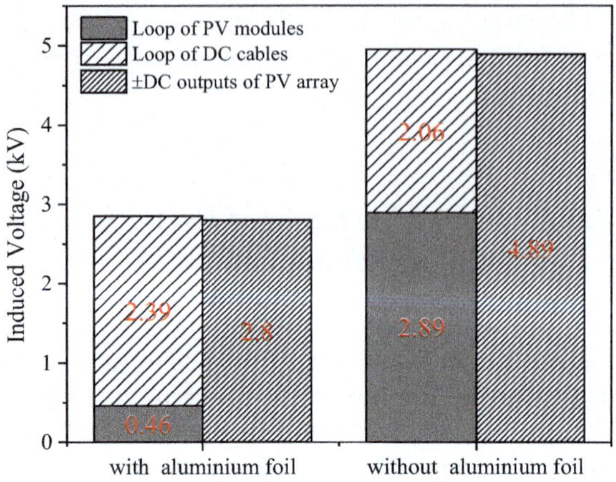

The DC loop of a PV array can be divided into two parts: the loop of DC cables and the loops inside the PV modules. The induced voltage at the output of PV array approximately equals to the sum of the induced voltages of PV modules and the induced voltage on the DC cables, as shown in Fig. 6.14. The eddy current of aluminum foil has a great impact in the induced voltages of PV modules, and it changes from 2.89 kV without the foil to 0.46 kV with the foil. However, the effect on the induced voltage on the DC cables is not remarkable. Since the aluminum foils are directly located below the PV modules, the loop of PV modules is a zone of the strongest field reduction. The loop of DC cables is not covered by the foils, and the eddy current has less effect on it.

Considering a lossy ground, the conductivity of soil is set to be 0.01 S/m. The lightning-induced voltages on PV arrays are calculated by the accurate numerical solutions of Sommerfeld integrals, as plotted in Fig. 6.15. The voltage on the lossy soil is larger than that on a perfectly-conducting ground plane. Since the magnetic field is affected by the soil, the magnetic field on the lossy soil is greater. However, it can be revealed that the lossy ground has the same effect on PV modules with and without Al foils on the backs. The conclusions regarding the effect of the eddy currents for PV modules on the induced voltages are still valid.

Moreover, in order to anylyze the effect of the transient ground potential rises, the grounding resistance is used to model the grounding structure of PV array. The induced voltage is calculated under various grounding resistances, as listed in Table 6.2. The results indicate that the DM induced voltage rises as the grounding resistances.

Principally, the lightning EM field will generate the magnetic flux passing through the loop of PV bracket. Meanwhile, the induced current of the loop would produce additional magnetic flux in opposite to the lightning magnetic flux according to the Lenz's law. Hence, the currents induced on the bracket can reduce the induced voltage. Since a high grounding resistance reduces the induced current of the bracket, the influence of the bracket becomes weakened and the voltage is greater. Clearly, the

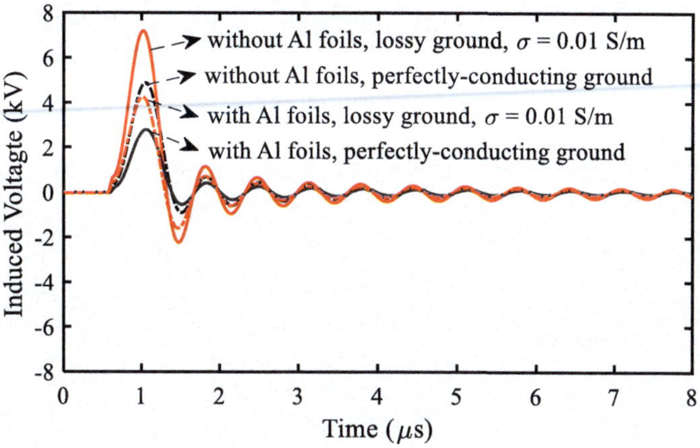

Fig. 6.15 Induced voltage of PV array on a lossy ground [1]

Table 6.2 Induced voltage for different grounding resistances (kV) [1]

Grounding resistance (Ω)	0	3	10	30
Without aluminum foil	4.89	4.93	5.21	5.59
With aluminum foil	2.80	2.78	2.88	3.08

aluminum foil cored in PV backsheet can reduce the induced voltage considering the effect of the transient ground potential rises. The grounding resistance plays a key role in the case of direct lightning strike. However, in the case of indirect lightning strike, the effect of a low grounding resistance on the induced voltage is minor.

6.3.2 Eddy Current Distribution of Photovoltaic Array

The induced voltages of PV modules are strongly dependent on the distance away from the lightning channel. The PV modules closer to the lightning channel, like P1 and P6, have larger voltages. The induced voltages of the upside PV panels, such as P6–P10, are greater. Meanwhile, the eddy current density on P6–P10 is the strongest, as shown in Fig. 6.16. Consequently, the drop of the induced voltage for P6 reaches up to 0.50 kV, which is the greatest. The eddy current on the aluminum foil is affected not only by the frame but also the crossbars of the bracket for PV array. Figure 6.16 shows that the eddy current is divided into two zones, the demarcation line locates the crossbar of PV bracket.

Considering the eddy current, the EM transient of PV array at different lightning-strike locations is evaluated. When S_b is struck by lightning, the presence of aluminum

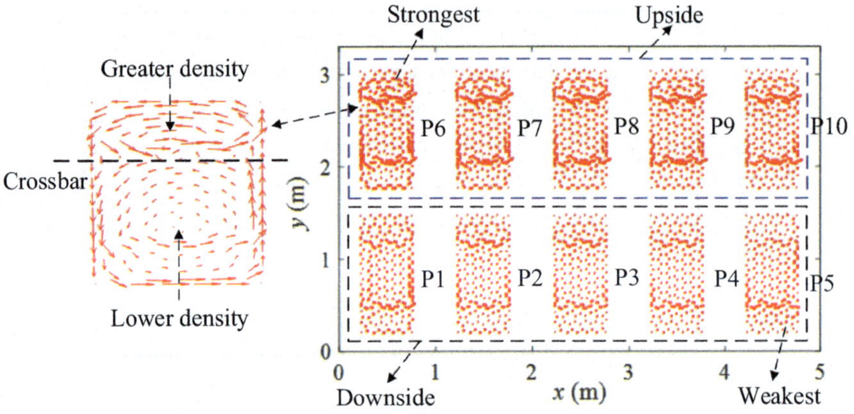

Fig. 6.16 Eddy current on PV array [1]

foils can still reduce the induced voltages of PV modules and DC cables, as shown in Fig. 6.17a, and the bypass diodes will hardly be broken down.

Since the magnetic flux passing through the PV array is almost offset for lightning striking S_c, the induced voltages of PV array are at low level, which is smaller than the maximum input voltage of the inverter. Nevertheless, the DM voltages of PV modules with aluminum foils is minimized, and the reduction is up to 275 V for P10, as shown in Fig. 6.17b. The lightning EM field passing through the PV array varies with the strike location. No matter how the lightning strike location changes, the eddy current on the aluminum foil still produces an EM field in the opposite to the original lightning EM field. The EM field near the aluminum foil is always weakened by the eddy current. Hence, the effect of the eddy current on the induced voltages is hardly affected by lightning strike points.

6.4 Design of Foil Embedded in Photovoltaic Backsheet Against Lightning Induced Voltage

6.4.1 Foil Material

The eddy current density of the foil is influenced by the foil material, and the effect of the eddy current for various foils is discussed. Three foils, including the aluminum foil, the copper foil and the graphite foil, have different conductivity: the conductivity of copper foil is 5.81×10^7 S/m, 5.71×10^7 S/m for the aluminum foil and 7×10^4 S/m for the graphite foil. The induced voltage peaks of PV modules with various foils are listed in Table 6.3. The induced voltages between ± DC cables for various foils are plotted in Fig. 6.18.

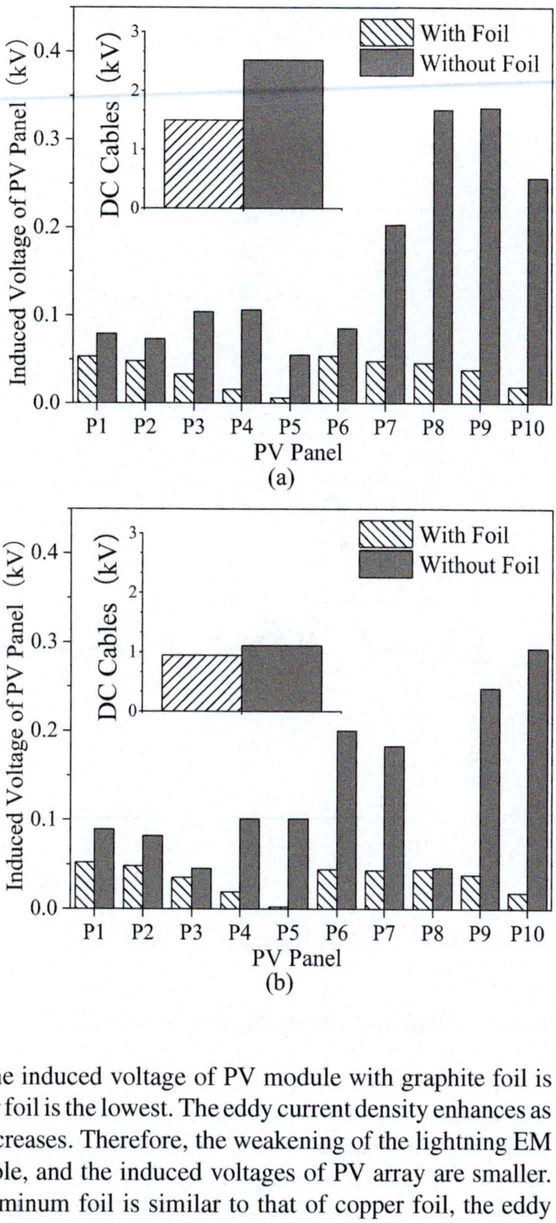

Fig. 6.17 Induced voltage at different lightning striking points. **a** Lightning strikes at S_b. **b** Lightning strikes at S_c [1]

The results indicate that the induced voltage of PV module with graphite foil is greatest, and that of the copper foil is the lowest. The eddy current density enhances as the conductivity of the foil increases. Therefore, the weakening of the lightning EM field becomes more remarkable, and the induced voltages of PV array are smaller. Since the conductivity of aluminum foil is similar to that of copper foil, the eddy current effect of aluminum foil is close to that of copper foil and hence the induced voltage for the PV array.

Table 6.3 Induced voltage peak of PV modules with various foils (V) [1]

PV module	Al foil	Copper foil	Graphite foil
P1	90	81	243
P2	55	49	183
P3	33	30	116
P4	15	13	57
P5	7	7	5
P6	81	72	365
P7	63	56	300
P8	53	47	246
P9	41	37	194
P10	24	22	122

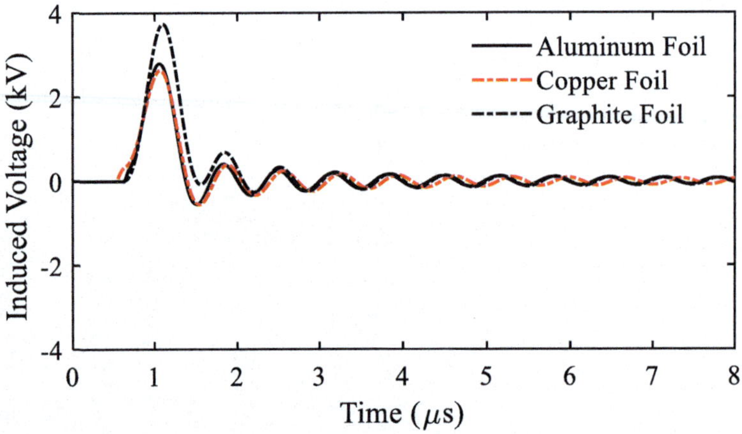

Fig. 6.18 Induced voltage between ± DC cables [1]

6.4.2 *Distance from Foil to Solar Cell*

Changing the distance *d* between the foil and the solar cells (see Fig. 6.7) to analyse the induced voltage between ± DC cables at the output of the PV array, as given in Table 6.4. The induced voltage is raised by 0.14–0.20 kV as the distance *d* increases from 0.5 to 2.0 cm for the three foils. However, the influence of the distance on the induced voltage is much smaller than the maximum open-circuit voltage of the inverter.

In terms of the PV modules in the PV array, as shown in Fig. 6.19, the DM induced voltages will increase when the aluminum foil is closer to the solar cells. The impact of the distance *d* on the induced voltages of PV modules becomes more significant when the lightning EM field is larger. Such as the PV modules P1 and P6, they are

Table 6.4 Induced voltage peak of DC cables at various distances (kV) [1]

d (cm)	Al foil	Copper foil	Graphite foil
0.5	2.80	2.64	3.74
1.0	2.88	2.71	3.79
1.5	2.95	2.78	3.84
2.0	3.00	2.82	3.88

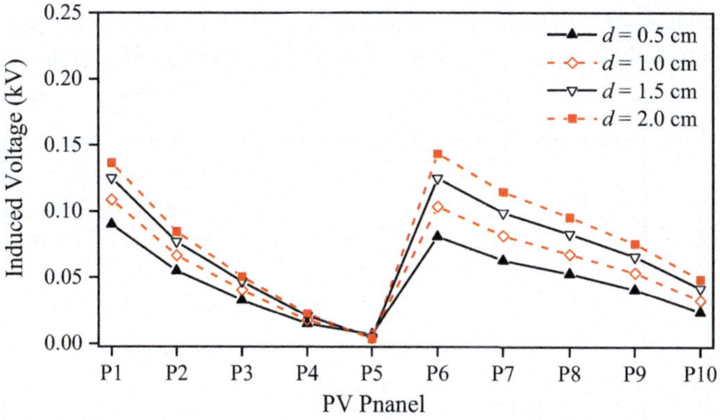

Fig. 6.19 Induced voltage peak of PV module with various distance *d* [1]

located in zone with the greatest EM field, and the differences of the voltages under various distances are obvious. In contrast, the zone with the weakest EM field, such as P5, the induced voltages for various distances are all close to zero. The EM field generated by the eddy current on the aluminum foil is reduced as the distance *d* increases. Hence, the effect of the eddy current on the lightning EM field is weaker, and the induced voltages will rise.

6.4.3 Foil Thickness

The induced voltages of the PV module with the foils for various thicknesses are presented in Figs. 6.20 and 6.21. It is clear that the voltage is hardly affected by the thickness of aluminum foil when the thickness is greater than 0.01 mm. The thickness changes from 0.01 to 0.001 mm, and the rise of voltage is about 278 V. The induced voltage is up to 1.22 kV when the thickness is 0.0001 mm. The eddy currents are not only distributed on top and base of the foil surface but also appear on the side of the foil surface. The greater the foil thickness is, the larger the eddy currents on the sides of the foil surface. The eddy current on the side also produces an EM field to weaken the lightning EM field, and can reduce the induced voltages of PV modules. The results indicate that the appropriate range of the aluminum foil thickness is 0.01–0.5 mm to effectively reduce the induced voltage and meet the manufacturing requirements for PV modules.

Figure 6.21 shows that the sensitivity of the induced voltage to foil thickness is relevant to the foil material. The induced voltage is significantly influenced by the thickness of graphite foil. If the PV backsheet is coated with graphite, the range of thickness is 5–10 mm, which is much greater than that of aluminum foil and copper foil. The thickness of PV backsheet foil needs to be thin as possible, otherwise the leakage currents inside PV modules will increase. In addition, the conductivity of graphite foil is much smaller than that of aluminum foil. Consequently, the aluminum foil is more suitable for the PV module to minimize the lightning induced voltage.

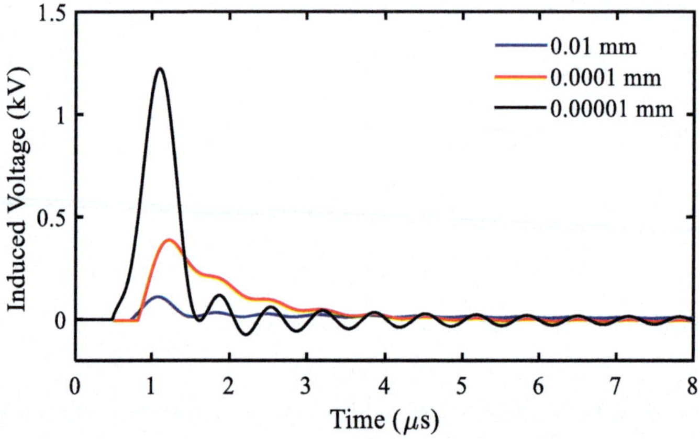

Fig. 6.20 Induced voltage of PV module with different thicknesses of Al foil [1]

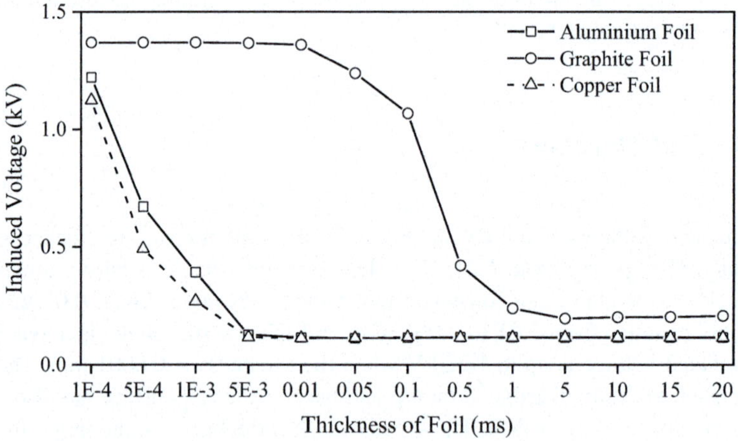

Fig. 6.21 Induced voltage peak of PV module with various thicknesses of three foils [1]

6.5 Conclusion

The mechanism of the eddy current generated by lightning EM field is studied, and the eddy currents on AL-BSF solar cells and the aluminum foil are calculated. The eddy current produces an EM field to weaken the lightning EM field. Hence, the DM induced voltage of PV panel is reduced. The internal loops of PV panels are directly located over the foil, and the lightning EM field can be effectively weakened.

The eddy current is mainly distributed on the outside of both the aluminum foil and Al-BSF cells, while its density on the center is the weakest. Since the induced currents on the metal frame also weaken the lightning EM field, the eddy current becomes less. The results indicate that the PV panels with aluminum foils can also minimize the DM induced voltages of DC cables in PV arrays.

The influence of foil material is discussed. The eddy current intensity rises as the conductivity of foil increases, and the effect of the eddy current for aluminum foil on the induced voltage is close to that of the copper foil but stronger than that of the graphite. The optimal range of the thickness for the aluminum foil is suggested 0.01–0.5 mm.

References

1. Zhong, X, Sun Q. Q, et al.: Effect of eddy current for photovoltaic module on lightning-induced overvoltage caused by electromagnetic coupling. IEEE Trans. Electromagn. Compat. 65(3), 850–861 (2023).
2. Rao, S. M. Wilton, D. R. and Glisson, A. W.: Electromagnetic scattering by surfaces of arbitrary shape. IEEE Trans. Antennas Propag. 30(4), 157−164 (1978).
3. Jin, J. M. Theory and Computation of Electromagnetic Fields. New York: John Wiley & Sons, Inc, 2015.
4. Ramarao, G. Chandrasekaran, K.: Lightning Channel-Base-Current Estimation Using Engineering Return-Stroke Models from Measured Magnetic Field Based on Deconvolution Method. IEEE Trans. Electromagn. Compat. 62(3), 798–806 (2020).
5. Wang, D. Takagi, N. Watanabe, T. et al.: Observed leader and return-stroke propagation characteristics in the bottom 400 m of a rocket-triggered lightning channel. J. Geophysical Res., Atmos. 104, 14369–14376 (1999).
6. Nucci, C. A. Rachidi, F. "Experimental validation of a modification to the transmission line model for LEMP calculations," Proceedings of 8th Symposium and Technical Exhibition on Electromagnetic Compatibility, Zurich, Switzerland, 389–394 (1989).
7. Ratnamahilan, P. Hoode, P. et al.: Computing transient electromagnetic fields radiated from lightning. J. Appl. Phys. 61(8), 525–536 (1987).
8. Yang, C. S. Zhou, B. H.: Calculation methods of electromagnetic fields very close to lightning. IEEE Trans. Electromagn. Compat. 46(1), 327–338 (2004).
9. Kaminski, A. Vandelle, B. and Fave, A. et al.: Aluminium BSF in silicon solar cells. Sol. Energy Mater Sol. Cells. 72, 373−379 (2002).
10. Visacro, S.: A representative curve for lightning current waveshape of first negative stroke. Geophys. Res. Lett. 31, L07112 (2004).

Part II
Lightning Electromagnetic (EM) Transients and Protection for Wind Turbines

Chapter 7
Introduction to Wind Power

7.1 Components of Wind Turbine

Wind power has become the most mature and scalable form of renewable energy generation in the world today, with significant prospects for commercial development [1]. Vigorous development of wind power is not only strategically important for energy security and diversification but will also play an increasingly crucial role in addressing climate change and promoting energy efficiency. Due to the locations of wind farms often being in windy areas such as hills, ridges, plains, or shallow sea regions, which are also prone to lightning, wind turbines, as tall structures in these areas, are highly susceptible to atmospheric lightning strikes [2].

According to observation data from 508 turbines over 5 years in wind farms in Texas, Kansas, and Illinois, if we calculate based on a turbine lifespan of 20 years, each turbine is expected to suffer damage from lightning strikes 2–3 times during its normal service life[3]. Data from the Nikaho wind farm in Japan recorded 128 in-stances of direct lightning strikes hitting turbines in a year, with multiple turbines being struck multiple times, up to 20 times[4]. In China, data from the Dongfang wind farm in Hainan shows a high lightning damage rate for turbine blades, reaching 5.56 blades per hundred blades per year [5].

Assessing the extent of damage to wind turbine components from lightning strikes is essential [6]. Statistics from Denmark on wind turbines reveal that for newer models (wind turbines above 450 kW), the proportion of control system and electrical system failures due to lightning strikes has remained relatively low, thanks to advancements in equipment withstand voltage and insulation technology. However, as turbine capacity increases and blades become longer, coupled with a lack of theoretical re-search on blade lightning protection systems, the probability of blade damage from lightning strikes is on the rise. This is illustrated in Fig. 7.1, where blade damage accounts for over 40% of total lightning strike damage, with most being irreparable. The high cost of turbine blades, coupled with expenses for maintenance, replacement, and installation, significantly impacts the economic stability

Q. Sun et al., *Lightning Transients and Protection for Renewable Energy System*,
https://doi.org/10.1007/978-981-97-7267-4_7

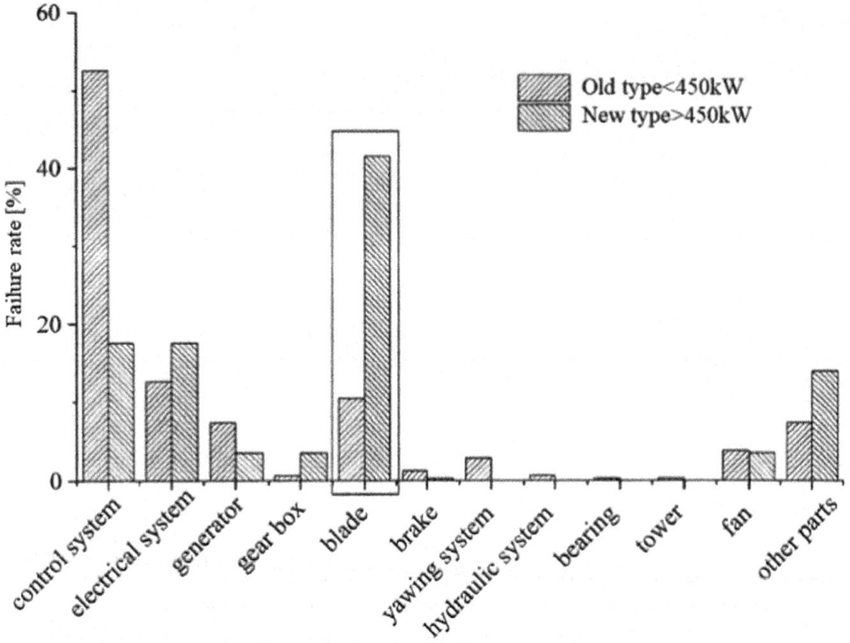

Fig. 7.1 Statistics on components damaged by lightning (Denmark) [6]

of wind farms. According to reports on the cost of replacing and repairing various turbine components in Germany, blade damage repair costs are currently the highest (as shown in Fig. 7.2). With increasing turbine capacity and the rise of offshore wind power, the cost of blade maintenance is expected to continue rising [7].

After being damaged by lightning strikes, wind turbines need to undergo repairs, during which they stop working. This downtime can result in a loss of electricity generation, which in turn affects the economic benefits of wind farms and can even impact the stability of the power grid. Figure 7.3 shows the average downtime for different components of wind turbines due to faults. Figure 7.2 and Fig. 7.3 reveal that blade damage not only incurs high maintenance costs but also results in longer downtime.

The lightning protection of wind turbine blades usually involves installing surge arresters on both sides of the blade tip or blade body to ensure that the blades are not damaged by lightning strikes. The successful arrester operation of the blade is also the foundation for preventing other components of the wind turbine from being damaged by lightning strikes. Most existing wind turbine lightning protection system designs refer to the IEC61400-24 technical standard, and usually install blade tip arresters at the blade tip. For longer blades, blade tip arresters cannot effectively protect the blades, and aluminum blade body arresters should be installed equidistantly on the blade surface from the blade tip, and connected to the ground through a down conductor as shown in Fig. 7.4 [8].

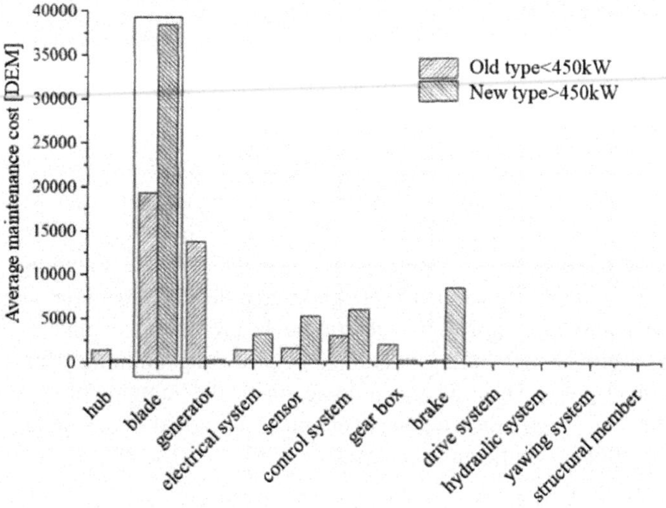

Fig. 7.2 Maintenance costs for different components of wind turbines (Germany)

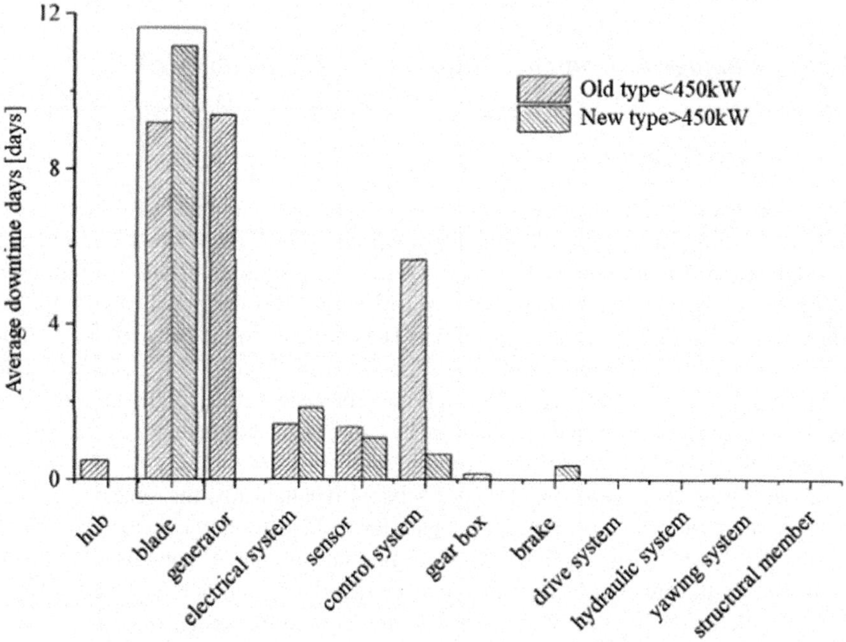

Fig. 7.3 Average downtime for repair of wind turbine components of different powers (Germany)

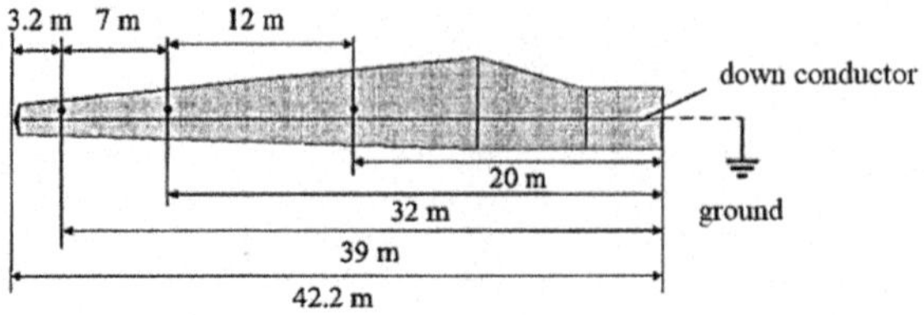

Fig. 7.4 Wind turbine blade lightning protection system structure

Therefore, conducting in-depth research on the physical lightning protection process of wind turbines, establishing a lightning protection model for wind turbines, providing a quantitative approach to determine the probability of blade arrester operation, and revealing the lightning shielding effects among multiple turbines in a wind farm are crucial steps. These efforts aim to enhance the lightning protection level of wind turbine generators, ensuring the economic, safety, and stable operation of wind farms, which holds significant importance.

7.2 Lightning Damage to Wind Turbine System

7.2.1 Damage to Propeller Blades from Lightning Strike

(a) **Damage Mechanism**

Propeller blades are the components most susceptible to lightning strikes on wind turbine systems. They account for the highest proportion of repair costs for lightning damage across the entire turbine assembly. Propeller blades are primarily composed of materials such as glass fiber reinforced plastic (GRP), carbon fiber reinforced plastic (CFRP), wood, steel, and aluminum. Their structure often consists of a combination of a shell and internal beams, as depicted in Fig. 7.5 for a typical blade structure. While the resistivity of the shell materials is generally high, making them non-conductive under normal circumstances, they can create conductive paths under the influence of intense lightning currents.

From the perspective of insulation discharge mechanisms, the surface of solid insulation presents the most susceptible path for surface discharge flashovers. When a propeller blade is struck by lightning, three possible discharge channels may appear on the blade: the outer surface of the blade shell, the inner surface, and the interface between the layers of shell material. Lightning current can either directly conduct along the surface discharge channel on the outer surface of the shell or create a thermally melted hole at the lightning strike point, penetrating through the hole to conduct along the inner surface or interface between the layers of shell material. After the lightning current penetrates the hole, it conducts along internal paths of the shell. After conducting a certain distance, it may jump out from the internal path of the shell to the outer surface of the shell for conduction. In some cases, such

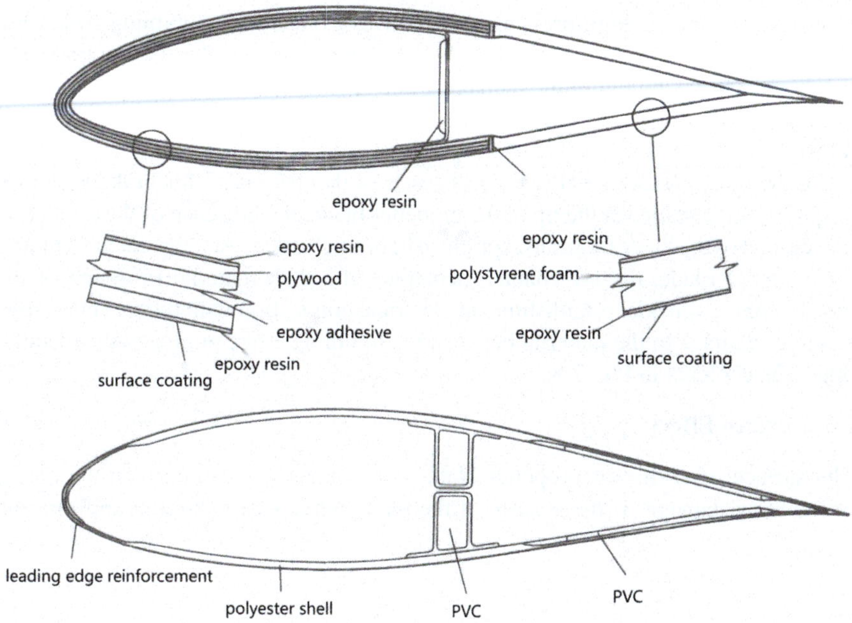

Fig. 7.5 Typical blade structure [9]

transitions may occur repeatedly. When the lightning current conducts along internal paths of the shell, arcs often occur, with arc temperatures reaching up to 30,000 K. Such high temperatures can severely damage the shell material near the arc path and generate high-pressure shock waves inside the shell, causing mechanical damage to the propeller blade shell. This damage, combined with the burning effect of high temperatures along the arc channel, often results in cracks on the blade after being struck by lightning, as shown in Fig. 7.6, and in severe cases, can lead to blade fracture [9].

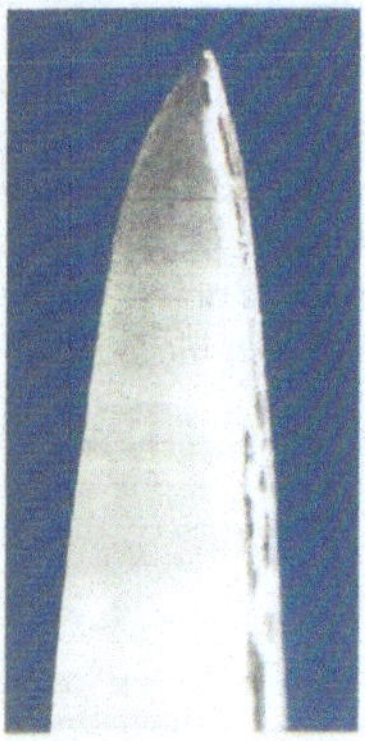

Fig. 7.6 Lightning strike-induced cracks on propeller blades [9]

In some cases, the high-pressure shock waves generated by lightning strikes not only damage the directly struck blade but also propagate to other blades that were not directly struck by lightning, damaging one or two of them, as shown in Fig. 7.7. Such severe lightning accidents have been recorded multiple times in foreign wind farms.

Furthermore, when metallic objects (such as dampers used for braking control purposes) are installed at the tip of the propeller blade, the presence of these metallic objects makes the blade more susceptible to lightning strikes. At the point of lightning strike on the blade, the arc channel penetrates the shell material in search of the metallic objects inside, establishing an electrical connection with them. This results in severe damage to the material between the lightning strike point and the metallic object, as depicted in Fig. 7.8.

(b) **Surface Effect**

The surface condition of propeller blades has a significant impact on the extent of lightning damage. If the surface of the blade remains smooth and uniform, the

Fig. 7.7 Both lightning-struck and unstruck propeller blades damaged simultaneously [9]

Fig. 7.8 The damage marks on the material between the lightning strike point and the metallic object on the propeller blade [9]

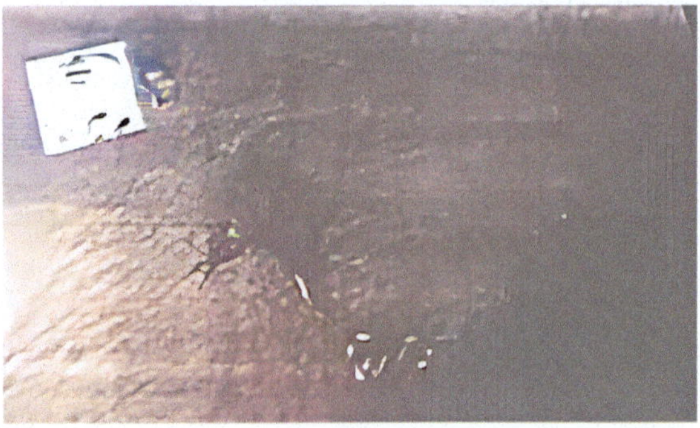

development of surface discharge flashovers caused by lightning strikes is relatively less likely. Due to the smooth and uniform surface of the blade, when struck by lightning, there won't be areas of distorted concentrated electric field on the surface. According to the theory of high-voltage discharge, the more uniformly distributed the electric field is on the interface between solid and gas mediums, the higher the voltage for surface flashover, and thus, the stronger its resistance to surface flashover.

However, in actual operation, it's often challenging to maintain a smooth and uniform surface on the blades. For instance, during blade manufacturing, there may be protruding fibers or certain rough spots on the surface due to weathering and erosion. These imperfections not only directly cause distortion in the electric field distribution along the blade surface during lightning strikes but also tend to accumulate moisture and atmospheric contaminants, forming a water film at these defect sites. The ions dissolved in the water film move along the surface under the influence of the electric field, further disturbing the uniformity of the electric field distribution along the surface. This results in a decrease in surface flashover voltage, making it easier for discharge breakdown to occur. The high temperature generated by these discharge paths will burn the surface material of the blade.

In addition to surface imperfections such as roughness and unevenness, surface contamination is another significant factor affecting the lightning damage condition of propeller blades. Due to prolonged exposure to the atmospheric environment, wind turbine systems are subject to pollution from industrial pollutants, natural salts, alkalis, and airborne dust, forming a layer of contaminants on the surface of the blades. In dry conditions, this surface contamination layer has high resistance and generally has little impact on surface flashover voltage. However, when atmospheric humidity is high, the surface contamination layer becomes wet, leading to a significant decrease in resistance and causing a decrease in surface flashover voltage. Consequently, when propeller blades are struck by lightning, surface flashovers are more likely to occur. This surface contamination effect is more pronounced for offshore wind turbine systems and coastal wind turbine systems.

Figure 7.9 depicts a schematic diagram of comparative tests investigating the influence of surface contamination on surface discharge flashover for propeller blades. The test specimens are made of Glass Fiber Reinforced Plastic (GRP), and their surface contamination level is set at a salt deposition density of 0.1 g/cm^2, equivalent to the salt contamination level near beaches.

In the lightning simulation tests, three types of specimens were used: uncontaminated propeller blades, contaminated propeller blades, and metal tubes of equal height. Multiple lightning impulse tests were conducted on each of these three types of specimens, and the respective 50% flashover voltages ($U_{50\%}$) were measured and statistically analyzed. The corresponding measurement and statistical results are shown in Fig. 7.10. From the graph, it can be observed that the flashover voltage of uncontaminated propeller blades is approximately 10% higher than that of contaminated propeller blades, while the flashover voltage of contaminated propeller blades is only slightly higher than that of the metal tubes.

In the impact tests on uncontaminated propeller blades, it is rare to observe obvious surface discharge channels, as shown in Fig. 7.11a. The discharge paths mainly go

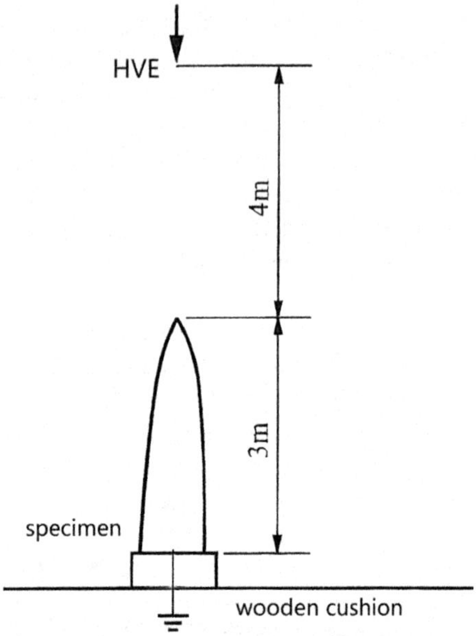

Fig. 7.9 The schematic diagram of lightning simulation tests on surface-contaminated propeller blades [9]

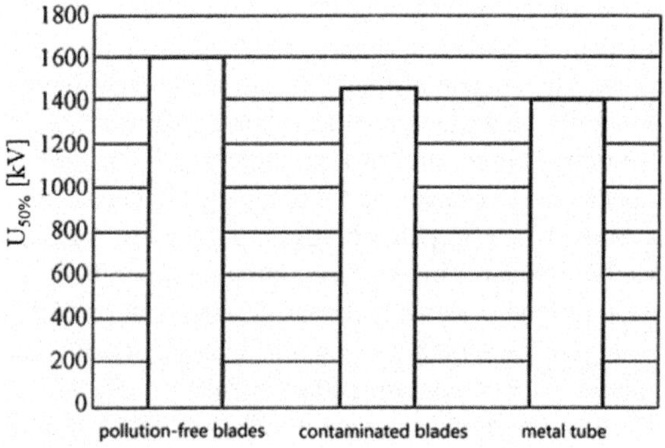

Fig. 7.10 The measurement and statistical results of the 50% flashover voltage [9]

directly from the high-voltage electrode to the ground, with only a few cases of surface flashover, as shown in Fig. 7.11b. However, in the impact tests on contaminated propeller blades, obvious surface flashover phenomena can be observed every time, with different flashover paths, as shown in Fig. 7.12. The flashover paths often penetrate the surface into the interior of the blade and then return to the surface, continuing along the surface, this form of discharge flashover is quite severe and leads to significant damage to the blades.

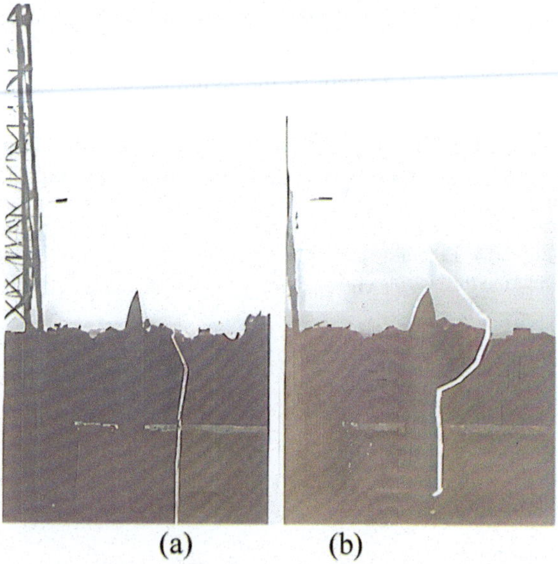

Fig. 7.11 The discharge path on uncontaminated propeller blades [9]

Fig. 7.12 Surface flashover on contaminated propeller blades [9]

7.2.2 *Damage to Bearings by Lightning Current*

(a) **Damage Mechanism**

When wind turbines are struck by lightning, bearings are another component prone to damage by lightning current within the system. As depicted in Fig. 7.13, after the blades are struck by lightning and the discharge occurs, a portion of the lightning current will conduct through the main shaft, and some of this current will flow into the bearings. The path of the lightning current carried by the main shaft through various major components inside the nacelle is illustrated in Fig. 7.14. After passing through the bearings, the lightning current will damage the rollers and races inside the bearings, thereby jeopardizing the operational reliability of the bearings.

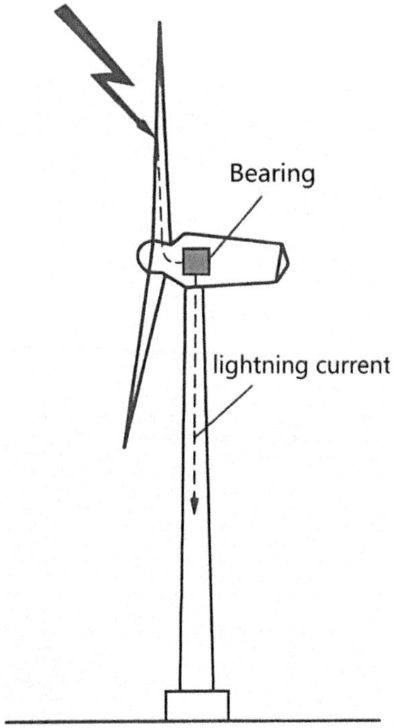

Fig. 7.13 Lightning current entering the bearings [9]

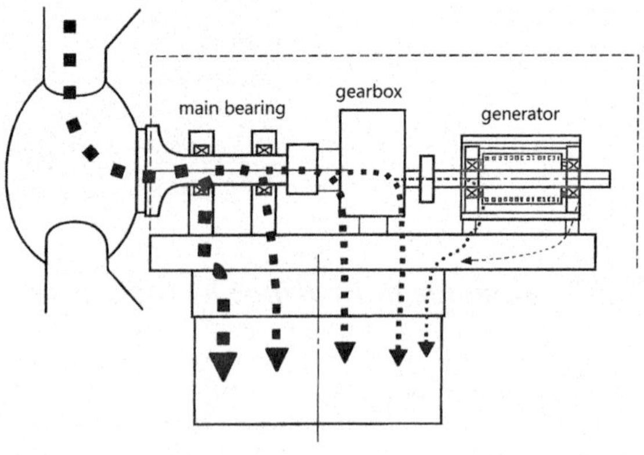

Fig. 7.14 The current path of lightning inside the nacelle [9]

Since lightning damage inside bearings often doesn't immediately result in operational failure, wind turbine systems typically don't undergo specialized bearing inspections immediately after a lightning strike. It may take several years from the initial occurrence of lightning damage inside the bearings to determine severe damage leading to operational failure. During this extended period, the damage caused by lightning current continues to develop and worsen under mechanical loads. The confusion between lightning damage and long-term mechanical wear makes it difficult to differentiate and assess the severity of damage caused by lightning current infiltration when inspecting bearings that have failed during operation. This potential damage results in increased noise during operation and a significantly shortened working life of the bearings. In recent years, this issue has drawn attention in some offshore wind farms abroad. Maintenance personnel in Danish wind farms found severe damage to the bearings in three previously lightning-struck offshore turbines during routine maintenance. They also conducted immediate inspections of the main bearings in a 90 kW small turbine after its blades were struck by lightning and fell off. The inspection revealed that out of a total of 46 rollers inside the main bearing, 12 were damaged, showing significant pitting and grooving on the surface. The average width of the grooves was 3 mm. In terms of maintenance costs, the repair costs for bearings are relatively high, especially for offshore wind turbine systems, where the maintenance costs of bearings can be even higher.

The damage caused by lightning current to bearings is closely related to the paths it takes inside the bearings. There are primarily two conduction paths: one formed by the contact between rollers and races in the loaded section along the circumference of the bearing, and the other formed by the short air gaps between rollers and races in the unloaded section along the circumference, after being pierced by an arc. For a bearing carrying a mechanical load, the number of rollers in contact with the inner race and their contact area along the circumference depend on the magnitude of the load and increase with the load. In the unloaded section, the breakdown characteristics of the short air gaps between rollers and inner race depend on the insulation strength of the lubricating oil filled inside the bearing. Furthermore, the rotational motion of the bearing also affects the conduction performance of lightning current on these two conduction paths and the extent of damage it causes.

(b) **Experimental Inquiry**

In order to investigate the damage mechanism of lightning current to bearings in wind turbine generators, conducting impulse current flow tests on bearing models is an effective research approach. The experimental setup is illustrated in Fig. 7.15, where the lightning impulse current is injected from the outer race of the bearing into the rollers, inner race, and shaft, and then flows back into the ground. The tested bearing selected here is a double-row self-aligning bearing with inner and outer diameters of 9 and 160 mm respectively. Both rollers and races are made of high-chromium steel. The experimental investigation is carried out using a complete bearing model.

The complete bearing model is composed of filling all the rollers originally present in the test bearing between the inner and outer races. The bearing is first subjected to mechanical load under static and unlubricated conditions, and then injected with

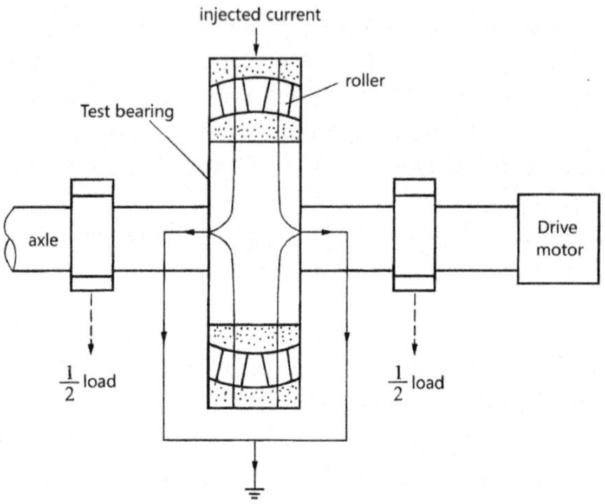

Fig. 7.15 Bearing model injected with lightning current experiment [9]

an impulse current, with an amplitude of 100 kA. During the experiment, it was observed that there were numerous damaged pits on both the rollers and races in the unloaded section along the circumference of the bearing, while in the loaded section, only small and sparse damaged pits were observed. The main reason for this difference in damage is that there are differences in contact conditions between the unloaded and loaded sections along the circumference of the bearing. In the unloaded section, poor contact between the rollers and races can create many air gaps. Due to the complex conduction paths inside the complete bearing model, these gaps may form open loops with nearby conduction paths. When the impulse current with a steep wavefront (di/dt) is injected into the bearing, it can induce a sufficiently high voltage in these gaps to break them down, generating arcs and burning the surfaces of the rollers and races, thus forming damaged pits.

The distribution of the maximum damage area on the surface of the rollers around the circumference of the bearing is shown in Fig. 7.16. When the bearing rotates at its rated speed, the observed damage phenomenon is more severe compared to when it is stationary, with a large number of scars found on the surfaces of both the inner and outer races. This is likely due to the metal melted by the high-temperature arc at the gap being crushed and adhered to the damaged area as the rollers rotate. When lubricating oil is added and the bearing rotates at its rated speed, it returns to the typical operating state of wind turbine bearings. Compared to the situation without lubricating oil, the introduction of lubricating oil causes the majority of damage pits to appear in the non-loaded sections adjacent to the load-bearing sections on the circumference of the bearing. In contrast, the gaps formed in the non-loaded sections further away from the load-bearing sections are relatively longer. The thinner lubricating oil layer filling shorter gaps is easily pierced and forms an arc, resulting in high-temperature damage. Consequently, more damage pits are observed near the load-bearing sections. Conversely, the thicker lubricating oil layer filling longer gaps has sufficient insulation strength, making it difficult for the gaps to be pierced and arcs

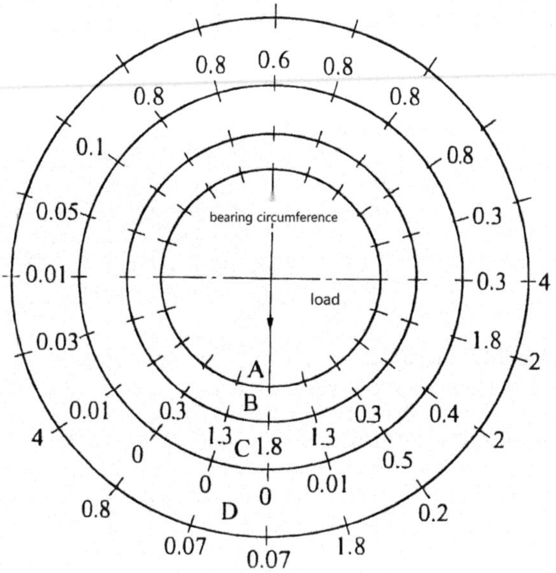

Fig. 7.16 Distribution of maximum damage area on roller surface along bearing circumference [9]

to form. Therefore, fewer damage pits are observed in the non-loaded sections further away from the load-bearing sections. Additionally, in the load-bearing sections, the introduction of lubricating oil worsens the direct metal-to-metal contact between the rollers and the races, leading to a decrease in electrical conductivity. Consequently, the thermal effects are intensified, resulting in a larger damage area in the load-bearing sections compared to when no lubricating oil is present.

7.3 Lightning Protection for Wind Turbine System

7.3.1 Lightning Protection of Blade

(a) **Lightning Protection**

The traditional blade lightning protection device is mainly composed of a lightning receiver and a down conductor. The lightning receiver is usually made into a disc shape, which is embedded in the tip of the blade, and the disc surface is flush with the blade surface, as shown in Fig. 7.17. The lightning receiver is electrically connected with the down conductor set inside the blade body and span the full length of the blade, as shown in Fig. 7.18. When the blade tip is struck by lightning, the lightning current is introduced into the down conductor by the lightning arrester, and then the down conductor introduces the lightning current into the hub of the blade root, the low-speed shaft and the tower, and finally discharges into the earth.

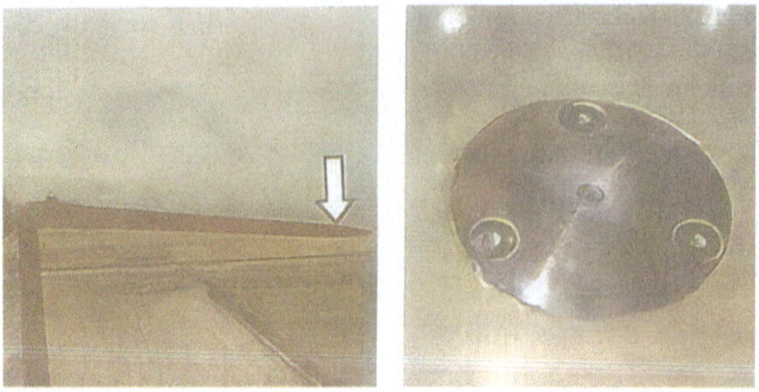

Fig. 7.17 Lightning protection equipment for lightning arrester at blade tip [9]

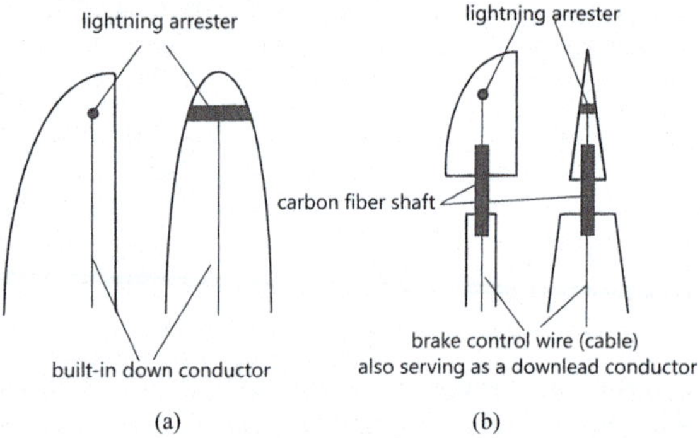

Fig. 7.18 Lightning protection down conductor in blade. **a** No tip damper. **b** Blade tip damper [9]

This lightning protection device with a single lightning receiver and a built-in down conductor is relatively easy to implement in manufacturing, but field operation experience shows that it is only effective for blade protection with a length of less than 20 m. The main disadvantage lies in the fact that there is only one lightning arrester capable of attracting and intercepting lightning strikes, and the area of the arrester is significantly small compared to the entire blade surface. Therefore, it is difficult to ensure that the arrester is the only point of lightning strike on the blade. There may be instances where a portion of the downward leading stroke of the lightning occurs at non-arrester areas on the blade surface, specifically at locations below the blade tip. When these areas below the tip are struck by lightning and a strike point is formed, the lightning current will follow the surface flashover path on the blade to reach either the arrester at the tip or the root of the blade. Additionally, there is a risk that the blade material itself may be punctured, allowing the lightning current to directly propagate to the built-in down conductor of the blade. As illustrated in Fig. 7.19, regardless of the path along which the lightning current travels, it will cause damage to the blade material.

In order to overcome this shortcoming, some manufacturers have embedded a metal strip on the blade surface, which can be set by spraying a metal layer on the blade surface or embedding a metal fiber braided mesh, as shown in Fig. 7.20. Compared with only installing the lightning arrester at the tip of the blade, the method of embedding metal mesh belt on the blade surface can effectively enhance

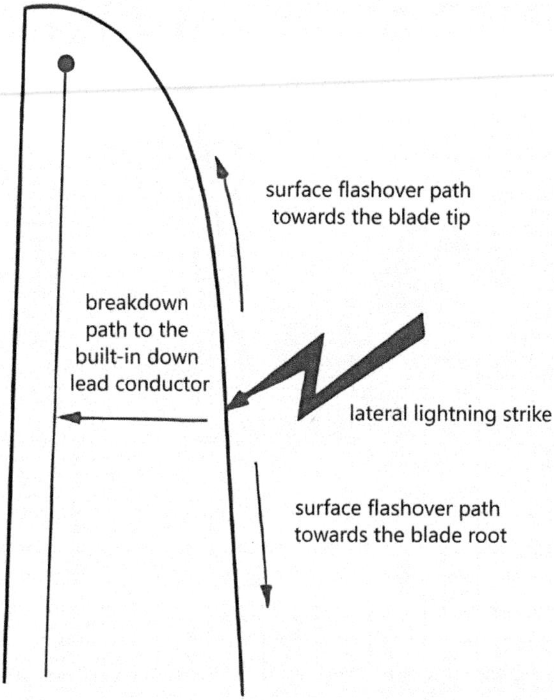

Fig. 7.19 Flashover path of non-flashover parts on the blade when struck by lightning [9]

the interception efficiency of the whole lightning protection device to the lightning downward leader and reduce the probability of blade material being damaged by lightning strike. However, it is difficult to ensure that the metal mesh belt on the blade surface maintains close electrical contact and firm adhesion with the surface throughout the blade length range, especially for those long blades after long-term operation. Due to the influence of temperature changes and ice and snow, this problem may be further aggravated. The lightning current transmitted on the metal mesh belt will cause damage to the blade material at the detached part where the contact is not close.

With the increase of wind turbine unit capacity, the length of the blade is also increasing. For the blade with a length of more than 20 m, the method of using a single lightning receiver and a built-in conductor usually cannot meet the requirements of lightning protection reliability. There will be a large number of lightning strike points at non-flashover parts, which will significantly increase the chance of blade material being damaged by lightning. At the same time, the metal mesh belt embedded along the blade surface will also cause technical and economic problems in the firmness of the metal mesh belt bonding. To this end, a more practical approach is to set up multiple lightning receivers on the long blade, each of which is electrically connected to the built-in down conductor, as shown in Fig. 7.21. This can greatly improve the interception performance of the lightning protection device to the lightning downward leader, and reduce the probability of lightning strikes on

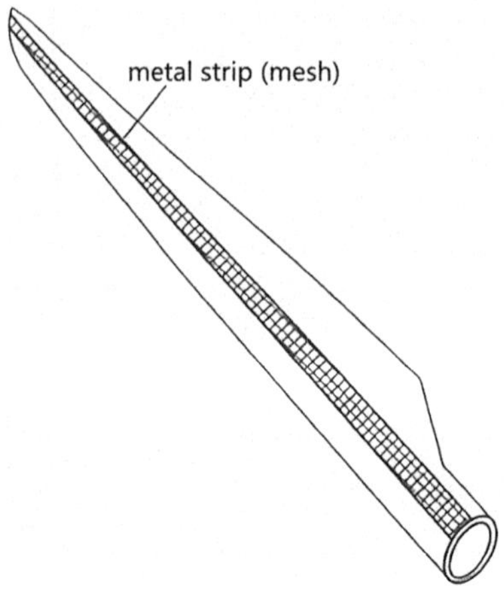

Fig. 7.20 The surface of the blade is embedded with a metal mesh belt for lightning protection [9]

the non-flashover parts of the blade surface. At present, this method has been put into practical application on the blades of megawatt-class units.

In order to more reliably protect the blades from lightning damage, some manufacturers have installed multiple metal layers along the full length of the front, rear edges and the center of the two sides of the long blades, as shown in Fig. 7.22. The blackened part in the diagram is the metal layer, so as to conduct the lightning diversion in an all-round way and reduce the severity of the lightning damage to the blade material.

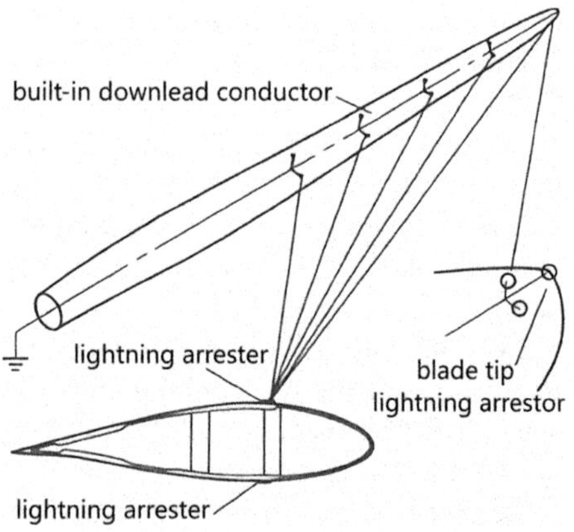

Fig. 7.21 Multiple lightning arresters are installed on the blade [9]

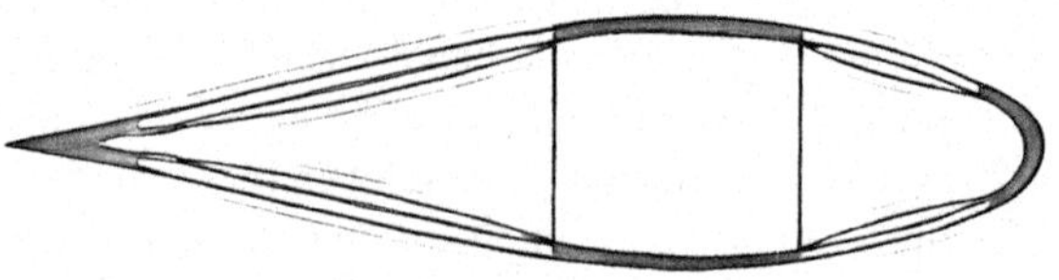

Fig. 7.22 Multiple metal layers are installed on the blade [9]

(b) Dimensions of Materials Used in Lightning Protection Devices

The material used in the lightning protection device on the blade should be able to withstand its thermal and mechanical effects when passing through the expected lightning current without being damaged. In general, the minimum material size used by the lightning arrester and the guide body should not be less than the data given in Table 7.1. After further considering various possible random factors and electrical and mechanical effects, the material size used by the lightning protection device can be selected according to the data recommended in Table 7.2.

Table 7.1 The minimum material size used in lightning protection devices

Material	Lightning arrester (mm^2)	Down conductor (mm^2)
Copper	35	16
Aluminum	70	25
Steel	50	50

Table 7.2 Recommended minimum material size for lightning protection devices

Material	Structural style	Minimum cross-section (mm^2)
Copper	Flat strap	50
	Cylindrical rod	50
	Fine wire woven bag	50
	Cylindrical rod for lightning arrester	200
Aluminum	Flat strap	70
	Cylindrical rod	50
	Fine wire woven bag	50
Aluminum alloy	Flat strap	50
	Cylindrical rod	50
	Fine wire woven bag	50
	Cylindrical rod for lightning arrester	200
Hot-dip galvanized steel	Flat strap	50
	Cylindrical rod	50
	Fine wire woven bag	50
	Cylindrical rod for lightning arrester	200
Stainless steel	Flat strap	60
	Cylindrical rod	78
	Fine wire woven bag	70
	Cylindrical rod for lightning arrester	200

(c) Estimation of Material Temperature Rise

When designing lightning protection devices for blades, a key consideration is the temperature rise caused by the thermal effect of the lightning current on the lightning arrester and the down lead. After being hit by lightning, the temperature of the lightning arrester and down lead may rise to a level that can burn the nearby blade material. When the temperature reaches the melting point of the lightning arrester material, holes and pits may form on the arrester. These lightning strike scars can further exacerbate corrosion in the atmospheric environment. For the down lead carrying the lightning current, the temperature rise may reduce its mechanical strength, making it prone to breakage. This hazard is particularly significant in blade lightning protection devices with tip dampers. The blade brake control wire, which also serves as a down lead, is already under mechanical tension before conducting the lightning current. The temperature rise caused by the lightning current may further reduce its mechanical strength and make it easier to snap. Such accidents have actually occurred several times in some wind turbines abroad, with some steel wires being broken by lightning strikes and having a cross-sectional area of up to 78 mm^2 (10 mm in diameter). The temperature rise caused by the thermal effect of the lightning current in metal materials can be estimated using the following formula:

$$\Delta T = \frac{1}{\alpha}\{\exp[\frac{(W/R)\alpha\rho_0}{\lambda S^2 C_W}]^{-1}\} \quad (7.2)$$

where, ΔT is the temperature rise of the conductor, K; α is the temperature coefficient of resistance, 1/K; *W/R* is the unit energy of the lightning current, J/Ω; ρ_0 is the resistivity of the conductor at ambient temperature, Ω·m; S is the cross-sectional area of the conductor, m^2; λ is the density of the conductor material, kg/m^3; C_w is the specific heat capacity of the conductor material, J/[kg· K]. Each quantity in the above formula can be valued according to Table 7.3 and Table 7.4.

Table 7.3 Physical parameters of typical materials for lightning protection devices

Parameter	Material			
	Copper	Aluminum	Low carbon steel	Stainless steel
ρ_0 (m)	17.8×10^{-9}	29×10^{-9}	120×10^{-9}	0.7×10^{-6}
A (1/K)	3.92×10^{-3}	4.0×10^{-3}	6.5×10^{-3}	0.8×10^{-3}
γ (kg/m^3)	8920	2700	7700	8000
θ_S (melting point) (°C)	1080	658	1530	1500
C_S (J/kg)	209×10^3	397×10^3	272×10^3	–
C_w [J/(kg·K)]	385	908	469	500

Note A Austenite is non-magnetic

Table 7.4 The relationship between temperature rise (K) and unit energy W/R of typical materials

	Material											
	Copper			Aluminum			Low carbon steel			Stainless steel		
	W/R (MJ/Ω)			W/R (MJ/Ω)			W/R (MJ/Ω)			W/R (MJ/Ω)		
	2.5	5.6	10	2.5	5.6	10	2.5	5.6	10	2.5	5.6	10
4	–	–	–	–	–	–	–	–	–	–	–	–
10	564	–	–	–	–	–	169	542	–	–	–	–
16	146	454	–	1120	–	–	56	143	309	–	–	–
25	52	132	283	211	913	–	22	51	98	940	–	–
50	12	28	52	37	96	211	5	12	22	190	460	940
100	3	7	12	9	20	37	1	3	5	45	100	190

7.3.2 Lightning Protection of Bearings

The main way of bearing lightning protection is to set a parallel low resistance channel at the front end of the bearing, and to implement bypass shunt for the lightning current transmitted along the shaft, so that the lightning current flows through the bearing as little as possible. In order to achieve this goal, conductor slip rings, brushes and discharge appliances are commonly used. Manufacturers have used carbon brushes to set bypass shunt channels, but carbon brushes will generate arcs on them when conducting lightning currents in frictional contact, aggravating their wear. The degree increases its contact resistance, weakens the bypass shunt effect, and deteriorates the lightning protection performance of the bearing. Wear will also make the carbon brush need to be replaced more frequently, which is not conducive to the normal operation of the unit, especially for offshore units, it is difficult to implement. At the same time, the carbon dust generated by friction will also cause pollution to the operating environment in the nacelle.

In order to make up for this defect, the carbon brush is replaced by copper brush with better wear resistance in some units. As shown in Fig. 7.23, the use of this copper brush can improve the wear and diversion conditions to a certain extent, but the corresponding cost will also be increased. If this brush is simply used for bypass shunt, because its contact resistance is maintained at a considerable value, only a part of the lightning current can be bypass shunted, and a part of the lightning current will still pass through the bearing. For this reason, a combination of bypass shunt and blocking isolation can be used for comprehensive management, as shown in Fig. 7.24. Installing insulating pads between the main bearing gearbox and the nacelle floor can effectively block the flow of lightning current through these paths. Additionally, insulating couplings should be installed between the gearbox and the generator to prevent the lightning current from entering the generator through the high-speed shaft. This approach can significantly force the lightning current to be diverted through the bypass channel of the front-end slip ring, guiding it towards

Fig. 7.23 Copper brush for lightning protection of bearings [9]

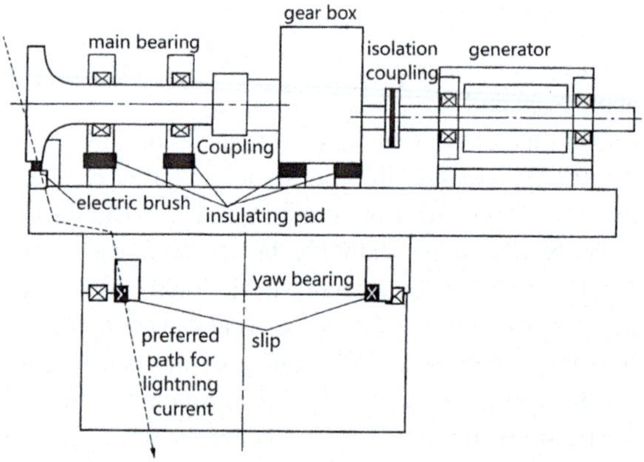

Fig. 7.24 Shunt and insulation blocking measures [9]

the nacelle floor and tower. It should be noted that during a lightning strike or electrical failure within the nacelle, such insulation methods can create a high potential difference, known as a transient overvoltage, between the insulated equipment and the nacelle floor. This transient overvoltage poses a hazard to the safety of maintenance personnel and the reliable operation of the equipment. Therefore, additional overvoltage suppression measures need to be taken at these locations.

7.3.3 Lightning Protection for Nacelle

After taking lightning protection measures on the rotor blades, a certain level of lightning protection can also be provided for the nacelle. Typically, lightning arrestors and down conductors installed on the rotor blades can effectively intercept downward

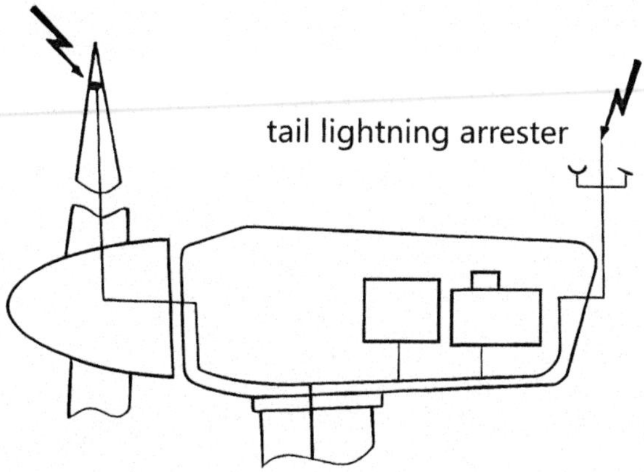

Fig. 7.25 Lightning Rod Installed at the Tail of the Nacelle [9]

leaders of lightning coming from the front and above the nacelle. However, they may not be able to intercept downward leaders approaching from behind the nacelle. Therefore, it is necessary to install lightning rods at the tail of the nacelle, as shown in Fig. 7.25. The lightning rod installed at the tail of the nacelle can effectively protect the anemometer and wind direction indicator at the rear of the nacelle, while also safeguarding the tail nacelle cover from direct lightning strikes. If no lightning protection measures are taken on the blades, it is necessary to install lightning rods at both the front and rear ends of the nacelle. When necessary, metal strips and metal mesh can be arranged on the surface of the nacelle cover to enhance the effectiveness of lightning protection. Some nacelle covers are made of metal materials, which act as Faraday cages, providing protection from lightning strikes and shielding the operating equipment inside the nacelle. However, it is still necessary to install a lightning rod at the tail of the nacelle to protect the anemometer and wind direction indicator.

All equipment inside the nacelle, except for those that require insulation isolation, should be electrically connected to the nacelle floor to achieve equipotentiality and prevent excessive transient potential differences between various equipment and components during lightning strikes, which could lead to counterattacks. Figure 7.26 shows an example of the actual equipotential connection between the generator and the floor inside the nacelle.

7.3.4 Divert and Conduct the Lightning Current

When struck by lightning, in order to reduce the harm of lightning current, it is always hoped that the lightning current will be introduced into the tower from the lightning protection device on the blade or the lightning rod at the tail of the nacelle through the brush, the bottom plate of the nacelle and the slip ring of the yaw system,

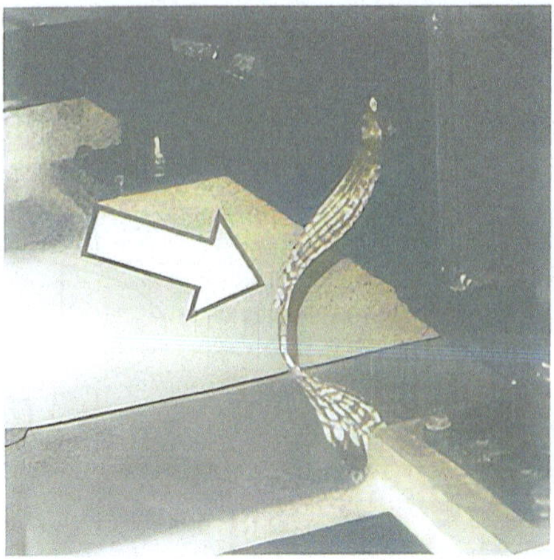

Fig. 7.26 Equipotential Connection between the Generator and the Nacelle Floor [9]

and then scattered into the earth through the grounding body of the unit, as shown in Fig. 7.27.

If this path can be maintained to conduct lightning current smoothly into the ground, the degree of harm caused by lightning can be significantly reduced, so as to achieve good lightning protection effect. However, on this path, there are two contact parts that affect the smooth conduction of lightning current, one is the moving friction contact part, and the other is the static contact part. The friction contact part refers to the part where the brush or slip ring is located. The lightning current passes through the friction contact surface where the two parts have relative motion. On the contact surface, the contact resistance will increase with the wear of the brush or slip ring surface. In addition, the tower of large capacity unit is relatively high, which is connected by several sections into a whole. If the contact between each two sections and the contact between the tower bottom and the grounding body is not close enough, the contact resistance value of these static contact parts is considerable.

From the circuit principle, the lightning current always needs to find the path with the lowest resistance to conduct into the ground. When the expected contact resistance on this path is large, it hinders the lightning current from conducting into the ground through this path. Then it will find other unwanted paths in the unit to conduct into the ground, such as through the high and low speed shafts in the nacelle and related equipment and the lines in the tower to conduct into the ground, which will seriously damage the equipment in the unit and endanger the safe and reliable operation of the unit. It can be seen that measures must be taken to dredge this desired diversion path to ensure that the lightning current can smoothly enter the ground along this path. For brushes and slip rings, materials with good electrical conductivity and high wear resistance should be used, and the contact area should be increased as much as possible when conditions permit, so as to increase the current capacity and reduce the contact resistance. Reliable electrical connections should be made between each two sections of the tower, and multiple crossover conductors should be arranged equidistantly on the circumference of the cylinder section, as shown in Fig. 7.28.

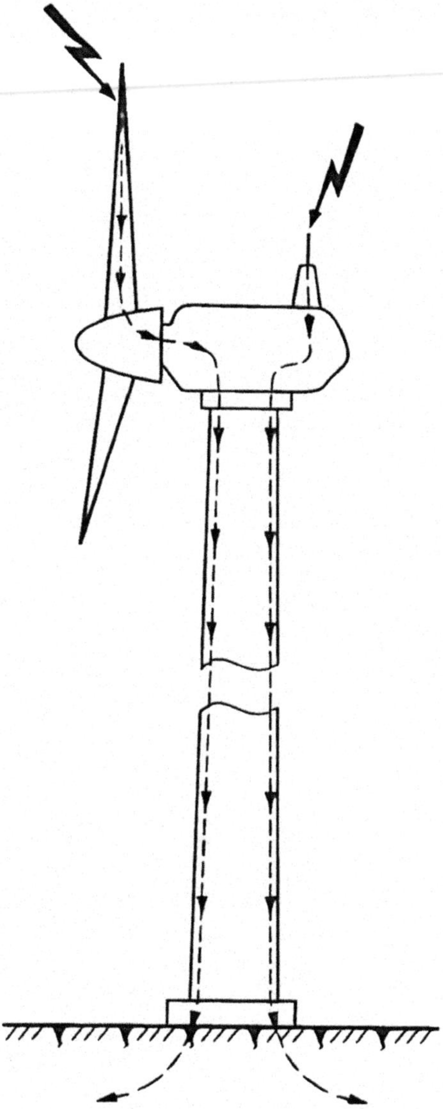

Fig. 7.27 Lightning current path of wind turbines [9]

The electrical connection between the tower and the grounding body should also be set up in many places, each with a dedicated connecting conductor, and ensure a strong and close connection, as shown in Fig. 7.29.

Fig. 7.28 Specialized jumper conductors between each two sections of the tower [9]

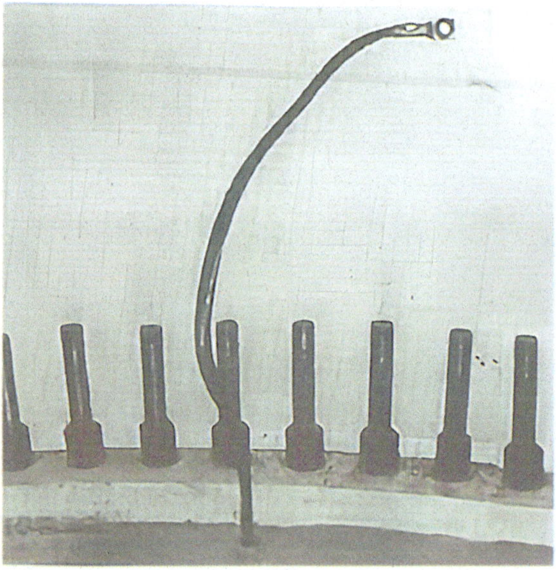

Fig. 7.29 Specialized connecting conductors between the bottom of the tower and the grounding body [9]

References

1. Global Wind Energy Council (GWEC). Global Wind 2014 Report. Brussels, Belgium, (2015).
2. Zhao, H., Wang, X.: Lightning Strike Mechanism and Lightning Protection Technology of Wind Turbine. Power Grid Technology. 27(7), 12–15 (2003).
3. Garolera, A. C., Madsen, S. F., et al.: Lightning Damage to Wind Turbine Blades from Wind Farms in U.S. IEEE Transactions on Power Delivery. (99), 1 (2014).
4. Hachiya, K., Hyodo, H., Honjo, N.: Lightning Strike Aspect from Wind Turbines in Winter Season - Observation Results at the Nikaho Kougen Wind Park in Japan. Proceedings of the

2008 International Conference on Lightning Protection (ICLP), Uppsala, Sweden. (9a-4), 1–7 (2008).
5. Guo, Z., Li, Q., Yan, J., et al.: Summary of Research on Physical Evolution Mechanism of Lightning Discharge of Offshore Wind Farms. Journal of Electrical Engineering. 10(5), 10–19(2015).
6. Wang, G.: Research on improved electro-geometric model of wind turbine blades based on lightning physics [D].Shandong University.(2018).
7. Wang, G., Zhang, L., Wu, H., et al.: Integrated Electromagnetic Transient Model of Offshore Wind Turbines and Research on Lightning Transient Overvoltage. Electric Power Automation Equipment. 37(11), 32–38 (2017).
8. IEC 61000–24: 2010 Wind Turbine Generator Systems - Part 24: Lightning Protection. IEC, (2010).
9. Zhang, X.: Wind Turbine Generator Set Lightning Protection and Grounding. China Electric Power Press. (2009).

Chapter 8
Evaluation of Lightning Strike Distance of Wind Turbine Blades

8.1 Electrical Geometric Model

The electrical geometric method establishes a connection between lightning current parameters and the geometric structure of the object under study. It utilizes the concept of striking distance to describe the object's ability to attract lightning from the ground. According to this method, the lightning leader that reaches the striking distance of a particular object first will discharge towards that object. The physical significance of striking distance is the critical breakdown distance from the head of the lightning leader channel developing from the thundercloud to the object being struck on the ground [1, 2].

In traditional electrical geometric methods for striking distance calculations, the striking distance is mainly determined based on operational data statistics or field strength around the ground object, which significantly deviates from the actual lightning process. Currently applied striking distance formulas are solely functions of lightning current [3, 4], while the rotating structure of actual wind turbine blades involves the embedding of one or more air terminals on the blades. Consequently, traditional striking distance calculation methods fail to account for the different striking distances between the air terminals on wind turbine blades and the blade body (without air terminals). Furthermore, they do not consider the impact of blade rotation angles on striking distance, rendering them unsuitable for lightning protection calculations for wind turbine blades [5, 6].

To establish an electrical geometric model for wind turbine blades, this paper introduces the concept of dynamic striking distance on wind turbine blades [7]. The term "dynamic" primarily encompasses two layers of meaning: first, different positions on wind turbine blades (such as air terminals and blade bodies) have varying striking distances; second, changes in the blade's position (such as angle and height) also lead to variations in striking distance. The dynamic striking distance of wind turbine blades exhibits a distribution characteristic, with striking distances not only present at the air terminals but also corresponding striking distances at the blade

Q. Sun et al., *Lightning Transients and Protection for Renewable Energy System*,
https://doi.org/10.1007/978-981-97-7267-4_8

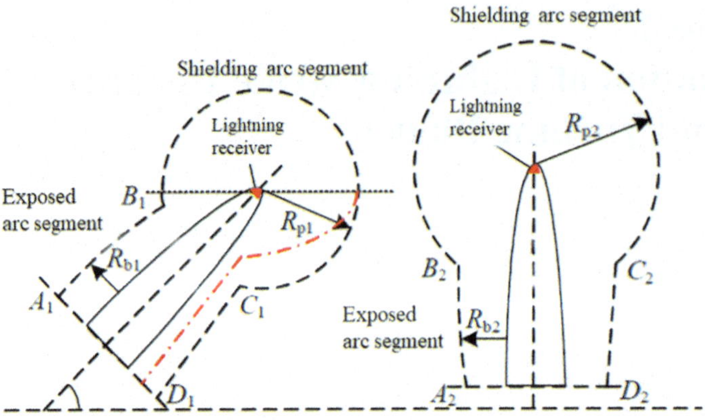

Fig. 8.1 Dynamic striking distance of turbine blades. **a** $0° < \theta < 90°$. **b** $\theta = 90°$[7]

body, albeit shorter. At the blade's plane, the striking distance at the blade body can be considered as a non-circular arc located at a certain distance from the blade surface with a smaller curvature. Figure 8.1 illustrates the variations in striking distance at different angles and positions on the blade. When the blade is not vertically oriented upwards, the striking distances on both sides are not strictly symmetrical; the striking distance on the lower side (as indicated by the dashed line in Fig. 8.1a lower part) is slightly smaller than on the upper side. Assuming vertical downward lightning development, the striking distance on the lower side does not affect the air terminal. Considering the rotational symmetry of the blade, the striking distance on the upper side of the blade is symmetric about the blade axis to the lower side, facilitating numerical calculations for the model.

8.2 Lightning Protection Model

Based on the dynamic striking distance concept of wind turbine blades as described above, the electrical geometric model for lightning protection of wind turbine blades can be derived, as shown in Fig. 8.2. It is assumed that the lightning leader initially develops vertically downward. Here, R_p represents the striking distance at the blade's air terminal, R_b is the striking distance at the blade body, and θ is the angle between the wind turbine blade and the horizontal direction, with counterclockwise direction considered positive. When the lightning current amplitude is I, the blade's striking distance corresponds to the arc ABCD, where arcs AB and CD are exposure arcs, and arc BC is the shielding arc. Point A marks the starting point of the envelope curve of the striking distance for the studied blade section (which can be a part of the blade). When the lightning leader reaches the exposure arc AB, it indicates that the blade body is struck by lightning, and the lightning protection system has failed to attract the lightning discharge.

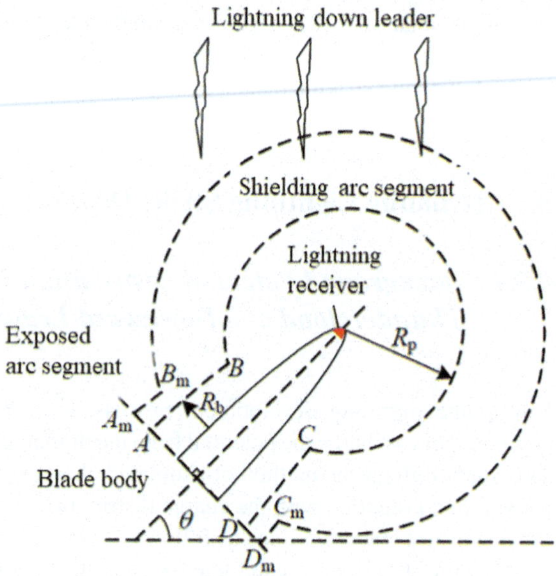

Fig. 8.2 EGM of blade lightning protection

With the increase of lightning current, R_p and R_b also increase accordingly. When the lightning current increases to I_m, the striking distance range extends to the arc $A_mB_mC_mD_m$. It can be observed that the exposure arcs A_mB_m and C_mD_m are significantly smaller compared to the exposure arcs AB and CD at lightning current I. As the lightning current magnitude continues to increase, the exposure arcs will decrease until they disappear, leading to complete shielding of the blade. At this point, the corresponding lightning current magnitude is the maximum lightning current I_{max} at which the blade may fail to attract the lightning discharge. When the lightning current magnitude exceeds I_{max}, the lightning protection system of the blade can provide complete shielding. I_{max} is primarily dependent on the blade's position and the design of the blade's lightning protection system. For a pre-designed wind turbine blade lightning protection system, the range of lightning current magnitudes that the system cannot effectively shield is $I_{min} < I < I_{max}$, where I_{min} represents the maximum lightning current magnitude that the material can withstand without damage when the blade is penetrated, and it is mainly related to the material and processing techniques used for the blade. If the lightning protection capability of the blade's lightning protection system is strong enough to ensure $I_{max} < I_{min}$, then the system can effectively shield against lightning currents of all magnitudes.

The electrical geometric model of wind turbine blades describes the lightning attraction capability of various positions on the blade through striking distances. Dynamic striking distance is a characteristic parameter in the electrical geometric model of wind turbine blades and serves as the fundamental quantitative representation of the lightning protection efficiency of the blades. To characterize the changing properties of dynamic striking distance, it is necessary to consider the distortion of spatial potentials at different positions on the blade and further incorporate the physical mechanisms of lightning leader development, thereby addressing the calculation issue of dynamic striking distance.

8.3 Dynamic Lightning Strike Distance

8.3.1 Background Potential Distribution Under Thundercloud and Downward Leader

During the lightning development process, the background potential is mainly composed of two parts: the potential from the thundercloud and the downward leader. This study utilizes the method of simulated charges to determine the distribution of background potential, with the charge distribution models for the thundercloud and the downward leader outlined as follows.

In nature, lightning predominantly consists of negative polarity lightning, with thunderclouds typically having 2–3 charge centers. The uppermost layer accumulates a large amount of positive charge, the lower layer carries negative charge, and sometimes there are one or more weak positive charge regions at the bottom. Amorouso and Lattarulo [8] found that when the charge density inside the thundercloud is relatively uniform, a charged disk model can simulate thundercloud charges, providing a realistic depiction of thundercloud charge distribution and spatial potential characteristics, as shown in Fig. 8.3. The model assumes typical values for various parameters: $\rho_p = 3$ nC/m3, $\rho_n =$ -0.9 nC/m3, $h_p = 10$ km, $h_b = 5.5$ km, $h_n = 1.5$ km.

The lightning stepped leader adopts the charge distribution model proposed by Cooray [9], which shows good agreement with measured values. The charge density in the leader is:

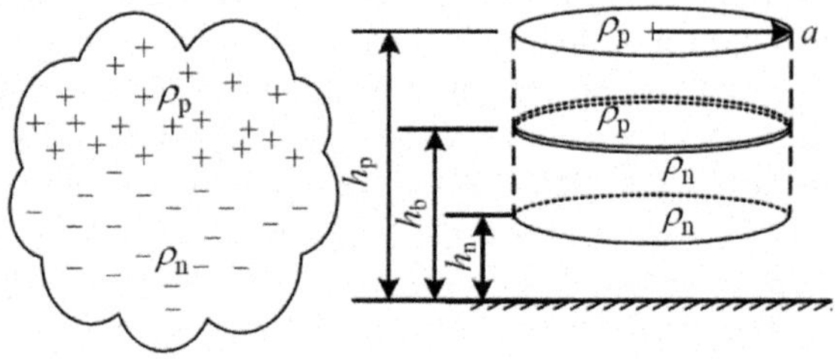

Fig. 8.3 Charge distribution of thundercloud

$$\begin{cases} r(z) = a_0\left(1 - \frac{z}{H_c - H_i}\right)G(H_i)I_p + \frac{I_p(a+bz)}{1+cz+dz^2}L(H_i) \\ G(z_0) = 1 - \frac{H_i}{H_c} \\ L(H_1) = 0.3a + 0.7b \\ a = e^{\frac{H_1 - 10}{75}} \\ b = 1 - \frac{H_i}{H_c} \end{cases} \tag{8.1}$$

where, $\rho(\zeta)$ represents the charge density in the stepped leader channel, in C/m; ζ denotes the distance from a point in the leader channel to the leader tip, in meters; I_p is the peak lightning current, in kA; H_c is the height of the thundercloud above ground, $H_c = h_n = 1500$ m; H_i is the height of the leader tip above ground, in meters; the coefficients are $a_0 = 1.476 \times 10^{-5}$, $a = 4.857 \times 10^{-5}$, $b = 3.9097 \times 10^{-6}$, $c = 0.522$, $d = 3.73 \times 10^{-3}$.

8.3.2 The Physical Mechanism of Positive Upward Leader Initiation

As the negative stepped leader progresses downward, the electric field on the surface of ground objects increases. Due to collision ionization and attachment effects, a positive polarity space charge region is generated. When the positive space charge exceeds a certain value, initial corona discharge occurs [10].

$$\exp\left(\int_{R_0}^{R} a(E) - h(E)dx\right) > N_{cri} \tag{8.2}$$

where, α and η represent the collision ionization coefficient and attachment coefficient, respectively; R_0 is the radius of curvature of the electrode; R is the boundary of the collision ionization region; x is the distance from the electron avalanche head to the electrode; N_{cri} can be taken as 0.55×10^8 [11].

After the initial corona discharge, a streamer zone is formed in the discharge gap. The free electrons generated in the streamer zone gather at the base of the streamer. These electrons collide with gas molecules at the base, transferring energy to the gas molecules (translational energy, rotational energy, electronic excitation energy, vibrational energy, etc.). This energy is converted into translational energy H_t (T_h) within 10^{-5} to 10^{-3} s, where T_h is the translational energy temperature of the gas molecules. Assuming that the mass of gas molecules in the streamer zone remains constant, the translational energy equation is satisfied [12]:

$$\begin{cases} H_t(T_h) = 3.5kn_hT_h\pi r_{stem}^2 \\ \frac{dH_t(T_h)}{dt} = (f_t + f_r + f_e)EI + \frac{W_v(T_v) - W_v(T_h)}{\tau_{vt}} \end{cases} \tag{8.3}$$

where, f_t, f_r, f_e represent the conversion coefficients of translational energy, rotational energy, and electronic excitation energy, respectively, where $f_t + f_r + f_e = 0.07$; k is the Boltzmann constant; n_h is the gas molecule density; r_{stem}, E, I represent the radius, field strength, and current passing through the base of the streamer; T_v is the temperature of molecular vibrational energy; τ_{vt} is the time constant for the conversion of vibrational energy to translational energy; W_v(T) represents the vibrational energy of gas molecules. The equilibrium equation for vibrational energy can be expressed as:

$$\begin{cases} W_v(T_v) = n_h \pi r_{stem}^2 \frac{\varepsilon_v}{\exp(\frac{\varepsilon_v}{kT_v})-1} \\ \frac{dW_v(T_v)}{dt} = f_v EI - \frac{W_v(T_v)-W_v(T_h)}{\tau_{vt}} \end{cases} \tag{8.4}$$

where, ε_v is the excitation energy of the nitrogen molecule's vibrational ground state, approximately 0.28 eV; f_v is the conversion coefficient for vibrational energy, where $f_t + f_r + f_e + f_v = 1$. When the temperature at the base of the streamer reaches 1500 K, the conductivity significantly increases, and the leader begins to form. The temperature at the base of the streamer is related to the number of electrons generated in the streamer region. According to the kinetic energy balance equation of gas molecules mentioned above, when the spatial charge in the streamer region exceeds 1 μC [11], it can be considered to meet the temperature conditions for the streamer to leader transition.

8.3.3 The Physical Mechanism of Positive Polarity Upward Leader Development

After the initiation of the upward leader, it is considered that the downward leader develops downward in a step-like manner at a certain speed, corresponding to the development process of the upward leader as shown in Fig. 8.4. The potential at the head of the upward leader is [13]:

$$U_{tip}^i = -l^i E_\infty - x_0 E_\infty \ln\left(\frac{E_{str}}{E_\infty} - \frac{E_{str} - E_\infty}{E_\infty} e^{-\frac{l^i}{x_0}}\right) \tag{8.5}$$

where, E_∞ is the steady-state value of the quasi-static leader field strength, 3×10^4 V/m; E_{str} is the approximate constant value of the electric field in the streamer region, 4.5×10^5 V/m; x_0 is the length coefficient, 0.75 m.

When the downward leader advances to the i-1 step without considering spatial charges in the streamer region, the background potential is $U_1{}^{i-1}$. Taking into account the spatial charges in the streamer region, the distorted potential becomes $U_2{}^{i-1}$. When the downward leader advances to the i step, assuming that the upward leader has not developed, the background potential becomes U_i. At this point, considering

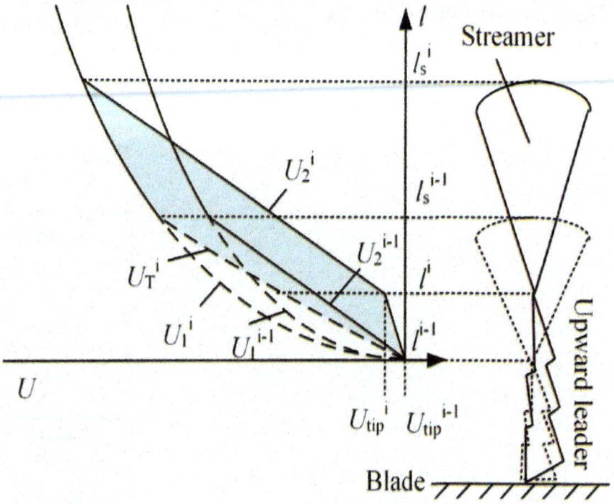

Fig. 8.4 Upward leader developing model

the spatial charges from the (i-1) step, the potential distortion becomes $U_T{}^i$. During the process of the leader's head advancing from l^{i-1} to l^i, with the streamer head advancing to $l_s{}^i$, the potential in the streamer region becomes $U_2{}^i$. The shaded area in Fig. 8.4 represents the newly generated charge ΔQ^i in this process [14], which can be approximated by the following equation for calculation:

$$\Delta Q^i = K_Q \int_{l^i}^{l_s^i} U_2^i - U_T^i dl = K_Q \int_{l^i}^{l_s^i} (U_2^i - U_2^{i-1} - U_1^i + U_1^{i+1}) dl \tag{8.6}$$

where, K_Q is the environmental factor and takes a value of 3.5×10^{-11} C/(V m). The increment of the next step leader can be expressed as:

$$\Delta l^i = \frac{\Delta Q^i}{q_L} \tag{8.7}$$

where, q_L represents the amount of charge per unit length of the leader required for the streamer to transition to the leader. For natural lightning, this value is approximately 65×10^{-6} C/m. Therefore, the position of the next step of the upward leader's head would be:

$$l^{i+1} = l^i + \Delta l^i \tag{8.8}$$

Assuming that the lightning strike occurs when DL and UL develop to the nth step and their streamers encounter one another. At this time, $l_s{}^n$ is the longest distance the UL and its streamer reach, which represents the lighting attraction range of tip receptor. This distance is determined to be R_p, the striking distance of the tip receptor, which can be calculated by:

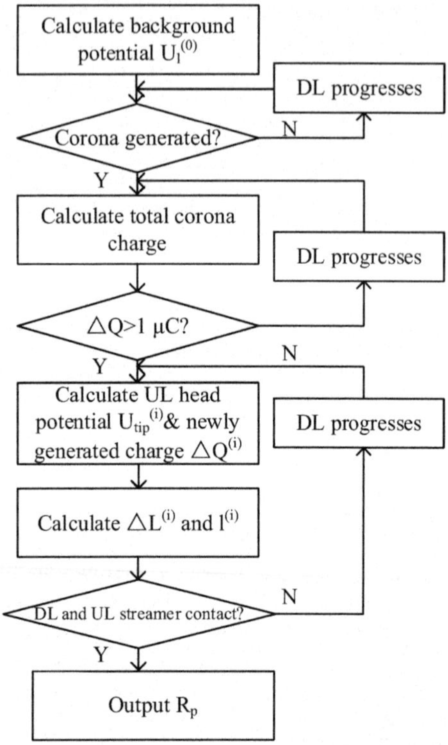

Fig. 8.5 Flowchart of the R_p calculation

$$R_p = l^n + \left(U_{tip}^n - U_d^n\right)/E_{str} \tag{8.9}$$

where, U_d represents the potential at the head of the downward leader; the calculation process of R_p is shown in Fig. 8.5.

8.3.4 Calculation of Dynamic Striking Distance of Fan Blade

Based on the described lightning strike process and the definition of dynamic striking distance, the dynamic striking distance can be determined using a leader model to better align with the actual lightning strike process. By employing a leader model, it becomes possible to calculate the height of the upward leader's streamer head relative to a specific point when the downward leader is directly above that point, thereby ascertaining the dynamic striking distance. This approach, which considers the development of leaders and the dynamic nature of lightning strikes, offers a more realistic assessment of lightning risk, making it more suitable for inclusion in English-language publications.

(1) Based on the thundercloud and the downward leader position, determine the background potential around the wind turbine. Use Eq. (8.2) to calculate the charge generated by spatial ionization and determine the onset of corona discharge. If corona discharge does not initiate, the downward leader continues to develop further downwards.
(2) Once corona discharge initiates, calculate the corona region charge ΔQ based on the potential curve around the wind turbine. When $\Delta Q > 1$ μC, the discharge transitions towards the leader. Otherwise, the downward leader continues to develop further downwards.
(3) After the leader initiates, assuming an initial leader length of 0.05 m, calculate the upward leader development process based on the potential distortion curve using Eqs. (8.5)–(8.8). When the average field strength between the heads of the upward and downward leaders reaches E_{str}, determine the lightning stroke, ultimately obtaining the development height of the upward leader and the streamer.

In the calculations, the parameters of the wind turbine model are as follows: tower height of 70 m, blade length of 40 m, radius of the conductor inside the blade of 0.01 m, and tower radius of 2.5 m.

To reduce computational complexity, this study assumes that an upward leader length greater than 2 m can form a stable upward leader [15]. The overall speed ratio of the development of upward and downward leaders varies between 1:4 and 1:1 [16, 17]. For natural lightning discharges with long air gaps, such as lightning strikes, calculations show that the overall speed ratio is closer to 1:4 [18]. By considering the height of the downward leader above ground when the upward leader stabilizes, combined with the speed ratio, an approximate value for the dynamic striking distance can be determined. Based on the leader model, for a lightning current of 30 kA and a downward leader step length of 10 m, with a speed ratio of 1:4, the dynamic striking distance curve for wind turbine blades is shown in Fig. 8.5. It can be observed that when the tower height and lightning current amplitude are fixed, the striking distance of the blades is influenced by their position and rotational angle.

The dynamic striking distance curve of wind turbine blades can be divided into three regions. In the tip effect region, there is a noticeable upward curl, which is caused by the tip effect of the blade tip arrester. Severe spatial potential distortion around the arrester makes it easier to generate upward leaders, resulting in a significantly greater striking distance compared to the blade body. The striking distance in the blade body region decreases as the distance from the blade tip increases, and can be approximated by a sloping line representing the blade body striking distance $R_b(l)$, where l is the distance from the blade tip, as shown by the dashed line in Fig. 8.6. The angle between the dashed line and the horizontal direction is denoted as θ', known as the blade body dynamic striking distance fitting angle, which is a negative value. In the shielding effect region, due to the shielding effect of the wind turbine tower, the rate of decrease in striking distance accelerates, a phenomenon that becomes more pronounced as the blade angle approaches horizontal. The striking distance formula recommended by IEEE based on extensive field operation data is widely used in lightning protection for ground objects. However, it can only provide the striking distance of the blade tip arrester when the wind turbine blade is oriented vertically upwards.

The traditional formula of striking distance has been widely used for the lightning protection of buildings with lightning rods. The upright blade is similar to a lightning rod. In this paper, the striking distance of the upright blade was calculated and compared with the result obtained with the traditional formulas (Fig. 8.7). The values obtained with this study are smaller than those of Armstrong and Whitehead as well as the results given by IEEE [19], and are closer to the results of the formula recommended by IEEE. The calculation method used in this paper takes into account the effect of space charge that attenuates the electric field intensity around the blade tip receptor and making the generation of an UL more difficult, which leads to the decrease of R_p. In addition, this method is more consistent with the physical development of lightning and considers the special structure of the wind turbine, unlike the traditional formula which makes it unsuitable for wind turbine lightning protection.

This chapter establishes an electrical geometric model for lightning protection of wind turbine blades. Based on the concept of dynamic striking distance for wind turbine blades, this model combines electrical geometric methods with the physical processes of lightning leader development, analyzing the mechanism of lightning protection for blades and providing a theoretical basis for evaluating blade lightning protection systems.

The concept and analysis method of dynamic striking distance of fan blades are proposed. By utilizing a physical development model of lightning upward leaders, the distribution characteristics of dynamic striking distance for blades are presented. This analysis method offers insights for lightning protection of ground objects with complex structures.

The simulated charge method is used to calculate the background potential distribution in the lower space of negative polarity thunderclouds and under downward

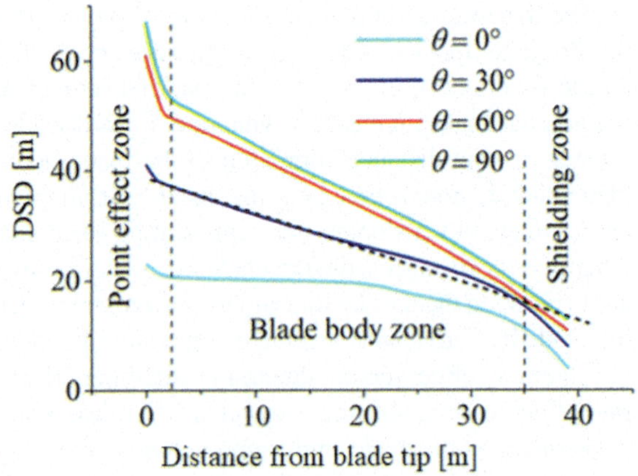

Fig. 8.6 Dynamic striking distance curve in I = 30 kA

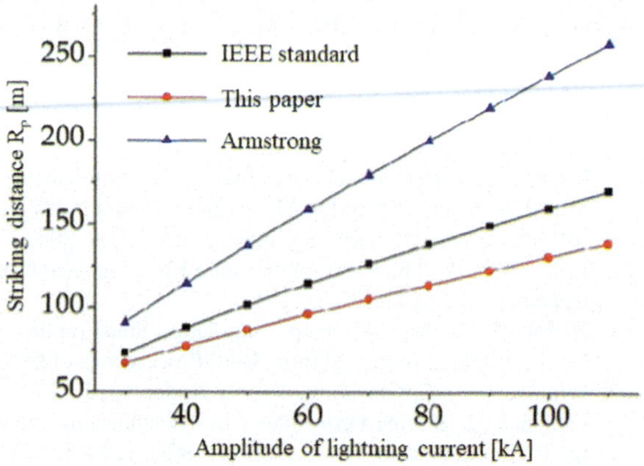

Fig. 8.7 Striking distances obtained by different methods

leaders. By incorporating the motion equations of spatial particles and considering the distortion effect of spatial charges around wind turbine blades on the background electric field, a numerical calculation method for dynamic striking distance is derived based on the step-like development model of lightning leaders.

References

1. Armstrong, H. R., Whitehead, E. R.: Field and Analytical Studies of Transmission Line Shielding. IEEE Transactions on Power Apparatus and Systems. 87(1), 270–281 (1968).
2. Brow, G. W., Whitehead, E. R.: Field and Analytical Studies of Transmission Line Shielding. IEEE Transactions on Power Apparatus and Systems. 88(5), 617–626 (1969).
3. Cooray, V., Rakov, V., Theethayi, N.: The Lightning Striking Distance Revisited. Journal of Electrostatics. 65(5), 296–306 (2007).
4. IEEE Std 998–2012 IEEE Guide for Direct Lightning Stroke Shielding of Substations. (2013).
5. Wang, G., Zhang, L., Guo, Z., et al.: Dynamic Striking Distance and Electrogeometric Model of Wind Turbine Blade Based on Lightning Physics. Proceedings of the CSEE. (21), 6427–6436 (2017).
6. Zhang, L., Wang, G., Zhang, W., et al.: An Electro-Geometric Model for Lightning Shielding of Multiple Wind Turbines. Energies. 10(9), 1272 (2017).
7. Wang, G.: Research on improved electro-geometric model of wind turbine blades based on lightning physics [D].Shandong University.(2018).
8. Amoruso, V., Lattarlo, F.: Thundercloud Pre-Stroke Electrostatic Modeling. Journal of Electrostatics. 56(2), 255–276 (2002).
9. Cooray, V., Rakov, V., Theethayi, N.: The Lightning Striking Distance-Revisited. Journal of Electrostatics. 65(5–6), 296–306 (2007).
10. Abdel-Salam, M., Turky, A. A., Hashem, A. A.: The Onset Voltage of Coronas on Bare and Coated Conductors. Journal of Physics D: Applied Physics. 31(19), 2550 (1998).
11. Gallimberti, I., Bacchiega, G., Bondiou-Clergerie, A., et al.: Fundamental Processes in Long Air Gap Discharges. Comptes Rendus Physique. 3(10), 1335–1359 (2002).
12. Gallimberti, I.: The Mechanism of the Long Spark Formation. Le Journal de Physique Colloques. 40(C7), C7–193–C7–250 (1979).
13. Rizk, F. A. M.: A Model for Switching Impulse Leader Inception and Breakdown of Long Air-Gaps. IEEE Transactions on Power Delivery. 4(1), 596–606 (1989).

14. Goelian, N., Lalande, P., Bondiou-Clergerie, A., et al.: A Simplified Model for the Simulation of Positive-Spark Development in Long Air Gaps. Journal of Physics D: Applied Physics. 30(17), 2441–2452 (1997).
15. Becerra, M., Cooray, V.: A Simplified Physical Model to Determine the Lightning Upward Connecting Leader Inception. IEEE Transactions on Power Delivery. 21(2), 897–908 (2006).
16. Dellera, L., Garbagnati, E.: Lightning Stroke Simulation by Means of the Leader Progression Model. I. Description of the Model and Evaluation of Exposure of Free-Standing Structures. IEEE Transactions on Power Delivery. 5(4), 2009–2022 (1990).
17. Eriksson, A. J.: The Lightning Ground Flash: An Engineering Study. Pretoria, South Africa: University of Natal. (1979).
18. Xie, S., He, J., Chen, W., et al.: Simulation Study on the Development Process of the Upward Leader Incepted from Lightning Rod. Proceedings of the CSEE. 32(10), 32–40 (2012).
19. Whitehead, J. T., Chisholm, W. A., Anderson, J. G., et al.: IEEE Working Group Report. Estimating Lightning Performance of Transmission Lines. II: Updates to Analytical Models. IEEE Transactions on Power Delivery. 8(3), 1254–1267 (1993).

Chapter 9
Lightning Protection Efficiency and Shielding for Wind Turbine Blades

9.1 Lightning Protection Efficiency for Wind Turbines

Nowadays, there are two main types of lightning arresters for wind turbine blades: one is to set the blade tip lightning arrester only at the tip of the blade, and the other is to set the blade tip lightning arrester and the blade body lightning arrester at the same time. However, there is no unified guidance plan for the selection of blade lightning arresters and the arrangement of lightning arresters by various blade manufacturers today. The influence mechanism of the arrangement of lightning arresters on the lightning protection efficiency of blades is still unclear. Therefore, finding a method to characterize the efficiency of the wind turbine blade lightning protection system is the theoretical basis for optimizing the wind turbine blade lightning protection system.

In this chapter, the numerical calculation method of the efficiency of the lightning protection system of the wind turbine blade is given according to the newly proposed electrical geometry model of the wind turbine blade. Then, the model is used to analyze the influencing factors of the efficiency of the lightning protection system of the actual size (1.5 MW) wind turbine blade, and the long gap breakdown test of the wind turbine blade is designed to verify the effectiveness of the electrical geometry model of the wind turbine blade.

9.1.1 Calculation Method for Lightning Protection Efficiency of Blades

(a) **Quantitative Characterization of Wind Turbine Blade Lightning Protection System Efficiency**

Q. Sun et al., *Lightning Transients and Protection for Renewable Energy System*,
https://doi.org/10.1007/978-981-97-7267-4_9

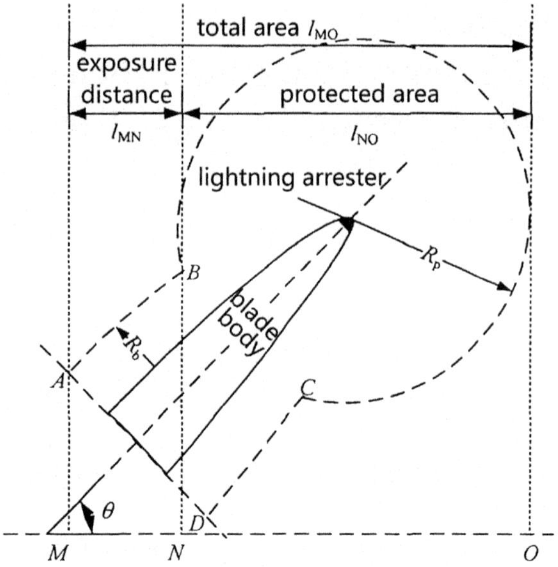

Fig. 9.1 Lightning interception efficiency analysis of blade LPS [2]

The electro-geometric model has been continuously improved since it was proposed. In 1993, based on long-term data statistics and research, the IEEE working group proposed a method of using the exposure distance to calculate the interception efficiency of the lightning protection system [1]. This method also assumes that the downward leader develops vertically downward, which is consistent with the previous assumptions made in this paper. The exposure distance refers to the horizontal distance between the two ends of the exposure arc, as shown in l_{MN} in Fig. 9.1.

The calculation formula of the efficiency of the wind turbine lightning protection system is given in IEC 61400-24, as shown in Eq. (9.1).

$$E = E_i \times E_s \tag{9.1}$$

where E_i is the interception efficiency of the wind turbine blade, and E_s is the classification efficiency of the blade lightning protection system.

When the lightning current amplitude is I and the blade is angled θ, the interception efficiency E_i of the blade can be obtained according to the exposure distance method as shown in Eq. (9.2).

$$E_i = \frac{l_{NO}}{l_{MO}} = \frac{l_{MO} - l_{MN}}{l_{MO}} \tag{9.2}$$

where, l_{MO} is the vertical projection distance of the blade part on the horizontal plane, l_{MN} is the exposure distance, and l_{NO} is the shielding distance.

According to the construction principle of the fan blade electrical geometry model, it can be seen that only when the lightning current amplitude is within the $I_{min} < I <$

I_{max} interval, the wind turbine blade may be damaged by lightning strike. According to Eqs. (9.1) and (9.2), when the wind turbine blade is at an angle θ with the horizontal direction, the efficiency of the blade lightning protection system is shown in Eq. (9.3).

$$E_\theta = \int_{I_{min}}^{I_{max}} \frac{l_{MO}(I) - l_{MN}(I)}{l_{MO}(I)} \times E_s \times f(I)dI \tag{9.3}$$

where, f(I) is the probability function of the lightning current amplitude[3]; for the fan blades with multiple lightning arresters, l_{MN} represents the sum of the exposure distances of each section.

In the actual operation of the fan, the fan blades are in a state of constant rotation. The distances l_{MO}, l_{MN}, and l_{NO} in Eq. (9.3) are not only a function of the lightning current I, but also a function of the rotation angle θ. Because the rotation angle of the wind turbine blade is in the upper half of the rotation circle at 0–180°, which is higher than the wind turbine tower, it is considered that the wind turbine is mainly struck by lightning within 0–180°. Therefore, the geometric mean value of Eq. (9.3) in 0–180° is taken as the efficiency of the lightning protection system of the fan blade during operation, as shown in Eq. (9.4).

$$E = \frac{1}{\pi} \int_0^{\pi} \int_{I_{min}}^{I_{min}} \frac{l_{MO}(I,\theta) - l_{MN}(I,\theta)}{l_{MO}(I,\theta)} \times E_s \times f(I)dId\theta \tag{9.4}$$

In this context, we believe that the discharge capacity of the lightning protection system is sufficiently large, and as long as the lightning arrester successfully intercepts the lightning strike, it will not cause substantial damage to the blade. Therefore, the value of Es is taken as 1.

(b) **Calculation of Exposure Distance**

The exposure distance l_{MN} is an important parameter to obtain the efficiency of wind turbine blade lightning protection system. This part will take the calculation of the exposure distance of the two lightning arrester blades as an example to illustrate the calculation method of the exposure distance of the fan blades. Because the rotation of the blade has rotational symmetry, the value of θ in the calculation is 0–90°.

When the blade is equipped with a side lightning arrester, the envelope range of the striking distance is indicated by the outer dashed line in Fig. 9.2. The values l_1 and l_2 represent the distances between the two lightning arresters (O_1O_2) and the starting point of the blade under study (O_1O), respectively. Shielding arcs B_1C_1 and D_1D_2 are circular arcs with the lightning arresters at the blade tip and blade body as the centers, and the striking distances R_{p1} and R_{p2} of the two lightning arresters as the radii. The vertical distance from the exposure arcs A_1B_1 and C_1D_1 to the blade body is the fitted striking distance $R_b(l)$ of the blade body. The exposure distance is the sum of the horizontal distances between A_1 and B_1, as well as between C_1 and D_1. At the starting point of the blade under study, the striking distance O of the blade body is $R_b(l_1 + l_2)$. Let the coordinates of point A_1 be (x_1, y_1) then the relationship

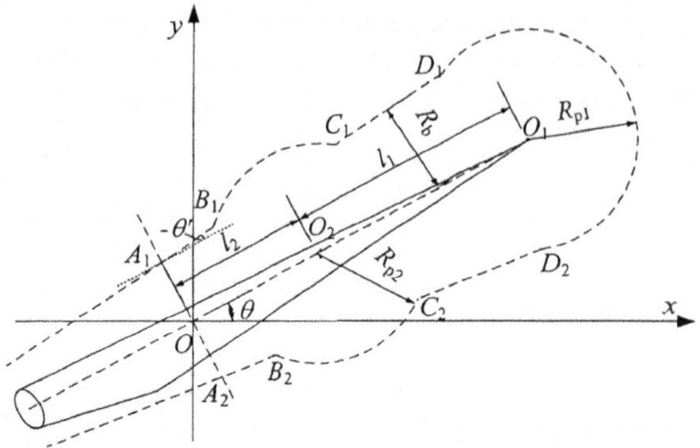

Fig. 9.2 Mathematical model of exposure distance

is shown in Eq. (9.5).

$$\begin{cases} x_1 = -R_b(l_1 + l_2)\sin\theta \\ y_1 = R_b(l_1 + l_2)\cos\theta \end{cases} \tag{9.5}$$

The exposed arcs (such as A_1B_1, C_1D_1) on the same side of the blade can be approximately considered as the numerical expression of the straight lines of A_1, B_1, C_1, and D_1 on the same straight line, as shown in Eq. (9.6).

$$\begin{aligned} y = {} & x\tan(\theta - \theta') + R_b(l_1 + l_2)\sin\theta\tan(\theta - \theta') + \cdots \\ & R_b(l_1 + l_2)\cos\theta \end{aligned} \tag{9.6}$$

where, θ' is the fitting angle of the dynamic striking distance of the blade body. The numerical expressions of O_1 and O_2 coordinates are shown in Eq. (9.7).

$$\begin{cases} O_1(x_{O_1}, y_{O_1}) = (l_1\cos\theta + l_2\cos\theta, l_1\sin\theta + l_2\sin\theta) \\ O_2(x_{O_2}, y_{O_2}) = (l_2\cos\theta, l_2\sin\theta) \end{cases} \tag{9.7}$$

Then the equation of the circle of O_1 and O_2 is:

$$\begin{cases} (x - l_1\cos\theta - l_2\cos\theta)^2 + (y - l_1\sin\theta - l_2\sin\theta)^2 = R_{p1}^2 \\ (x - l_2\cos\theta)^2 + (y - l_2\sin\theta)^2 = R_{p2}^2 \end{cases} \tag{9.8}$$

The values of coordinates $B_1(x_2, y_2)$, $C_1(x_3, y_3)$ and $D_1(x_4, y_4)$ can be obtained by combining Eqs. (9.6) and (9.8). According to the dynamic striking distance of the blade, the electrical geometric model diagram can be made to determine the position of each point. For the case of two exposed arcs shown in Fig. 9.2, a set of solutions with the largest x value needs to be discarded, and there is $x_2 < x_3 < x_4$. The exposure distance l_{MN} is:

$$l_{MN} = x_4 - x_3 + x_2 - x_1 \tag{9.9}$$

The total area l_{MO} corresponding to the studied leaf part is:

$$l_{MO} = (l_1 + l_2)\cos\theta + R_{pl} + \max\left[R_b(l_1 + l_2)\sin\theta, R_{p2} - l_2\cos\theta\right] \quad (9.10)$$

Substituting Eqs. (9.9) and (9.10) into Eq. (9.4) (or Eq. (9.3)), the operating lightning protection system efficiency E (or fixed angle lightning protection system efficiency E_θ) of the studied fan blade part can be obtained.

9.1.2 Sensitivity Analysis

(a) The Influence of Blade Rotation Angle on the Efficiency of Blade Lightning Protection System

According to the electrical geometric model of fan blades, the lightning protection system efficiency E_θ of blades with lightning current amplitude of 30 kA and only tip lightning arrester at typical angles of 0°, 30°, 60° and 90° can be obtained. In order to give a more intuitive picture of the lightning protection failure of the blade, this book draws the failure rate curve $(1 - E_\theta)$ of the lightning protection system within 0–180° according to the rotational symmetry of the blade, as shown in the 30 kA dotted line in Fig. 9.3. In the figure, the angle of the polar coordinate represents the rotation angle of the blade θ, and the radius represents the failure rate of the blade lightning protection system $1 - E_\theta$.

When the lightning current amplitude is 30 kA, the closer the blade rotation angle is to the level, the greater the failure rate of the lightning protection system is. When the angle is 0°, the failure rate is about 40%. When the blade angle is turned to 30°, the failure rate decreases rapidly to close to 0. On the one hand, due to the larger the blade angle, the more obvious the spatial potential distortion at the tip of the blade tip lightning receiver, the easier it is to start the upward leader, resulting in an increase in the striking distance of the lightning receiver and a rapid decrease in the length of the exposed arc. On the other hand, due to the increase of the angle, the exposure

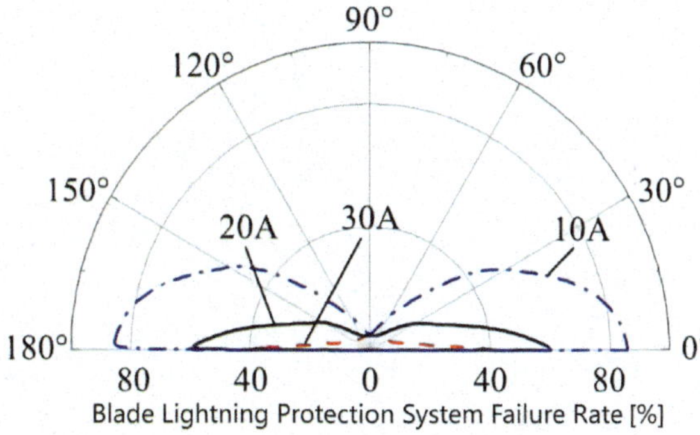

Fig. 9.3 Failure rate of blade LPS under different lightning current

distance obtained by the projection of the exposed arc in the horizontal direction decreases in a trigonometric function relationship. Therefore, with the increase of the blade rotation angle (0–90°), the blade will soon be within the protection range of the lightning receiver.

Through the study of the angle of the fan, it is suggested that when the fan is shut down in stormy weather, the fan blades should be avoided to be close to the horizontal level, so that the top blades can be vertically upward, which can effectively improve the efficiency of the blade lightning protection system.

(b) The Influence of Lightning Current Amplitude on the Efficiency of Blade Lightning Protection System

The influence of lightning current amplitude on the lightning protection efficiency of blades is studied by changing the lightning current amplitude (30 kA, 20 kA, 10 kA) and ensuring that other parameters of the fan remain unchanged. The failure rate of the blade lightning protection system under different lightning current amplitudes is shown in Fig. 9.3. When the lightning current is 20 kA, $\theta = 30°$, the failure rate is 17%, $\theta = 60°$, the failure rate has been reduced to 0; when the lightning current is 10 kA, $\theta = 30°$, the failure rate is 53%, $\theta = 60°$, the failure rate is 15%. It is found that with the decrease of lightning current amplitude, the lightning protection efficiency of the blade decreases obviously, and the angle range of complete protection of the blade also decreases. Therefore, the smaller the lightning current, the more vulnerable the blade to lightning.

This is because the change of lightning current affects the spatial potential distribution around the blade, which changes the striking distance and ultimately affects the lightning protection efficiency of the blade. When the lightning current is 20 kA, the dynamic striking distance curve of the blade is shown in Fig. 9.4. Compared with the above, it is found that the striking distance at the tip of the blade and the blade body is significantly reduced, and the tip effect of the lightning arrester is weakened, which makes the shielding arc range smaller, resulting in an increase in the probability of unsuccessful lightning of the blade.

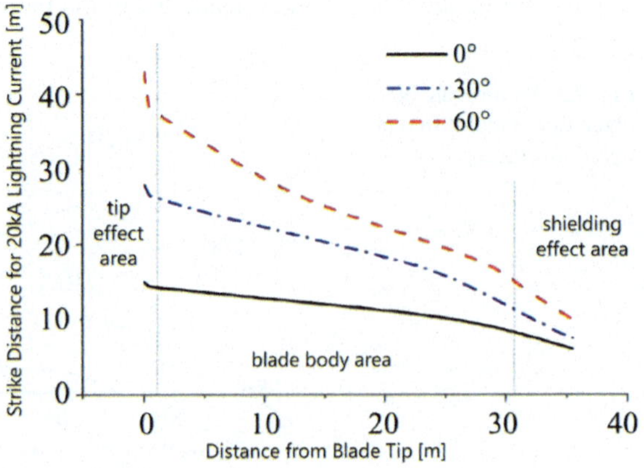

Fig. 9.4 Dynamic striking distance curve ($I = 20$ kA)

According to the analysis of the principle of the electrical geometric model, as long as the amplitude of the lightning current $I < I_{\min}$, even if the lightning flashover is not successful, due to its small energy, the blade will not cause structural damage. In order to obtain the relevant numerical range of $I_{\min}$, this book designs the impulse current test, the principle and the physical wiring are shown in Figs. 9.5 and 9.6 respectively.

The structure of the blade material is shown in the sandwich structure shown in Fig. 9.5. Both sides are glass fiber reinforced epoxy resin with high strength, and the middle interlayer is porous material (PVC, PET, light wood, etc.), which can effectively reduce the weight of the blade. The nickel–chromium wire for arc ignition is passed through the blade material, and the two ends of the nickel–chromium wire are connected to the high-voltage end and the low-voltage end of the impulse current generator through the wire. The fan blade sandwich material used in this book is PET, with a size of 10 cm × 10 cm. The damage under impulse current is shown in Fig. 9.7. When the current is 15 kA, the material around the lightning arc is seriously ablated and bursts. When the current is 7.5 kA, there is only carbonization in a small

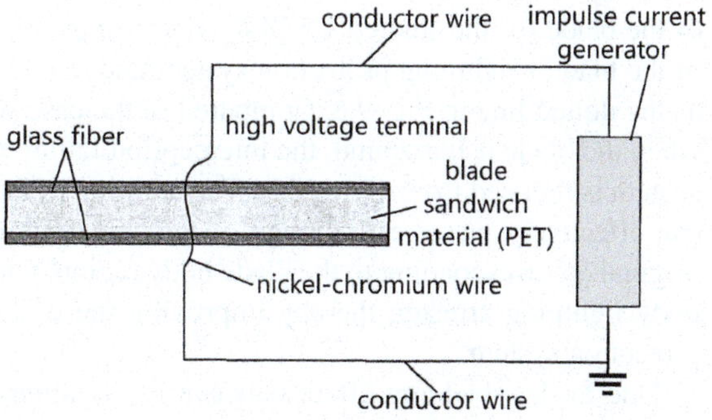

Fig. 9.5 Principle diagram of impulse current test

Fig. 9.6 Figure of test object

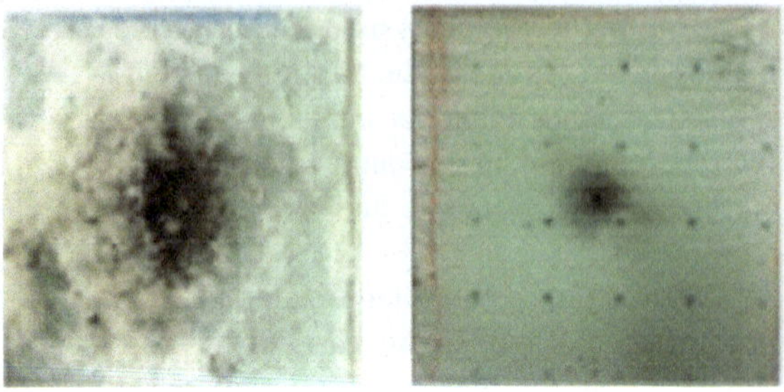

Fig. 9.7 Blade material damage under impact current. **a** 15 kA. **b** 7.5 kA

area around the arc, and the structure of the blade remains intact without bursting. Therefore, it is recommended that the value $I_{\min}$ should be ≤ 7.5 kA.

(c) The Influence of Blade Body Lightning Arrester on the Efficiency of Blade Lightning Protection System

To investigate the impact of the blade body lightning arrester on blade protection, this book introduces a blade body lightning arrester at a distance of 7 m from the tip of the blade for the studied 1.5 MW wind turbine blade. The calculated failure rate of the blade's lightning protection system under a 30 kA lightning current is shown as the dotted line in Fig. 9.8. Compared to the case with a single lightning arrester, when the blade is horizontal, the interception failure rate of the lightning protection system is reduced from 41 to 25%. The addition of the blade body lightning arrester can effectively increase the range of the shielding arc, replacing the exposed arc originally corresponding to the blade body region with the shielding arc of the blade body lightning arrester, thereby improving the efficiency of the blade's lightning protection system.

Due to the shielding effect between the lightning arresters, when the distance between two lightning arresters is close, the striking distance of the blade tip lightning arrester will decrease compared to its original striking distance. Calculations have found that when the blade body lightning arrester is approximately 1 m away from the blade tip lightning arrester, the striking distance of the blade tip lightning arrester

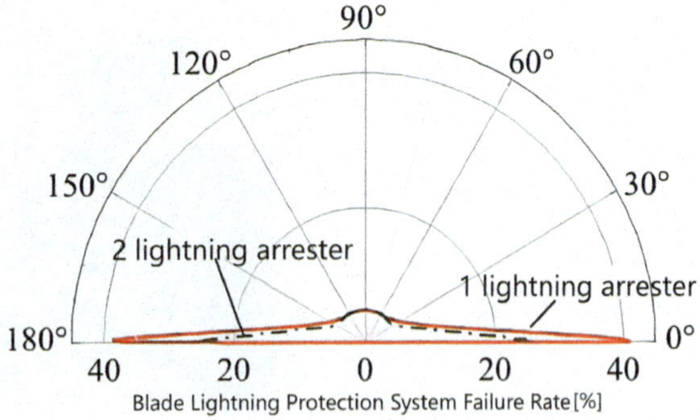

Fig. 9.8 Failure rate of blade LPS under different number of receptors (I = 30 kA)

decreases by approximately 2 m. Therefore, when arranging the positions of the lightning arresters, it is necessary to determine the spacing of the lightning arresters based on the electrical geometric model of the blade to meet the requirements for blade lightning protection efficiency. It is not advisable to simply increase the number of lightning arresters, as this may actually lead to a decrease in blade lightning protection efficiency.

9.1.3 Experiment Study

Defining the range of exposed arc and shielded arc is the basis of blade electrical geometry model calculation. In order to verify the validity of the model proposed in this book and simplify the process of actual lightning stroke on wind turbine blades, a long gap discharge experiment of solid wind turbine blades was carried out at Shahe Experimental Station of North China Electric Power Research Institute to find the boundary between the exposed arc and the shielding arc. The principle diagram of the experiment is shown in Fig. 9.9.

The fan blade adopts the first five meters of the Sinoma 45.2 B–1.5 MW blade (only the tip lightning receiver), and the angle θ between the ground and the ground is 30°. The rod-shaped upper electrode is used to simulate the lightning downward leader. Along the direction shown in the arrow in the figure, the step length of 0.75 m begins to move directly above the blade tip to find the position of the upper electrode at the 50% flashover rate. This position can be considered as the boundary between the exposed arc and the shielding arc. The impulse voltage is a 250/2500 μs negative operating wave generated by the impulse voltage generator. The digital camera uses a

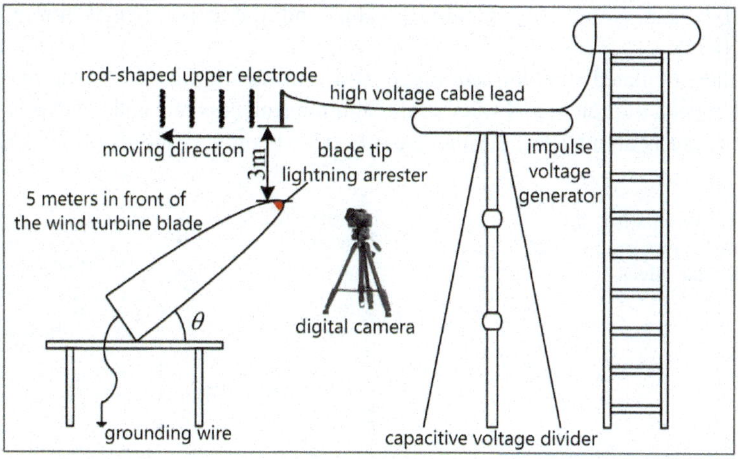

Fig. 9.9 Schematic diagram of experimental setup

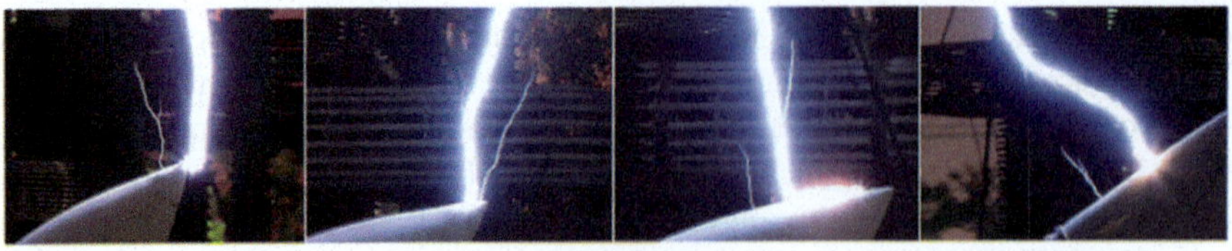

Fig. 9.10 Observation results of discharge path. **a** Successful lightning strike (hit the tip of the lightning arrester). **b** Successful lightning strike (hit the side of the lightning arrester). **c** Successful lightning strike (Lightning slide flash hits the lightning arrester). **d** Lightning protection failure (lightning strike on the blade body)

long exposure method to record the discharge path to determine the success or failure of the lightning protection system of the blade. The lightning success includes hitting the tip of the blade tip lightning receiver, the side of the lightning receiver and the sliding through the blade surface to hit the lightning receiver, as shown in Fig. 9.10a–c; the lightning failure is to hit the blade body and does not pass through the lightning receiver, as shown in Fig. 9.10d [4].

Twenty discharge experiments were carried out at each upper electrode position. The experimental data were compiled and a fitted curve was plotted, as shown by the green dashed line in Fig. 9.11. If the probability of 50% was taken as the junction of the shielding arc and the exposed arc, the value was about 2.3 m. Using the electrical geometric model of the fan blade introduced in this book, the boundary between the exposed arc and the shielding arc is obtained at 1.8125 m in this case. The relative error between the calculated value and the experimental value δ is shown in Eq. (9.11)

$$\delta = \left| l_c - l_e \right| / l_e \times 100\% \tag{9.11}$$

where, l_c and l_e are the calculated value and the experimental fitting value, respectively.

In the calculated shielding arc range (0–1.8 m), using Eq. (9.11), the maximum error between the calculated value of the lightning success rate and the experimental fitting curve is about 19%, but the average error is only about 3%. In the range of

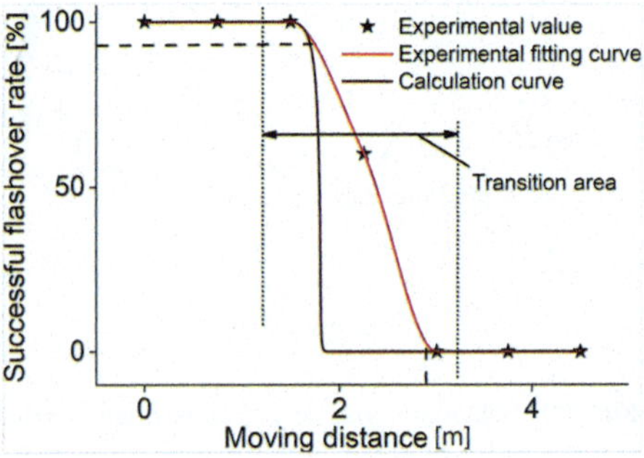

Fig. 9.11 Comparison of calculated values and experimental values

exposure arc (> 1.8 m), there is only a certain error in the transition area. This is because the model of this book equivalently regards the lightning success rate of the blade lightning protection system as 0 when the lightning success rate is lower than a certain value. It should be noted that this equivalence provides a certain degree of protection domain for the application of the blade electrical geometry model to a certain extent.

On the whole, the calculated curve is close to the shape of the experimental fitting curve, and has a certain degree of protection, only in the transition region there is a certain error. The experiment proves the effectiveness of the electrical geometry model of the fan blade in this book.

It should be noted that there are inevitable differences between the experimental conditions and the lightning environment of the real wind turbine. The equivalence between the experiment and the actual situation still needs to be verified, but the designed experiment proves the reliability of the method in this book to a certain extent.

9.2 Lightning Shielding and Protection for Multiple Wind Turbine Blades

9.2.1 Lightning Shielding for Multiple Wind Turbines

The electrical geometric method suggests that when the leading head of lightning reaches the critical breakdown distance (strike distance) to a target object, it discharges to that object. This method allows us to link the parameters of lightning current with the geometric structure of the object being studied. Through geometric drawing, we can obtain the exposure area of the object under different amplitudes of lightning current, and then calculate the situation of an object being struck by lightning bypassing it. Among them, the strike distance is the core parameter in the electrical geometric model that characterizes the ability of an object to attract lightning. Its physical meaning is the critical breakdown distance from the leading discharge channel of the thundercloud developing towards the ground to the struck object[5, 6]. With the large-scale application of wind power, the clustered development of wind turbines has become the mainstream.

Based on the electrical geometric model of single wind turbine blades proposed earlier, an analysis of the lightning shielding distance between two wind turbines is conducted to provide a theoretical basis for optimizing the lightning shielding layout of large-scale wind farms. As shown in Fig. 9.12, the colored circles in the figure represent the strike distance circles of the blade tip arrester at certain lightning current amplitudes I and rotation angles of 0°, 45°, and 90° respectively. The center of the strike distance circle is located on the circumference of the blade tip rotation, and the radius of the strike distance circle is the strike distance R_p of the blade tip

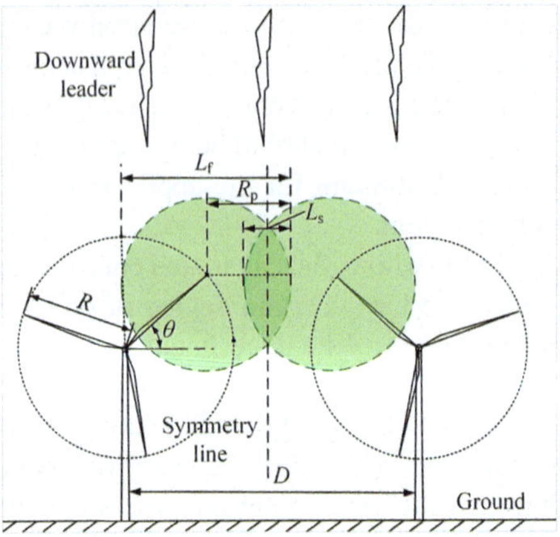

Fig. 9.12 Shield of multiple turbines

arrester when the lightning current amplitude is I; D represents the spacing between the two wind turbines, in meters; L_f represents the horizontal shielding distance of one of the wind turbines at a certain angle, in meters. According to the geometric relationship shown in the figure, the expression for L_f can be obtained as follows:

$$L_f = R\cos\theta + R_p \tag{9.12}$$

where, R represents the length of the wind turbine blade, in meters; θ represents the angle between the wind turbine blade and the horizontal direction.

Assuming that the lightning downward leader develops vertically downward, the two wind turbines as the research object operate axially symmetrically along the symmetry axis in Fig. 9.12. When the two wind turbines' blade tip lightning striking distance circles intersect, as shown in Fig. 9.12, the overlap distance of the two striking distance circles is L_s, and the number of lightning strikes dangerous times N of the two wind turbines at the blade rotation angle θ is represented by Eq. (9.13).

$$N = (4L_f - L_s) \times n \tag{9.13}$$

where, n represents the equivalent lightning strike density in the region where the wind turbines are located.

$$N_0 = 4L_f \times n \tag{9.14}$$

Comparing Eqs. (9.13) and (9.14), it can be observed that there is always $N < N_0$, indicating that when the strike distance circles of the two wind turbine blades intersect, it can result in mutual shielding between the two turbines, thereby reducing the overall lightning strike hazard occurrences for both turbines. This plays a positive role in lightning protection for the entire wind farm. When the spacing D between the two turbines satisfies the requirement of Eq. (9.15), the strike distance circles of the two turbines intersect, demonstrating the presence of lightning shielding effect.

$$D < D_{\max} = 2R\cos\theta + 2R_p \tag{9.15}$$

In the equation, $D_{\max}$ represents the maximum distance between two wind turbines where lightning shielding exists under the conditions of blade rotation angle θ and lightning current amplitude I. Equation (9.15) is referred to as the criterion formula for the existence of lightning shielding between multiple wind turbines in this book.

9.2.2 Discussion on Environmental Factors Impacting Lightning Shielding

The appropriate layout spacing for lightning shielding between two wind turbines can be derived by analyzing a series of curves depicting the variation of Dmax values, which are dependent on the amplitude of lightning current under various blade rotation angles. Therefore, this section simplifies the influence of environmental factors on the lightning shielding distance between turbines to analyzing the relationship between environmental factors and the maximum shielding distance $D_{\max}$ between two turbines. Since the value of $D_{\max}$ is related to the angle of blade rotation and the lightning current amplitude, in order to facilitate the analysis of the impact of environmental factors on the spacing between two turbines and avoid extreme cases of 0° and 90°, this book selects an intermediate value ($\theta = 45°$) for analysis to reduce the overall relative error. The lightning current amplitude is set to $I_p = 60$ kA, the turbine is a 1.5 MW turbine with a tower height of 70 m, blade length of 40 m, inner radius of down conductor in the blade of 0.01 m, tower radius of 2.5 m, and only blade tip arrester is present.

(a) **The Relationship Between $D_{\max}$ and Pressure, Temperature, and Humidity**

The electric field intensity E_{str} in the corona discharge region is an important parameter for determining whether the lightning upward leader is stable[7], but it is susceptible to environmental factors such as pressure, temperature, and humidity[8], as shown in Eq. (9.16).

$$\begin{cases} E_{str} = 425\delta^{1.5} + (4 + 5\delta)\gamma \\ \delta = \frac{P}{P_0}\frac{T_0}{T} \end{cases} \tag{9.16}$$

where, γ represents absolute humidity, in g/m^3; δ represents air relative density; P represents air pressure, in atm; T represents temperature, in K; P_0 and T_0 are respectively the pressure and temperature at sea level under standard atmospheric conditions, where $P_0 = 1$ atm and $T_0 = 293$ K.

According to the principle of lightning leader development, it can be seen that the size of E_{str} will affect the potential of the leader head and the new charge ΔQ

generated in the U_{tip} domain of the streamer region for each development step, and ultimately affect the start and development length of the upstream leader. According to the simulation process and Eq. (9.4), the variation of D_{max} with pressure, temperature and humidity is obtained when the lightning current amplitude is 60 kA, as shown in Figs. 9.13 and 9.14. It can be found that D_{max} increases with the increase of air pressure and absolute humidity, and decreases with the increase of temperature. Taking into account the relationship between the relative density of air and air pressure and absolute humidity (Eq. 9.16), the above relationship can be simplified to describe that D_{max} increases with the increase of relative air density a and absolute humidity b. For the wind farm with small air density and relative humidity, in order to make the mutual shielding effect between the wind turbines, it is necessary to reduce the distance between the two wind turbines as much as possible.

(b) The Relationship Between Dmax and Altitude

Due to the precise data of γ and δ requiring special instruments for measurement, Eq. (9.16) poses some inconvenience in practical engineering applications. However, since there exists a relationship between the altitude H_L and P, T, and γ as shown in Eqs. (9.17)–(9.19), it is possible to approximately calculate the maximum shielding

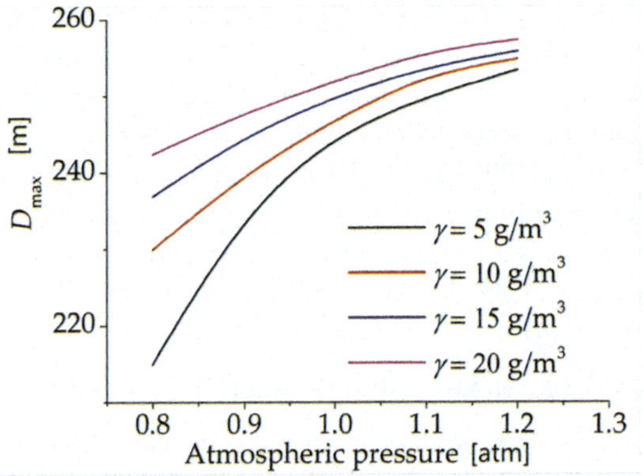

Fig. 9.13 Change of D_{max} with atmospheric pressure and absolute humidity of air

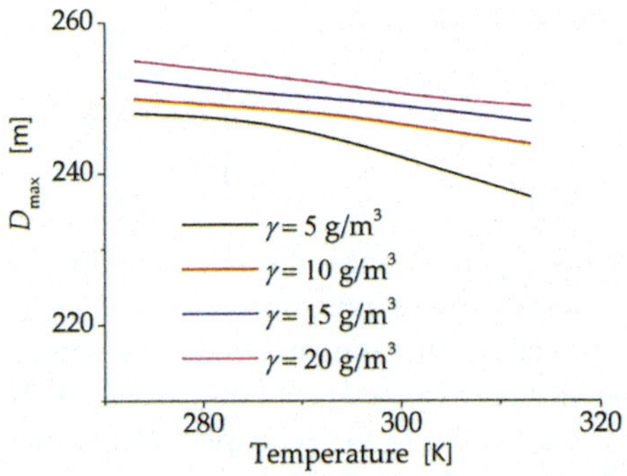

Fig. 9.14 Change of D_{max} with temperature and absolute humidity of air

distance between two turbines in the local wind farm based on the altitude of the area [9].

$$P = P_0 \times e^{-H_L/8} \tag{9.17}$$

$$T = T_0 - 6H_L \tag{9.18}$$

$$\gamma = \gamma_0 \times e^{-H_L/3} \tag{9.19}$$

where, H_L represents the altitude, measured in kilometers, and γ_0 represents the absolute humidity at sea level under standard atmospheric conditions, which is 11 g/m^3.

Substituting Eqs. (9.17)–(9.19) into Eq. (9.16), we can obtain the relationship between the electric field strength E_{str} in the flow region and the altitude H_L as shown in Eq. (9.20).

$$E_{str} = 9.18H_L^2 - 102.16H_L + 523.69 \tag{9.20}$$

After calculation, the relationship between the maximum shieldable distance D_{max} between two wind turbines at a lightning current of 60 kA and the altitude H_L is shown in Fig. 9.15. It can be observed that D_{max} decreases with increasing altitude. For wind farms in high-altitude areas, it is necessary to minimize the spacing between two turbines as much as possible to meet the lightning shielding requirements between turbines.

It should be noted that factors such as terrain, latitude, and season can also cause slight differences in P, T, and γ between two areas at the same altitude. Therefore, for a more accurate calculation, it is still necessary to measure the local δ and γ and calculate using Eq. (9.16).

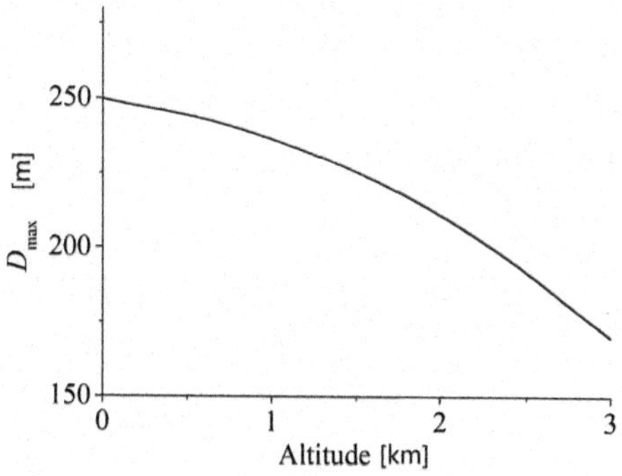

Fig. 9.15 Change of D_{max} with altitude

9.2.3 Spatial Optimization and Lightning Protection Strategies

In this section, a wind farm composed of 1.5 MW turbines is taken as an example to analyze the layout of large-scale wind farms using the electrical geometric method of turbine blades. By selecting the envelope curve of the lightning shielding distance at various blade angles (0°, 30°, 60°, 90°) with respect to the lightning current amplitude, the spacing between turbines is determined. The parameters of the turbine model are as described, with environmental conditions set to standard temperature $T = 298.15$ K, standard pressure $P = 1$ atm, and altitude $H = 0$ km. Using Eq. (9.12), the horizontal distance L_f of the envelope of the strike distance is calculated to analyze its relationship with the lightning current amplitude at different blade rotation angles, as shown in Fig. 9.16. If the distance between two turbines is $4R$ (where R is the length of the blade), when the lightning current $I < 26$ kA, there is no mutual shielding between the two turbines. When 26 kA $< I <$ 40 kA, there is a shielding effect between the two turbines at certain blade angles. When the lightning current $I >$ 40 kA, the two turbines can provide shielding effects at any angle from 0° to 90°. Therefore, the upper envelope curve represents the maximum possible shielding distance with respect to the variation of lightning current amplitude, while the lower envelope curve represents the minimum possible shielding distance with respect to the variation of lightning current amplitude. Two red lines divide the coordinate graph into three zones: Zone 1, Zone 2, and Zone 3. Zone 1 and Zone 3 respectively represent non-shielding and fully shielding regions, while Zone 2 represents the potentially shielding region, where shielding effects occur only at certain rotation angles within this zone.

Considering the prevailing wind direction, since the blades of the wind turbine are in the direction perpendicular to the prevailing wind direction most of the time, the lightning shielding between the wind turbines perpendicular to the prevailing wind direction is ensured, which contributes the most to the lightning protection safety of the entire wind farm. According to the existing research results, the recommended distance perpendicular to the prevailing wind direction is no more than 10R [10–12].

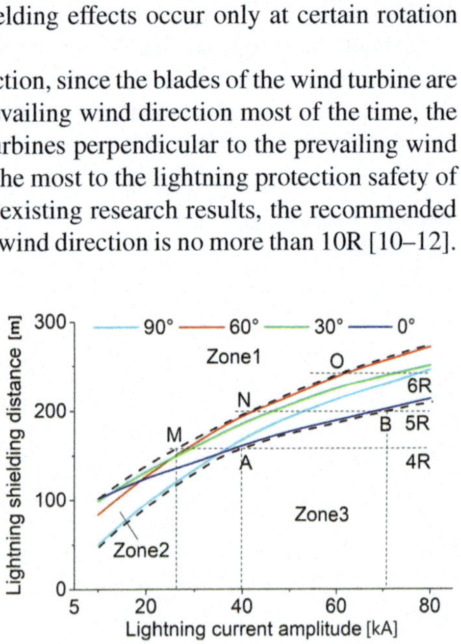

Fig. 9.16 Lightning shielding distance

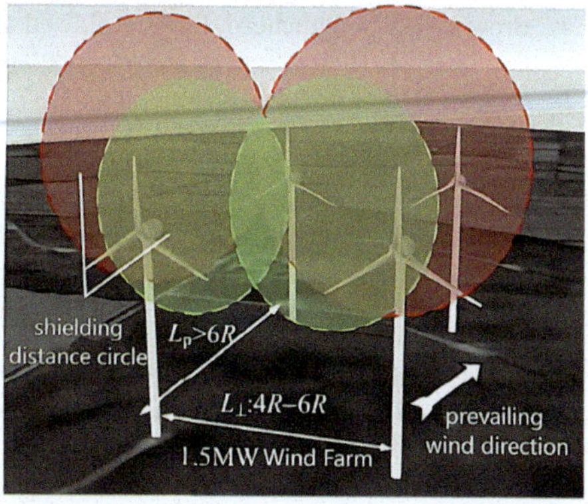

Fig. 9.17 Spatial allocation of a wind farm with 1.5 MW turbines with consideration for mutual lightning-damage shielding

According to the shielding curve between the two wind turbines (Fig. 9.16), and considering the influence of air flow on the two wind turbines, for the wind farm of 1.5 MW wind turbine, the arrangement spacing $L_{\perp}$ of the two wind turbines in the direction perpendicular to the prevailing wind direction is preferably 4R–6R. At this time, most of the lightning current amplitude (26–60 kA or more) has a certain shielding effect, which is a supplement to the traditional wind turbine arrangement method. The time when the blade is in the prevailing wind direction is relatively short. The time that turbine blades spend aligned with the prevailing wind direction is relatively short. Therefore, even if the spacing between turbines increases along the prevailing wind direction, the overall impact on the lightning safety of the wind farm will not be too significant. Therefore, the determination of the spacing of the wind farm in the prevailing wind direction should first ensure that the wake of the front row fan does not affect the rear row fan, so that the power generation efficiency of the wind farm is as large as possible, so the spacing can be greater than the maximum recommended spacing perpendicular to the prevailing wind direction, that is, $L_p > 6R$. The spatial layout of 1.5 MW large-scale wind farm considering lightning disaster shielding is shown in Fig. 9.17 [13].

References

1. Whitehead, J. T., Chisholm, W. A., Anderson, J. G., et al.: Estimating lightning performance of transmission lines. II: Updates to analytical models. IEEE Transactions on Power Delivery, 8(3), 1254–1267 (1993).
2. Wang, G. Z., Zhang, L., Guo, Z. X., et al.: Dynamic Striking Distance and Electrical Geometry Model of Wind Turbine Blades Based on Lightning Physics. Proceedings of the CSEE, 37(21): 6427–6436 (2017).
3. IEC. IEC 62305–1 Protection against lightning (part 1): General principles. [Standard] (2006).
4. Zhang, L., Zhang, Y., Wang, G.Z., et al.: Lightning Shielding of Multiple Wind Turbines Based on Lightning Physics and Optimized Spatial Allocation of Wind Farm. Proceedings of the CSEE, 2018, 38(18): 5335–5342.

5. Brown, G. W., Whitehead, E. R.: Field and analytical studies of transmission line shielding: Part II. IEEE Transactions on Power Apparatus and Systems, 88(5), 617–626 (1969).
6. Rakov, V. A., Uman, M. A.: Lightning: Physics and effects. Cambridge University Press. (2003).
7. Zhang, L., Wang, G., Zhang, W., et al.: An electro-geometric model for lightning shielding of multiple wind turbines. Energies, 10(9), 1272 (2017).
8. Eriksson, A. J., Roux, B. C. L., Geldenhuys, H. J., et al.: Study of airgap breakdown characteristics under ambient conditions of reduced air density. IEE Proceedings A—Physical Science, Measurement and Instrumentation, Management and Education - Reviews, 133(8), 485–492 (2008).
9. Petrov, N. I., Waters, R. T.: Determination of the striking distance of lightning to earthed structures. Proceedings of the Royal Society of London. Series A: Mathematical and Physical Sciences, 450(1940), 589–601 (1995).
10. Patel, M. R.: Wind and solar power systems—Design, analysis, and operation. Wind Engineering, 30(3), 265–266 (2006).
11. Mosetti, G., Poloni, C., Diviacco, B.: Optimization of wind turbine positioning in large windfarms by means of a genetic algorithm. Journal of Wind Engineering and Industrial Aerodynamics, 51(1), 105–116 (1994).
12. Muto, A., Suzuki, J., Ueda, T.: Performance comparison of wind turbine blade receptor for lightning protection. In Proceedings of the International Conference on Lightning Protection (ICLP). Cagliari, Italy: IEEE (2010).
13. Wang, G. Z.: Research on improved electrical geometry model of wind turbine blades based on lightning physics. Shandong University (2018).

Chapter 10
Damage Mechanism for Wind Turbine Blade Under the Effect of Electric Arc

10.1 Molecular Dynamics of Material Degradation

10.1.1 Introduction to Molecular Dynamics

Molecular dynamics simulation is an important research method for studying the structure and properties of molecules or molecular systems through computer simulation. It includes methods such as molecular mechanics, Monte Carlo simulation, and molecular dynamics simulation. These methods are based on classical mechanics models of molecules or molecular systems; or through optimizing the total energy of individual molecules to obtain stable configurations; or by repeatedly sampling the configuration space of molecular systems and calculating their total energy to obtain the most reliable configurations and thermodynamic equilibrium properties of the systems; or by numerically solving the classical mechanics equations of motion of molecular systems to obtain the trajectory of the systems, and statistically analyzing the structural features and properties of the systems [1]. Currently, thanks to the development of molecular simulation theory, methods, and computer technology, molecular simulation has become the third means of understanding and recognizing the world at the molecular level, following experimental and theoretical methods.

The most direct outcomes of molecular dynamics simulation are the structural characteristics of molecular systems [2]. These include coordination structures in solutions, configurations and morphologies of biological and synthetic polymers, interactions among high polymers and solvent molecules, adsorption of molecules on solid surfaces, and orientations and distributions of molecules in external fields such as gravitational and electromagnetic fields [3]. Additionally, molecular dynamics simulation can explore various thermodynamic properties of molecular systems, such as kinetic energy, potential energy, Gibbs free energy, enthalpy, heat capacity, as well as velocity autocorrelation functions, velocity cross-correlation functions, mean square displacement, and calculate transport properties such as diffusion coefficients.

Q. Sun et al., *Lightning Transients and Protection for Renewable Energy System*,
https://doi.org/10.1007/978-981-97-7267-4_10

It also provides insights into the relationships between density, pressure, volume, temperature, and other state equations related to the system's status.

The mechanical damage of wind turbine blades under arc action is mainly caused by material pyrolysis, gas product expansion, and internal air expansion [4]. Through molecular-level molecular reaction dynamics simulation, the damage mechanism of sandwich materials balsa wood and PVC light cedar in wind turbine blades under arc action can be studied [5].

10.1.2 Molecular Dynamics of Balsa Wood

(a) **Molecular Composition and Properties**

Balsa wood has a low density and high strength, achieving an ideal balance between weight and strength. Therefore, it is often used as a sandwich material for wind turbine blades to reduce their weight. However, its strength is not sufficient to withstand lightning strikes [6]. In cases where the lightning rod fails to discharge the lightning current from the wind turbine blades, balsa wood is vulnerable to lightning damage. To quantitatively study the micro-mechanism of lightning damage to balsa wood materials, we need to quantitatively calculate the content of each component inside the balsa wood material. The calculation process is as follows:

The basic density of balsa wood averages around 0.2 g/cm^3. The bulk density of wood, which refers to the density of wood substance or cell wall material, averages about 1.5 g/cm^3, approximately equal to the density of cellulose. After drying, the moisture content of balsa wood is 6%, resulting in a density under standard conditions of approximately 1 g/cm^3. The density of air in wood cavities is calculated to be 1.29 kg/m^3 under standard conditions (0 °C, 1 standard atmosphere pressure, 1 atm).

Assuming the total volume of balsa wood is 1, with the volume occupied by wood material denoted as x and the volume occupied by air denoted as y, the equation set can be formulated as follows:

$$\begin{cases} 1500x + 1.29y + 6\% * 200 = 200 \\ x + y + \frac{12}{1000} = 1 \end{cases} \tag{10.1}$$

Further calculation yields the volume and mass fractions of each component in balsa wood as in Table 10.1.

Table 10.1 Proportion of the components of basal wood

Component	Wood substance	Air	Water	Balsa wood
Molar mass (g/mol)	1500	1.29	100	200
Volume percentage (%)	12.46	86.34	1.2	100
Mass percentage (%)	93.45	0.56	6	100

The wood component consists of plant cell walls, forming the main structure of balsa wood, with cellulose being the primary constituent of the cell wall, accounting for approximately 40–50%. Analysis of balsa wood's minute details by the North American Advanced Materials Research Center (NORA) indicates that the strength of balsa wood comes from cellulose fibers. Therefore, when modeling the wood portion of balsa wood, cellulose is used to model the cell wall.

The simple molecular formula of cellulose is $(C_6H_{10}O_5)n$, where n is the degree of polymerization. The molar mass of a single cellulose monomer is approximately 162 g/mol, while the molar mass of a water molecule is about 18 g/mol, and the molar mass of dry air is 28.9634 g/mol. Calculations can be made to determine the proportions of cellulose, water, and air molecules in a given volume of balsa wood, as shown in Table 10.2.

(b) **Model Construction and Simulation**

In molecular dynamics simulation software, a cellulose model with a degree of polymerization of 16 is established. After calculation, the number of water molecules in the model is approximately 9. The molecular construction function is utilized to establish the water molecule model and to perform energy minimization optimization on the two types of molecular models. The two molecular models are then imported into the same unit cell, with periodic boundary conditions selected to eliminate boundary effects of the model, thus establishing a cellulose-water mixed system model. As shown in Fig. 10.1, gray represents carbon atoms, red represents oxygen atoms, and white represents hydrogen atoms.

The molecular dynamic reaction of the optimized model at 4000 K was carried out in NVT ensemble. The force field was ReaxFF in MS (Materials Studio) software, equilibration time was 1 ps, production time was 200 ps, time step was 0.1 fs, and frame output every 100 steps.

(c) **The Mechanism of Pyrolysis Process**

After multiple repeated simulations, it was found that under the action of thermal stress, the entire molecular chain begins to twist in the initial stage of the reaction. The binding effect of chemical bonds between atoms prevents immediate dissociation of each atom. However, as the reaction progresses, chemical bonds begin to break. Statistical analysis of the breaking of initial bonds in multiple experiments shows that the first step of cellulose thermal degradation can be viewed as a relatively regular depolymerization process. Glycosidic bonds between molecular monomers are more prone to breakage, accounting for approximately 51.43% of the total, as shown in the

Table 10.2 Molecular proportion of the components of balsa wood

Component	Cellulose (monomer)	Air	Water
Molar mass (g/mol)	162	28.96	18
Mass ratio	15.575	0.093	1
Molecular ratio	1	0.033	0.583

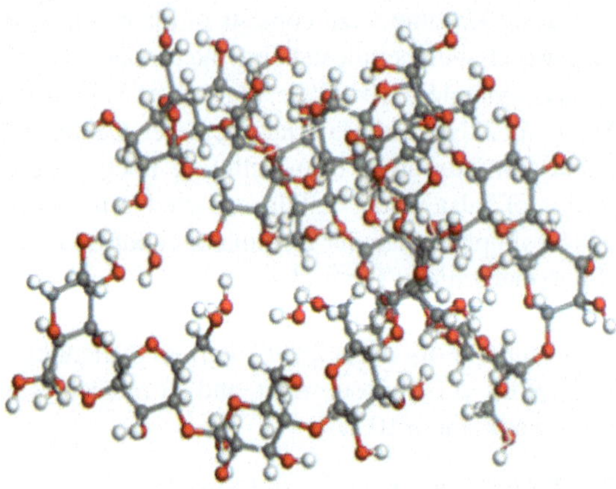

Fig. 10.1 Molecular model of balsa wood

Fig. 10.2. This is because the glycosidic bond in cellulose molecules has the lowest dissociation energy, which leads to its earlier rupture than other bonds under thermal conditions, causing cellulose molecules to decompose into glucose monomers and reducing their degree of polymerization. As the reaction proceeds, hydrogen atoms and hydroxyl radicals continue to detach, promoting the progress of the pyrolysis reaction. Opening of the D-glucose ring begins to occur, and the opening follows a certain regularity. The carbon–oxygen bond in the oxygen-containing bridge –O–C–C– within the glucose ring usually breaks earlier than other C–C bonds, generating carbonyl or aldehyde groups, as shown in the Fig. 10.3. This is because the dissociation energy of the carbon–oxygen bond in the oxygen-containing bridge is lower than that of other chemical bonds in the glucose ring.

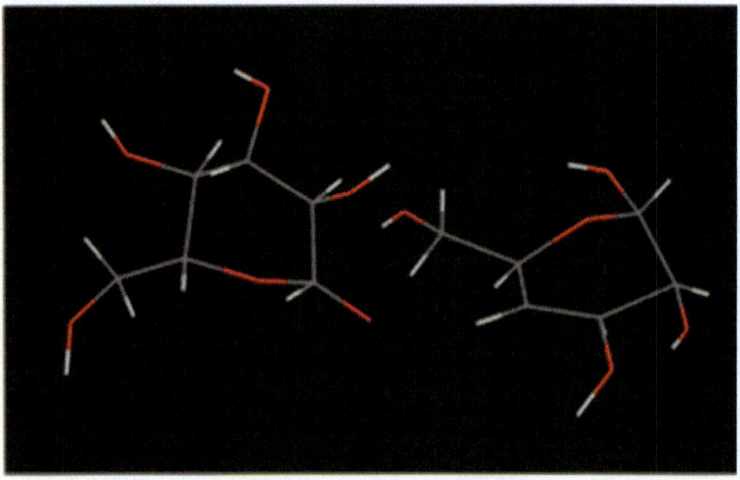

Fig. 10.2 The glycosidic bond break at the beginning of the reaction

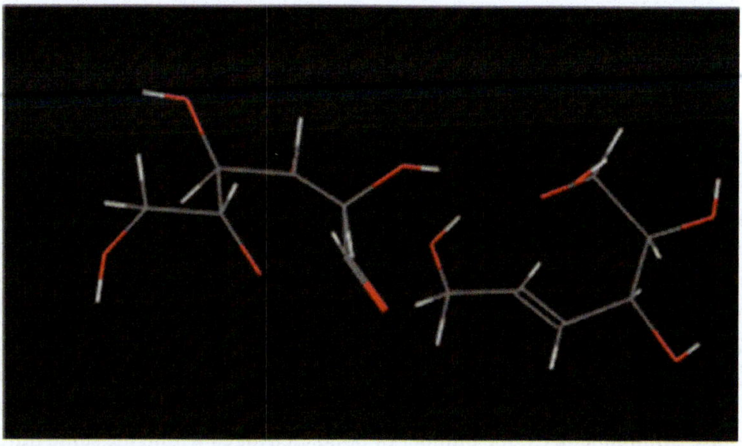

Fig. 10.3 The pyranoid rings break

The decomposition path of cellulose molecules is shown in Fig. 10.4. Subsequently, with the detachment and recombination of hydrogen and oxygen atoms, a large number of carbon atoms undergo continuous processes of recombination, fragmentation, recombination, and intramolecular ring closure, eventually forming carbon long chains and multi-ring carbon structures, manifesting macroscopically as material carbonization.

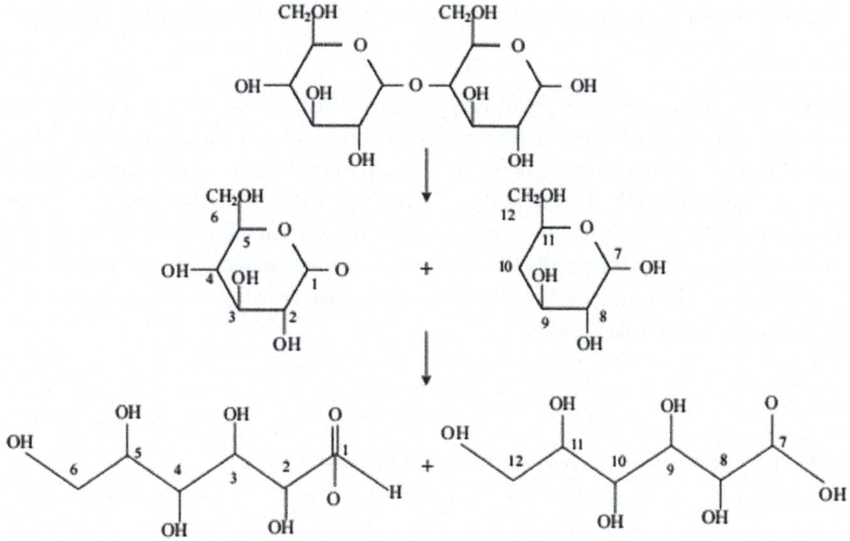

Fig. 10.4 Break process of cellulose

Performing thermal decomposition reactions of the balsa wood molecular model at different temperatures, it can be observed that the higher the temperature, the faster the reaction progression of cellulose molecules. However, it does not alter the sequence of bond cleavage; rather, an increase in temperature merely advances the time of bond cleavage. The cleavage times of characteristic bonds at various temperatures are tabulated in Table 10.3. Examination of reaction products at 100 ps reveals that higher temperatures result in deeper reaction extents and more small molecule products, while conversely, lower temperatures yield more large molecule products. Since temperature, as an external factor, does not alter the nature of molecular chemical bonds, the reaction sequence and mechanism remain unchanged at different temperatures. However, the elevation of temperature enables atoms to acquire more energy within the same timeframe, accelerating their motion rates, and facilitating chemical bonds to reach dissociation energy more readily, thereby advancing reaction times. Simultaneously, the number of bond cleavages within the same timeframe increases, leading to a deeper reaction extent.

(d) **The Main Pyrolysis Gas Products and the Generation Mechanism**

The rapid expansion of gas due to the instantaneous thermal effect of lightning arcs is a major factor causing damage to blade materials. Besides the small amount of moisture and minimal air contained within balsa wood itself, its thermal decomposition also generates a certain amount of gas products. Therefore, we conduct a detailed analysis of the gas products generated during the pyrolysis of balsa wood. The gas products of cellulose were very complex, including CO, CO_2, H_2O, H_2, and a variety of low carbon compounds. Selecting several main gas products, detailed research was conducted on their generation process, obtaining the following reaction pathways.

(1) **H_2O**

H_2O was the main product in cellulose pyrolysis. The hydrogen atoms and hydroxyls in cellulose molecules were active and likely to break off the main chain. These free hydrogen and hydroxyl radicals could easily combine to form water molecules under the influence of hydrogen bonds, as shown in Fig. 10.5. The simulation results showed that the water was not stable and further participated in the generation of other products during the pyrolysis reaction, resulting in a decrease in water content. However, new H_2O molecules were generated, thus showing a fluctuating trend in the increase of water molecules.

(2) **H_2**

Table 10.3 Break time of characteristic bonds at different temperature

Temperature/K	Glycosidic bond cleavage time/ps	First pyranose ring cleavage time/ps	Second pyranose ring cleavage time/ps
2000	3.6	3.9	4.8
3000	0.3	0.95	1.73
4000	0.05	0.1	0.35

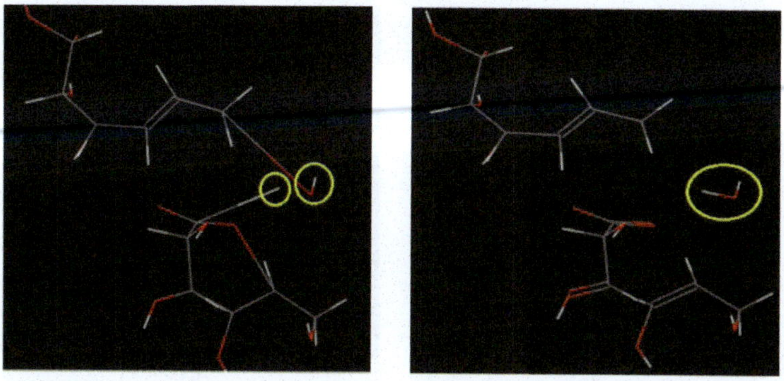

(a) (b)

Fig. 10.5 The generation process of H_2O. **a** Hydroxyl and hydrogen radicals break from carbon chain. **b** Formation of H_2O from dropped hydroxyl and hydrogen

As the reaction progresses, more and more hydrogen atoms are removed from the main chain at high temperature. Some of these highly reactive hydrogen radicals combined with hydroxyl radicals to give water, while some of them combined with each other to generate hydrogen. The others abstracted hydrogen atoms from other molecules to generate hydrogen and promote the reaction.

(3) **CO_2**

After the pyranoid ring broke, the C1 atom connecting with the oxygen atoms in pyranoid rings and the oxygen atoms in glucosidic bond easily fell off the main chain to form CO_2 while losing the hydrogen atom, as shown in Fig. 10.6. Additionally, the C_7 atom with the oxygen atoms in pyranoid rings and a hydroxyl group could drop off the main chain and generate a formic acid molecule, which would then turn into CO_2 by dehydrogenation. Due to the absence of oxygen in this simulation, unlike the actual environment, the production of CO_2 was lower than expected. In fact, carbon burning occurs when lightning strikes the blades, resulting in a larger amount of CO_2 production.

(4) **CO**

In the oxygen-free simulation environment, CO was one of the most common products in cellulose pyrolysis. Formic acid molecules easily lose the hydrogen and hydroxyl group to produce CO. Other groups such as aldehyde groups, carbonyl groups, and the carbon atoms with hydroxyl groups were likely to produce CO. Additionally, existing CO_2 could be reduced to CO under the strong reducing action of hydrogen radicals.

(5) **C_2H_2**

Acetylene was another main product in cellulose pyrolysis. After the hydrogen atoms and hydroxyl groups fell off, the saturation of the main chain and existing small

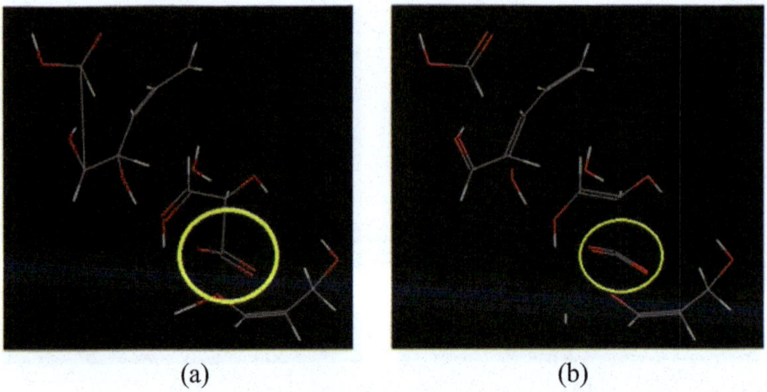

Fig. 10.6 The generation process of CO_2. **a** C–C bond break. **b** CO_2 generation

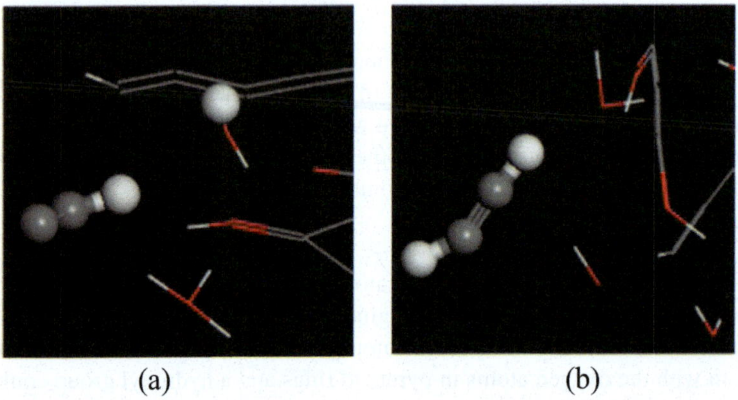

Fig. 10.7 The generation of C_2H_2. **a** Unsaturated low-carbon hydrocarbon compounds capturing an H atom forming H_2O. **b** C_2H_2 formation

molecules was very low, and many single bonds turned into double or triple bonds. Some double carbon intermediate products such as ethynol (HC≡COH), unsaturated C2 (C≡C), and unstable ethylene glycol $\left(\begin{matrix} H\backslash \\ HO/ \end{matrix} C = C^{/OH}_{\backslash H}\right)$ could change into C_2H_2 through the replacement of hydroxyls by H atoms (as shown in Fig. 10.7), hydrogen absorption, or hydroxylation.

10.1.3 Molecular Dynamics of PVC

(a) **Molecular Composition and Properties**

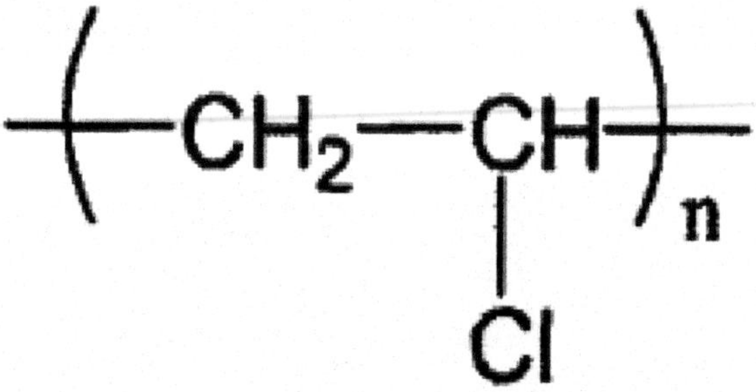

Fig. 10.8 The molecular formula of PVC

Table 10.4 Molecular proportion of the components of PVC

Component	Polyvinyl chloride (monomer)	Water
Molar mass (g/mol)	62	18
Mass ratio	98.83	1.1689
Molecular ratio	24.64	1

PVC is a polymer formed by the polymerization of vinyl chloride monomers under the action of oxidants or other initiators, or through polymerization induced by light or heat. Its simple molecular formula is $(C_2H_3Cl)_n$, where n represents the degree of polymerization, and its molecular formula model is shown in Fig. 10.8.

The molar mass of the monomer molecule is 62, with an initial moisture content of 1.1689%. The molar mass of a water molecule is 18. The air density is extremely low, and the air content can be ignored. The calculation of the volume and mass proportions of each component in PVC materials is similar to the calculation method of the components in the last section, and the calculation yields the proportions of various molecular components in the PVC material as shown in Table 10.4. Guided by the calculated molecular proportions of various components in PVC materials, molecular modeling of PVC materials is conducted.

(b) **Model Construction and Simulation**

To make the damage characteristics of basswood and PVC materials more comparable, this section establishes a PVC model with the same volume as the basswood model in the previous section. Through calculations, it is found that the PVC molecular system model with the same volume contains approximately 306 PVC monomers, corresponding to a water molecule content of 12. Since the polymerization degree does not affect the PVC pyrolysis reaction process, a PVC molecular model with a polymerization degree of 51 and water molecule model are established in molecular dynamics simulation software. The molecular models undergo spatial geometry optimization and energy minimization to obtain their rational structural models. The resulting PVC material system model is shown in Fig. 10.9, where gray represents carbon atoms, green represents chlorine atoms, white represents hydrogen atoms, and red represents oxygen atoms.

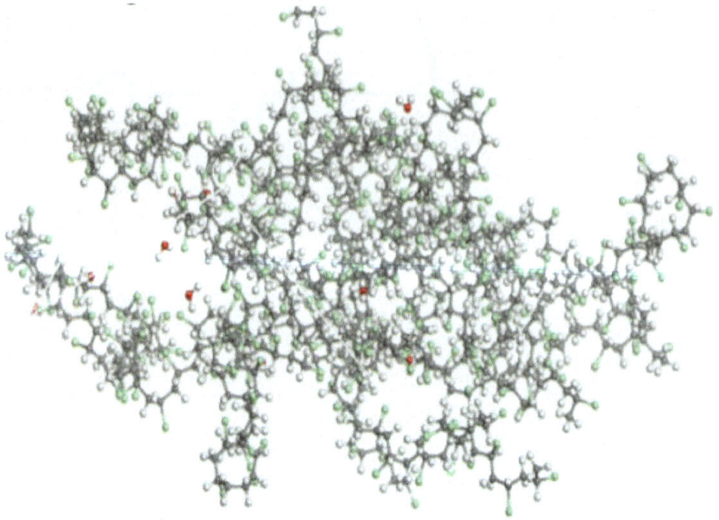

Fig. 10.9 Molecular model of PVC

The simulation experiment is set at a temperature of 4000 K with a reaction time of 200 ps. ReactFF force field is employed for molecular dynamics calculations of the constructed system. The ensemble chosen is NVT ensemble, with temperature control using the Nosé-Hoover extended system method, and pressure control using the Berendsen control method. The reaction temperature is set to 4000 K, temperature control parameters are set to 0.1 ps, reaction time is set to 200 ps, and the calculation time step is set to 0.1 fs. Output is generated every 100 steps. The PVC molecular pyrolysis process is observed, and the generation pattern of bond breaking, types of decomposition products, and their formation process are recorded.

(c) **The Mechanism of Pyrolysis Process**

As for PVC degradation process, the C_l atom substituents were the most unstable part and broke off first from the main chain, as shown in Fig. 10.10. The active chlorine free radicals abstracted H atoms from the PVC chain, and formed HCl. The adjacent two C atoms, which have lost C_l atoms and H atoms, easily generate double bonds or triple bonds, or the interphase C atoms connect to form a ring, as shown in Fig. 10.11. As the reaction progresses, the C–C bonds broke and low-carbon small molecules were produced, and the DP decreased. However, these low-carbon molecules were not stable, and they were easily reconnected with the main chain. In addition, the tail carbon atoms sometimes connected with other PVC molecular chains, which made the DP increase, as shown in Fig. 10.12. The C chains constantly broke and were restructured during the pyrolysis process, thus there was no regular pattern change in the DP of PVC.

Comparing the pyrolysis reactions of PVC at different temperatures, the higher the temperature, the faster the movement rate of atoms in PVC, leading to a more rapid

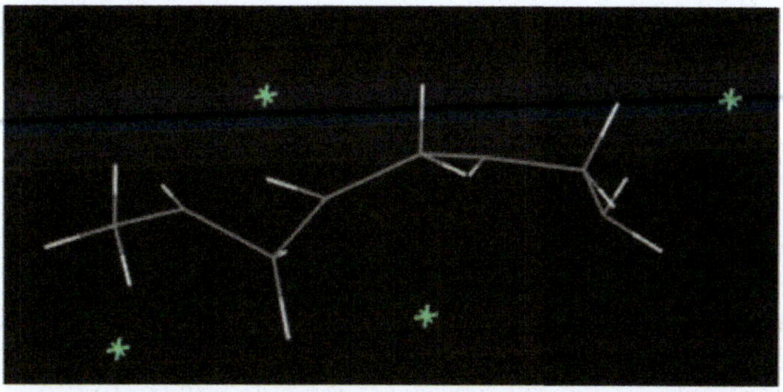

Fig. 10.10 The chlorine atoms falling off the main chain

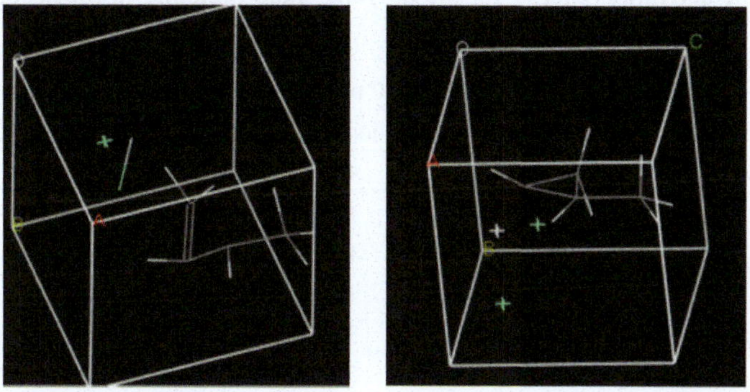

(a) (b)

Fig. 10.11 Formation of double bonds and ring structures. **a** The adjacent two C atoms generated double bond. **b** The interphase C atoms connected to form a ring

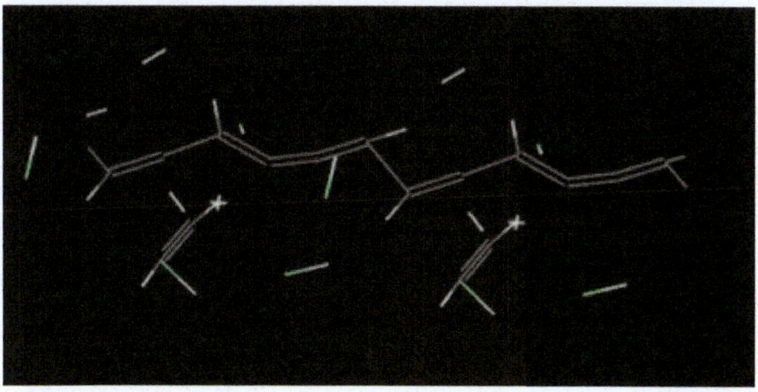

Fig. 10.12 The tail carbon atoms connection with other PVC molecular chain and DP increase

pyrolysis reaction. This reduces the time for bond breakage of various chemical bonds and increases the number of bond fractures. The statistical results of the changes in the number of bond fractures of various characteristic bonds as a percentage of the total number of bonds of that type over time are shown in Table 10.5. The first step of PVC pyrolysis can be rapidly completed at several temperatures to generate HCl. This is because the energy required for the first step reaction is low, and it is sufficient to rapidly meet the reaction conditions within a short time at 2000 K. The number of HCl generated under different pyrolysis temperatures changes over time, as shown in Fig. 10.13. It can be observed that increasing the temperature at the same reaction time increases the reaction rate.

From the breakage of C–H bonds and C–C bonds in Table 10.5, it can be seen that at 2000 K, the carbon chain in the PVC pyrolysis process did not react. This is because the activation energy for the polyene chain breaking in the second step of

Table 10.5 Broken bond proportion during PVC pyrolysis at different temperature %

Chemical bonds	Temperature/K	Reaction time /ps					
		0	40	80	120	160	200
C–Cl	2000	0	100	100	100	100	100
	3000	0	100	100	100	100	100
	4000	0	100	100	100	100	100
C–H	2000	0	33.3	33.3	33.3	33.3	33.3
	3000	0	58.3	58.3	54.2	58.3	58.3
	4000	0	75.0	70.8	66.7	62.5	60.4
C–C	2000	0	0	0	0	0	0
	3000	0	26.7	37.5	25	31.3	43.7
	4000	0	53.3	53.3	60.0	66.7	53.3

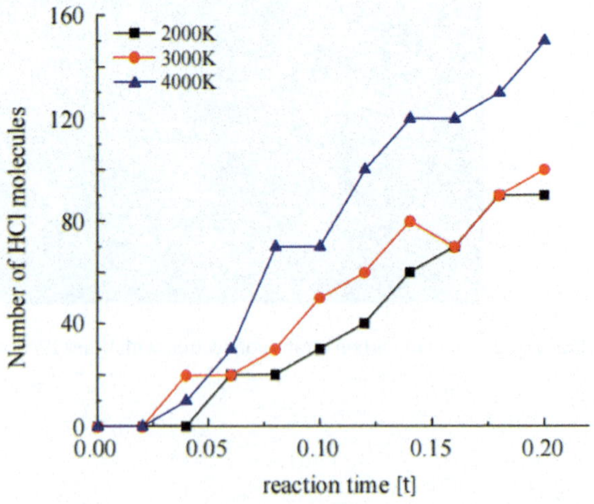

Fig. 10.13 HCl number change curve at different temperature

PVC pyrolysis is much higher than that of the first step, and the energy absorbed from the environment within 200 ps at 2000 K cannot meet the conditions. When the temperature increases to 3000 K, multiple fractures occur in the PVC carbon chain by 40 ps. At 4000 K, the reaction degree of PVC deepens further.

(d) The Main Pyrolysis Gas Products and the Generation Mechanism

The pyrolysis of PVC generates a certain amount of gas products. These small molecule products rapidly expand under the intense thermal effect of lightning arcs, and their expansion pressure is a major factor causing damage to blade materials. Therefore, we conduct a detailed analysis of the gas products from PVC pyrolysis. At high temperatures, the main pyrolysis gas product of PVC was HCl, H_2 and C_2H_2. There were also a few olefins and chlorohy drocarbons.

HCl was the main gaseous product in PVC pyrolysis because the chlorine atoms in PVC are unstable and are easily removed from the main chain. The active chlorine radicals promoted the removal of H atoms from the PVC chain, forming HCl, as shown in Fig. 10.14.

Ethylene was mostly derived from the head or tail two carbon atoms falling off the long chain. There were also some ethylene and acetylenes that transformed into each other. The number of ethylenes in the earlier period was higher than in the later period because many Cl atoms and H atoms were abstracted from the main chain, and the saturation level of the chain was low. Thus, the triple bond was more stable than the double bond, and some C_2H_4 changed to C_2H_2 or were added to the main chain, leading to the slight decrease of C_2H_4, as shown in Fig. 10.15.

The simulation results showed that the number of C_2H_2 molecules was nearly the same as that H_2 ones at 4000°C. The saturation of the PVC molecule was reduced after dechlorination and dehydrogenation, and double or triple bonds or ring structures were formed. As the reaction progressed, some C–C bonds broke and generated low carbon molecules, then the small molecules such as C_2H_4 were reduced to C_2H_2 and H_2 as shown in Fig. 10.16, or unstable intermediates C_2 obtained two hydrogen atoms

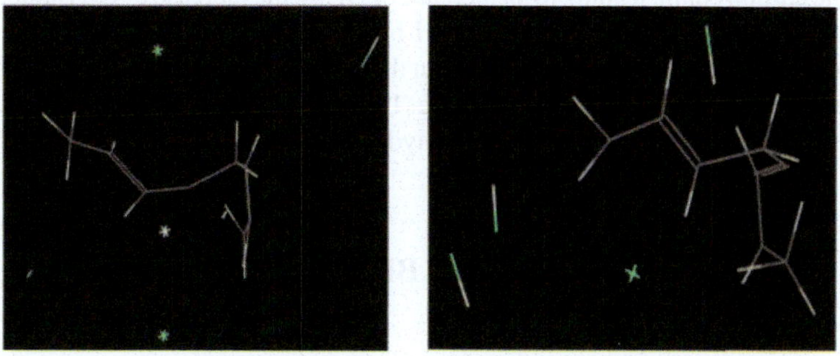

(a) (b)

Fig. 10.14 The generation of HCl. **a** Cl atoms dropped off the main chain. **b** Cl radicals captured H atoms and form

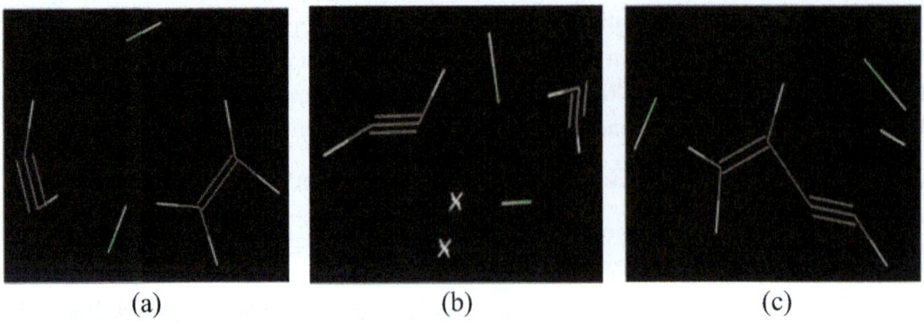

Fig. 10.15 The decrease of C_2H_2. **a** A C_2H_4 and a C_2H_2. **b** C_2H_4 lost two H atoms and were changed into C_2H_2. **c** C_2H_4 and C_2H_2 lost one H atom and combined together

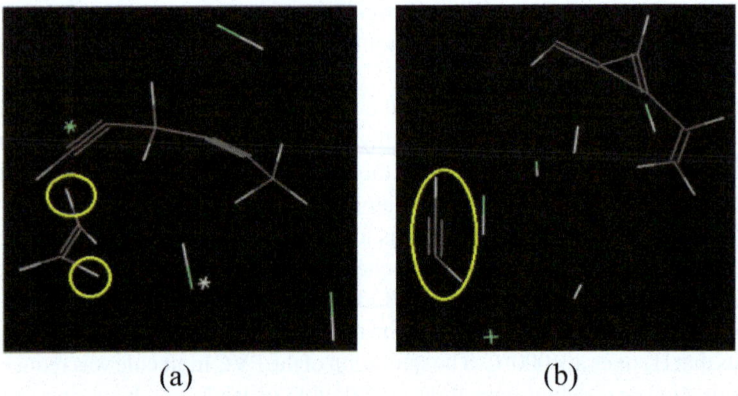

Fig. 10.16 The generation of C_2H_2. **a** C_2H_4 lost H atoms **b** C_2H_2 was generated

to produce C_2H_2, but C_2H_2 and other low carbon molecules would also reconnect with the carbon chain, leading to an increase of DP, so the increase of C_2H_2 fluctuated.

As the reaction progressed some C–H bonds broke at high temperature. These hydrogen radicals that were removed from the main chain were active and combined with each other to form H_2 as shown in Fig. 10.17, or abstracted H atoms from carbon chains or other small molecules to form hydrogen.

10.2 Damage of Wind Turbine Blade Under Lightning Strike

10.2.1 Damage of Balsa Bood and PVC Blade

(a) **Internal Arc Damage Experiment on Materials**

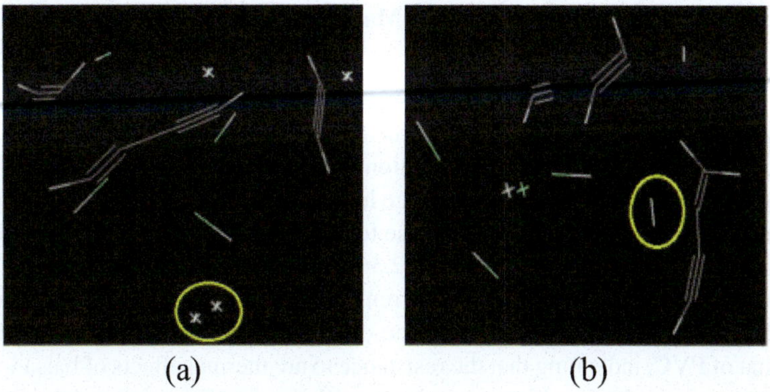

Fig. 10.17 The generation of H_2. **a** Uncombined H atoms. **b** Two H atoms combined to form H_2

The main types of damage from blade lightning strikes mainly include material thermal decomposition damage and structural damage [7]. To study the damage situation of wind turbine blades caused by lightning arcing within the material, large current impulse experiments were conducted on two materials, balsa wood and PVC.

The PVC and balsa wood of wind turbine blades were cut into 15 cm × 10 cm × 2 cm cuboids. In order to simulate the internal arc damage of material, two pieces of cuboid material were stacked together, and a nickel–chromium wire (d = 0.1 mm) was put between the two pieces of material. Then the specimen was fixed by insulating tape. One side of the nickel–chromium wire was connected to the high-voltage port of an impulse current generator and the other side of the wire was connected to the low-voltage port. When the wire was melted, the high temperature was used to simulate the thermal effect of the lightning arc [8]. A diagram is shown in Fig. 10.18.

Then the output current was set to an impact lightning current and the peak value was set to 1.45, 6.28, 12.5, and 21 kA, respectively. Then the impact currents with different peak values were added to the nickel–chromium wire to test the damage current of the two materials and observe their damage characteristics. The morphological damage features of the two materials were observed by scanning electron microscope (SEM), and the residual strength of the two materials was tested by a plastic pendulum bob shock testing machine.

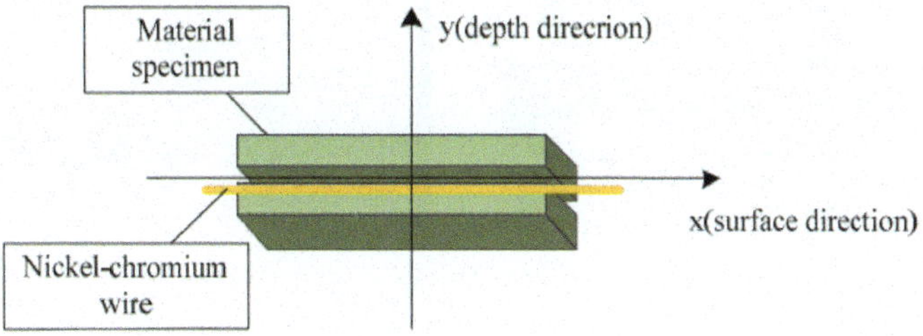

Fig. 10.18 The specimen diagrammatic sketch

(b) **Damage Morphology of the Two Materials**

The two pieces of specimen were unfolded, and the damage condition of the two materials is shown in Figs. 10.19 and 10.20. As we can see, there are carbonization traces along the nickel–chromium wire on the two pieces of material. The higher the peak current is, the bigger the carbonization areas are. There are obvious mechanical damages in both materials but the damage level is different. The upper layer of PVC was fractured when the peak current rose to 6.28 kA, and the two layers were both broken when the peak current rose to 12.56 kA. However, the upper layer of balsa wood was not broken until the peak current rose to 21 kA, and exhibited cracking, indicating severe damage. Therefore, the damage current of balsa wood is higher than that of PVC, indicating that the resistance to arc thermal effects of balsa wood is higher than that of PVC, which is consistent with the molecular dynamics simulation results mentioned earlier.

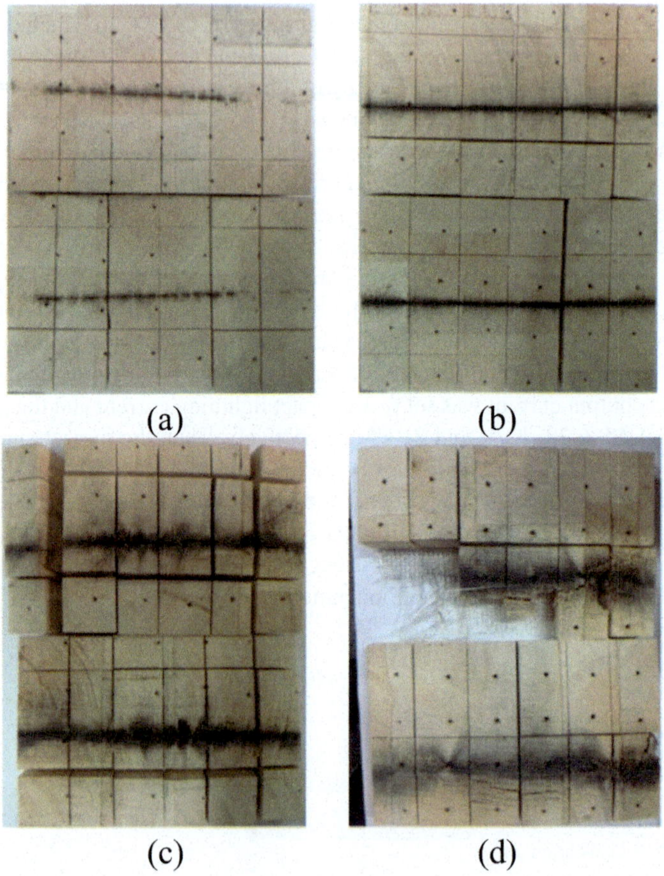

Fig. 10.19 The damage condition of balsa wood under different impulse currents. **a** 1.45 kA. **b** 6.28 kA. **c** 12.56 kA. **d** 21 kA

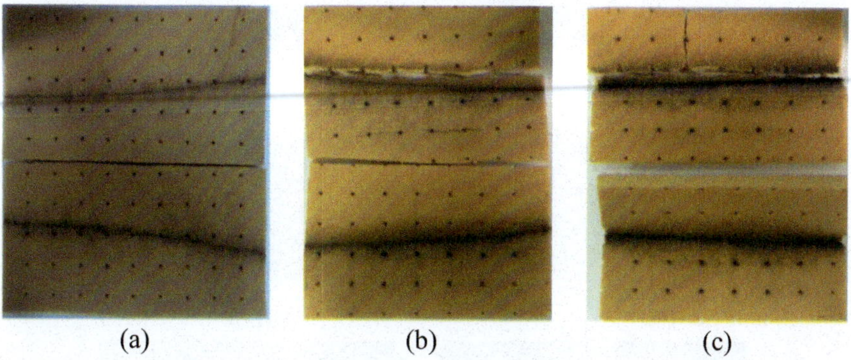

Fig. 10.20 The damage condition of PVC under different impulse currents. a 1.45kA. b 6.28kA. c 12.56kA

The damage morphological features observed by SEM (magnification was 200 times) are shown in Fig. 10.21. From the damage morphology we can see that the damage to PVC and balsa wood is different. The balsa wood damage showed itself as fractured fibers and missing material, and the PVC damage was mainly pore expansion. That was consistent with the simulation results: the pyrolysis results of cellulose consisted of main chain breakages, DP decrease and gas generation, which caused the fiber breakage and material defects.; the PVC chain broke and recombined in pyrolysis process and there was no obvious reduction in DP, so there was nearly no break in PVC damage, but PVC were softened locally under the effect of arc high temperature because the glass transition temperature of PVC was low, and the large amount of HCl and other gases produced in pyrolysis process could expand the gaps in the softened PVC. The experiment results were consistent with the simulation results.

(c) Damage Degree of the Two Materials

The sizes of the damage areas in the two materials were measured. The width size along the surface and the depth size of the two materials are shown in Tables 10.6 and 10.7, respectively. As we can see, the damage width was larger than the depth in balsa wood but the size of width and depth in PVC was not regular. The impact residual strength of the two materials is shown in Fig. 10.22. It was found that the impact residual strength of PVC and balsa wood decreased with the increase of peak current, and the fall in PVC was faster. In addition, PVC was easier to fracture along the depth direction than balsa wood because arc heat in PVC spread non-directionally, but in balsa wood it spread along the surface direction.

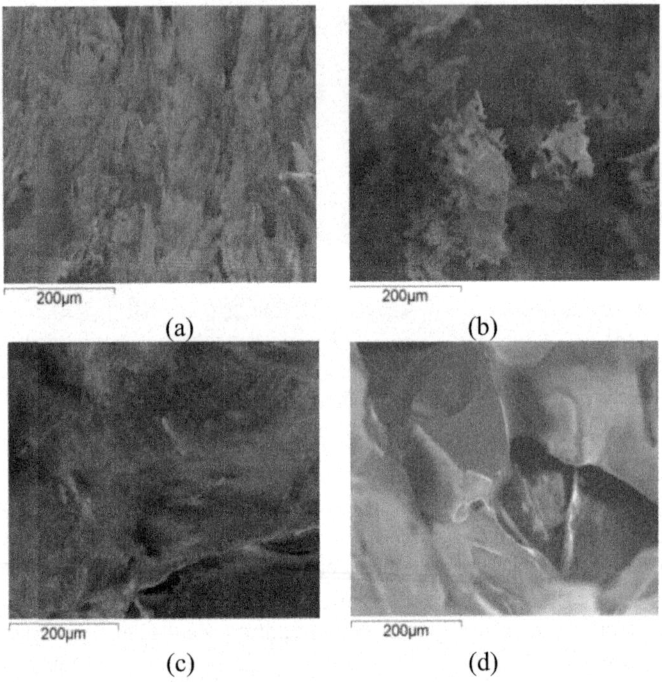

Fig. 10.21 The morphological features of balsa wood and PVC. **a** Undamaged balsa wood. **b** Damaged balsa wood (21 kA). **c** Undamaged PVC. **d** damaged PVC (12.56 kA)

Table 10.6 Damage width and depth of balsa wood

Peak value of current (cm)	1.45 kA	6.28 kA	12.56 kA	21 kA
Width size	0.4	0.7	3	4.5
Depth size	0.1	0.2	2	2

Table 10.7 Damage width and depth of PVC

Peak value of current (cm)	1.45 kA	6.28 kA	12.56 kA
Width size	0.35	1	2.5
Depth size	0.1	2	2

10.2.2 Damage in the Interlayer of Balsa Wood and PVC Blade

(a) **Arc Damage Experiment on Blade Interlayer Structure**

In order to investigate the structural damage of the interlayer section materials when subjected to lightning arcs striking wind turbine blades, balsa wood and PVC were

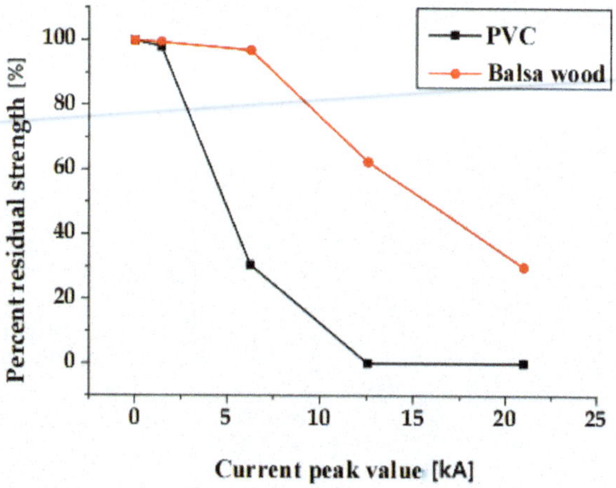

Fig. 10.22 Curve of residual strength versus current peak value

chosen as the interlayer materials. Blade samples were fabricated to simulate the structure of wind turbine blades, and were subjected to lightning currents of varying magnitudes in the interlayer section. The damage condition of the two interlayer materials was then analyzed.

The experimental model, mimicking the structure of wind turbine blades, is depicted in Fig. 10.23. The specifications of the fabricated blade shell section are 80 cm × 90 cm, with widths of 10, 42, 35, and 12 cm from top to bottom, as shown in the figure. The structural analysis of the U-section of the blade is illustrated in Fig. 10.24. The interlayer consists of three layers, with the outer two layers reinforced with 0.9 mm thick epoxy resin strengthened with glass fibers, and the middle layer made of either balsa wood or PVC with a thickness of 0.6 cm.

To simulate the arc path of lightning strikes on the material, multiple holes with a diameter of 0.18 cm were drilled at various locations in the interlayer section of the experimental samples. Nickel–chromium wires with a diameter of 0.01 cm were

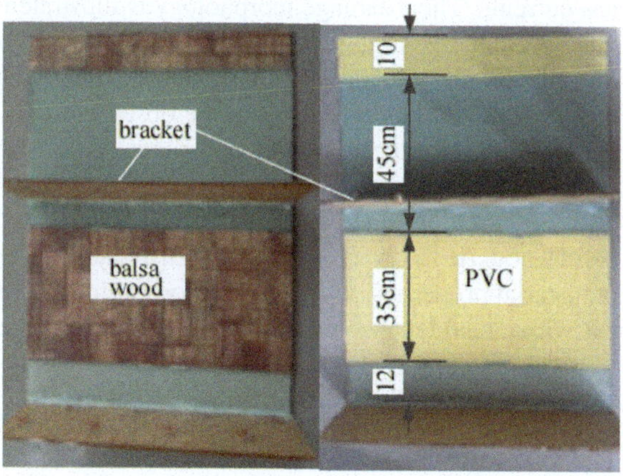

Fig. 10.23 Experimental sample model

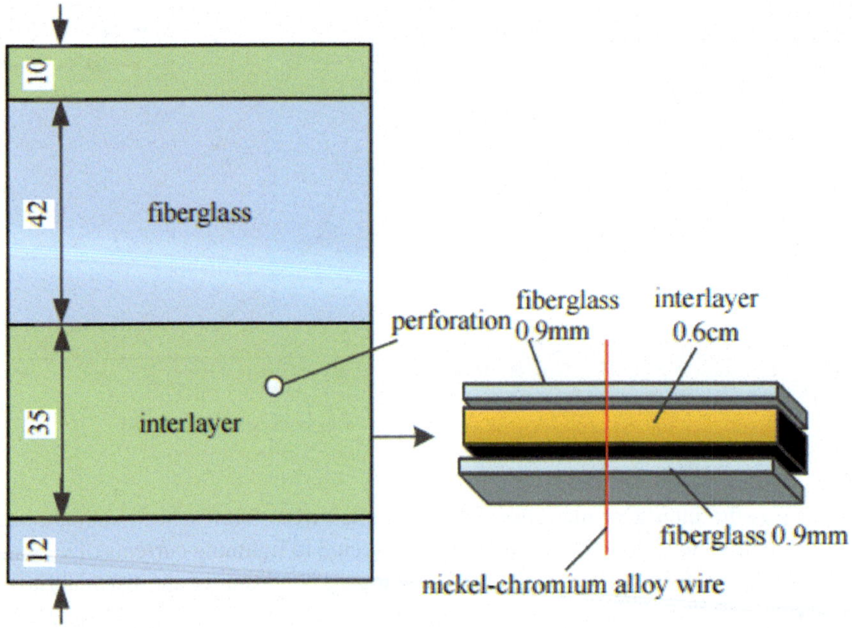

Fig. 10.24 Experimental sample structural analysis diagram

threaded through these holes, allowing currents of 7.5 kA, 7.7/18.1 μs, 15 kA, 6.7/15.5 μs, and 30 kA/5.1/13.1 μs to pass through. The high temperatures generated when large currents pass through the nickel–chromium alloy were used to simulate the arc path when the current penetrates the blade material at different intensities. The damage characteristics of the two materials were analyzed.

(b) **Damage Morphology of the Two Materials**

The two experimental materials experienced varying degrees of damage under several test currents. Their damage morphology is illustrated in Figs. 10.25 and 10.26. The black areas depict high-temperature pyrolysis of the materials, resulting in burn and carbonization, while the white areas indicate delamination between the sandwiching material and the fiberglass. It can be observed from the figure that at 7.5 kA, the damage to both materials is relatively minor, with no severe impairment. The diameter of damage in balsa wood is slightly larger than that in PVC, where PVC exhibits signs of scorching and blackening, indicating a lesser degree of charring compared to balsa wood. At 15 kA, PVC shows extensive scorching, with significant deterioration in material properties, while balsa wood exhibits a similar damage diameter to PVC but without evident scorching, and less pronounced carbonization. At 30 kA, balsa wood also displays scorching phenomena.

(c) **Damage Degree of the Two Materials**

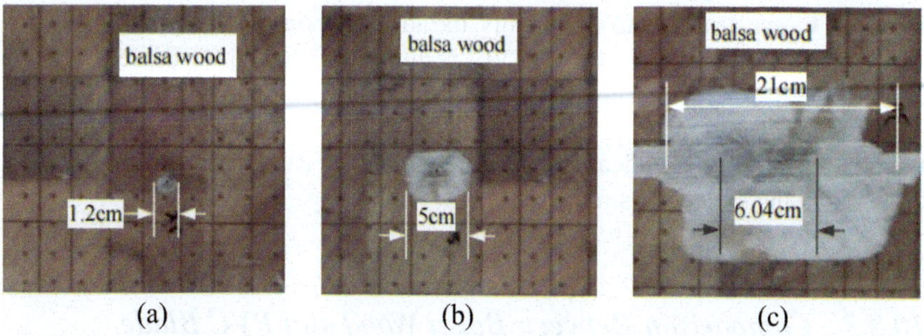

Fig. 10.25 The damage condition of balsa wood under different impulse currents. **a** 7.5 kA. **b** 15 kA. **c** 30kA

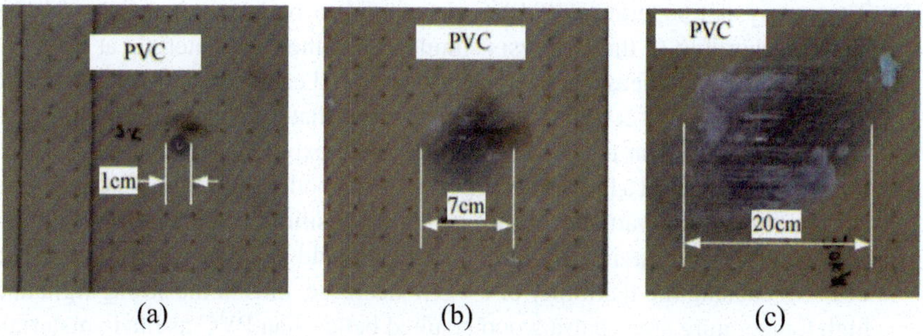

Fig. 10.26 The damage condition of PVC under different impulse currents. **a** 7.5 kA. **b** 15 kA. **c** 30kA

The diameters of damage and burn marks under different impulse currents for the two materials are presented in Table 10.8. The experimental results reveal that although there is not a significant difference in the damaged areas of the two materials under different currents, PVC exhibits more severe burning. This is attributed to the weaker thermal stability of PVC compared to balsa wood, requiring less energy for thermal damage, corroborating with the molecular simulation results.

Moreover, balsa wood exhibits more severe layered damage. Despite molecular simulations indicating a significantly lower volume of gas produced by balsa wood during thermal decomposition compared to PVC, the higher porosity of balsa wood

Table 10.8 The damage parameters under different impulse currents for the two materials

Experimental current (kA)		7.5	15	30
Damage diameter (cm)	Balsa wood	1.2	5	21
	PVC	1	7	20
Scorch diameter (cm)	Balsa wood	0.02	0.22	1.04
	PVC	0.22	1.74	4.62

leads to a looser structure. Consequently, the adhesive bond between balsa wood and epoxy resin is weaker than that of PVC, resulting in a higher air content between layers, which exacerbates layering upon heating. Additionally, PVC also demonstrates some degree of layering, attributed to unavoidable air gaps during material bonding and the significant gas production during PVC thermal decomposition observed in the molecular simulations.

10.2.3 Comparison Between Balsa Wood and PVC Blade

The damage characteristics of balsa wood and PVC were compared.

(1) The pyrolysis temperature of PVC was found to be lower than that of balsa wood. According to the pyrolysis simulation of the two materials at different temperatures, the release of HCl in PVC occurred earlier than glycosidic bond breaks in balsa wood at the same temperature, so that the heat resistance of balsa wood was better than that of PVC. The impact experiments also showed that PVC was damaged at a lower current than balsa wood. In addition, PVC softened easily at high temperatures because its glass transition temperature is low (78–81 °C), and the ultraviolet light and oxygen could also cause photooxidation of PVC, causing the flexibility of PVC to decrease, thus in the strong light and high temperature areas balsa wood resulted better than PVC as blade material.

(2) The rigidity of cellulose decreased greatly during pyrolysis. Cellulose has stronger rigidity compared with PVC. Firstly, the molecular polarity of cellulose made the interaction between the molecular chains stronger; secondly, the hexagonal pyranoid rings in the molecule made molecular chains stronger; thirdly, there were a lot of hydrogen bonds in the molecules or between molecules, especially intramolecular hydrogen bonds, making the glycosidic bonds unable to rotate, thus greatly increasing the rigidity of balsa wood. However, it could be seen from the pyrolysis of cellulose that the glycosidic bonds broke during the pyrolysis process, the DP decreased sharply, the pyranoid ring structure was broken, and the intramolecular hydrogen bonds were changed, as shown in Fig. 10.27, so the molecular chain structure was destroyed, and thus the rigidity of the cellulose was greatly reduced.

(3) The decomposition of cellulose was mainly due to the breakage of the chains, the decrease of DP, and formation of a lot of gaseous products, leading to macroscopic fiber fractures and missing materials. The decomposition of PVC was mainly shown in terms of release of HCl gas, and other small gas molecules would be generated in further reaction, and the molecular chains broke and reconnected constantly. Considering its low softening temperature, the damage of PVC was mainly shown as pore extension on a macroscopic level.

(4) The main air products of balsa wood in the simulation were CO, H_2O, H_2 and C_2H_2. Those of PVC were HCl, H_2, and C_2H_2. The gas product expansion accelerated the fiber fracture in balsa wood and the pore extension in PVC.

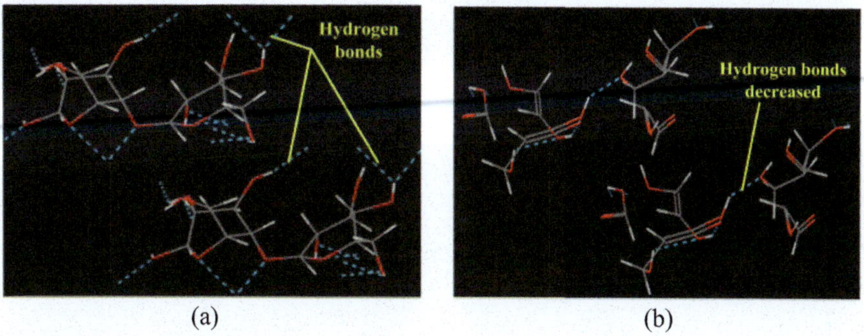

Fig. 10.27 The hydrogen bonds in cellulose. **a** The hydrogen bonds before pyrolysis. **b** The hydrogen bonds were destroyed during pyrolysis

According to the simulation results, the gas products of balsa wood were much more plentiful than those of PVC, so the material damage caused by gas products was more serious in balsa wood. In addition, because the cellulose in the balsa wood can generate hydrogen bonds with the water molecules in the surrounding environment, the moisture content of balsa wood is much higher than that of PVC. On the one hand, the water molecules could promote the pyrolysis of the cellulose; on the other hand, the combined water would evaporate at high temperature, which could also exacerbate balsa wood damage, so PVC was better than balsa wood as a turbine blade material in humid areas (such as those where offshore wind turbines are installed).

References

1. Adri, C.T. Siddharth, D. Francois, L. et al.: A reactive force field for hydrocarbons. J. Phys. Chem. A. 105, 9396–9409 (2001).
2. Kimberly, C. Adri, C.T. William, A.G. et al.: Reactive Force Field for Molecular Dynamics Simulations of Hydrocarbon Oxidation. J. Phys. Chem. A. 112, 1040–1053 (2008).
3. Adri, C.T. Yehuda, Z. Faina, D. et al.: Atomistic-Scale Simulations of the Initial Chemical Events in the Thermal Initiation of Triacetonetriperoxide. J. Am. Chem. Soc. 127, 11053–11062 (2005).
4. Zhang, L. Jiang, LY. Zhao, T. Zou, L. Microcosmic Mechanism Investigation on Lightning Arc Damage of Wind Turbine Blades Based on Molecular Reaction Dynamics and Impact Current Experiment. Energies, 2017, Vol.10(12): 2010.
5. Feraboli, P. Miller, M.: Damage resistance and tolerance of carbon epoxy composite coupons subjected. Compos. Part A. 40, 954–967 (2009).
6. Kithil, R.: Case Study of Lightning Damage to Wind Turbine Blade; National Lightning Safety Institute: Louisville, CO, USA, 2008.
7. Toshio, O. Yoshiyasu, H. Akinori, Y. Coupled thermal–electrical analysis for carbon fiber epoxy composites exposed. Compos. Part A. 41, 973–981 (2017).
8. Vasa, N. J. Naka, T. Yokoyama, S. et al.: Experimental Study on Lightning Attachment Manner considering Various Types of Lightning Protection Measures on Wind Turbine Blades. In Proceedings of the International Conference on Lightning Protection, Kanazawa, Japan, 17–21 September. pp. 1483–1487 (2006).

Chapter 11
Lightning Electromagnetic Interference to Electronics Inside Offshore Wind Turbines

11.1 Integrated Transient Response Analysis Model

11.1.1 Numerical Calculation Method for Grounding Resistance

The design of the offshore wind turbine grounding system is different from that of traditional onshore WTs. The pipe rack structure is used as the foundation of offshore WTs. Considering the economy and the ease of construction, the general engineering uses the foundation as the natural grounding body. At present, the basic forms of offshore wind turbine can be subdivided into single pile foundations, gravity foundations, suction foundations, multi-pile foundations and floating foundations [1]. Among them, the most widely used is the single pile foundation, and its structural diagram is shown in Fig. 11.1.

The grounding resistance is calculated by point current source Green function in the soil. The Laplace equation of cylindrical coordinate system is solved by Fourier transform and separation of variables [2]. The potential function of the midpoint current source is obtained, as shown in Eq. (11.1):

$$\varphi = \phi(\lambda)J_0(\lambda r)e^{-\lambda r} + \gamma(\lambda)J_0(\lambda r)e^{\lambda r} \tag{11.1}$$

In Eq. (11.1), $\phi(\lambda)$ and $\gamma(\lambda)$ are undetermined coefficients, and $J_0(\lambda r)$ is the first class of zeroth order Bessel functions, λ is an arbitrary constant. The ground environment of offshore wind turbine is divided into sea water layer and soil layer. The grounding body is divided into the upper and lower two sections, and the potential function should be calculated respectively [3]. The mathematical model is shown in Fig. 11.2.

where, l is the total length of grounding, a is based radius, h_l is the upper water depth, ρ_l is the upper seawater resistivity, ρ_2 is the lower resistivity of seabed sand.

Q. Sun et al., *Lightning Transients and Protection for Renewable Energy System*,
https://doi.org/10.1007/978-981-97-7267-4_11

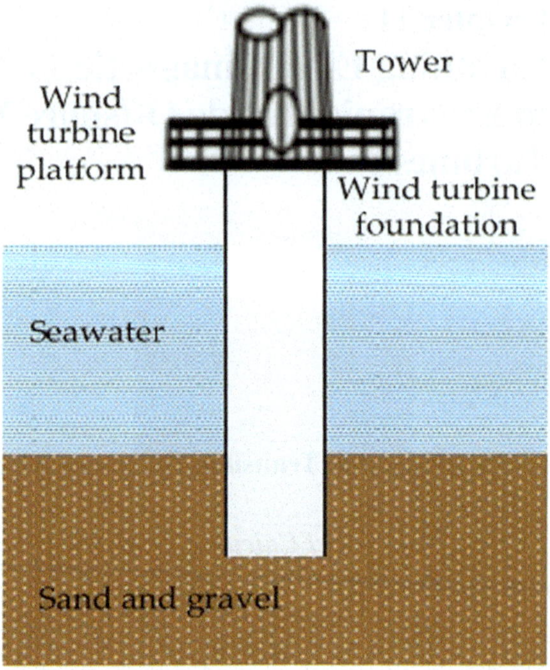

Fig. 11.1 Single foundation schematic

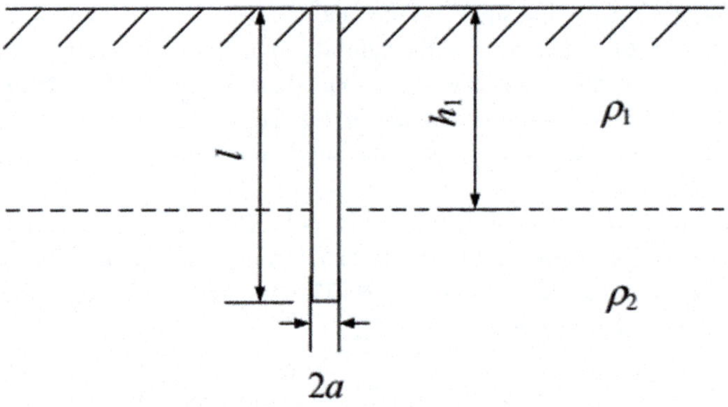

Fig. 11.2 Grounding model of offshore WT

The current density of the ground body in the upper seawater and the lower soil is δ_1 and δ_2 respectively, and the relation is as shown in Eq. (11.2):

$$\begin{cases} \delta_1 = \frac{\rho_2 I}{[\rho_2 h_1 + \rho_1 (l - h_1)]} \\ \delta_2 = \frac{\rho_1 I}{[\rho_2 h_1 + \rho_1 (l - h_1)]} \end{cases} \tag{11.2}$$

The potential functions of the upper seawater and the underlying soil $\varphi\delta_I$ and $\varphi\delta_2$ can be calculated by using Eqs. (11.1) and (11.2), respectively. Based on the superposition principle, the numerical expression of the impulse grounding resistance of the single pile wind turbine in the double deck grounding environment of the ocean can be obtained, as shown in Eq. (11.3):

$$R = \frac{\alpha\rho_1\rho_2}{2\pi[\rho_2 h_1 + \rho_1(l - h_1)]} \times \left[\ln\frac{l + \sqrt{a^2 + l^2}}{a} + \sum_{n=0}^{\infty} k^n \ln\frac{2nh_1 + l + \sqrt{a^2 + (2nh_1 + h_1)^2}}{2nh_1 - h_1 + \sqrt{a^2 + (2nh_1 + h_1)^2}}\right] \tag{11.3}$$

In Eq. (11.3), α is the impulse factor. Considering that the single pile foundation is a vertical tube, its radius is much larger than the radius of flat steel in the common grounding device. The spark discharge around the grounding body is not obvious [4]. The approximate value of α is 1.

11.1.2 Wave Impedance Model for Blades of Offshore Wind Turbines

The rotation of the blade will change the capacitance, inductance and other parameters of the blade. From the definition of wave impedance, it can be seen that the blade wave impedance changes with the increase of blade length. Therefore, in order to accurately establish the blade wave impedance model, the simultaneous effects of rotation angle, height of tower and blade length should be taken into account. Because of the good conductivity of sea water, the sea surface can be considered as a good conductor. Calculation of wind turbine and its mirror is shown in Fig. 11.3, where l_1, l_2, l_3, h respectively represent three blades and tower, l_1', l_2', l_3', h' respectively represent the mirror images of Energies the wind turbine blades and the tower.

The blade capacitance is calculated by the mean potential method. Considering that the height of the offshore wind turbine is much larger than the radius of the tower, the radius of the tower can be neglected. It is assumed that the blade and tower are uniformly charged and the charge is concentrated on the axis, and the charge linear density is τ. As shown in Fig. 11.3, the blade to ground capacitance C_g, the blade to column capacitor C_{lh} and the blade capacitance $C_{l1/2}$ can be obtained:

$$C_g = \frac{4\pi\varepsilon L^2}{\int\limits_{l_1}\left(\int\limits_{l_1}\frac{1}{D_{l_1l_1}}dl_1 - \int\limits_{l_1'}\frac{1}{D_{l_1l_1'}}dl_1'\right)dl_1} \tag{11.4}$$

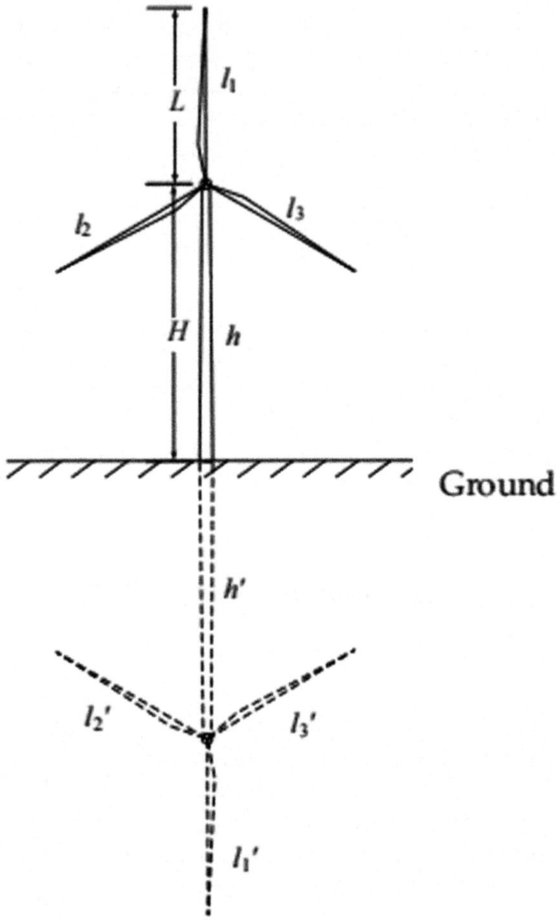

Fig. 11.3 Schematic of a WT and its mirror image

$$C_{lh} = \frac{4\pi \varepsilon L^2}{\int\limits_{l_1} \left(\int\limits_{h} \frac{1}{D_{l_1 h}} dh - \int\limits_{h'} \frac{1}{D_{l_1 h'}} dh' \right) dl_1} \tag{11.5}$$

$$C_{l1/2} = \frac{4\pi \varepsilon L^2}{\int\limits_{l_1} \left(\int\limits_{l_2} \frac{1}{D_{l_1 l_2}} dl_2 - \int\limits_{l_2'} \frac{1}{D_{l_1 l_2}} dl_2' \right) dl_1} \tag{11.6}$$

In Eqs. (11.4)–(11.6), L is the blade length, ε is the space conductivity, and D is the distance between the two micro units. The inductance of a conductor consists of two parts, the inner inductance and the outer inductance, but considering that lightning current changes rapidly and the lightning transient process time is short, it is generally recognized that the internal inductance is negligible in the calculation of lightning electric parameters, and only the external inductance should be considered [5]. For space conductors, the inductance is the sum of inductance and mutual inductance. At present, Neumann integral method in electromagnetic field theory is widely used to calculate self-inductance and mutual inductance in domestic and foreign scholars [6]. According to the geometrical relationship of the blade, the self-inductance L_g, the mutual inductance L_{lh} and the mutual inductance between blades $L_{l1/2}$ can be obtained:

$$L_g = \frac{\mu \cos\theta_1}{4\pi \int\limits_{l_1} \left(\int\limits_{l_1} \frac{1}{D_{l_1 l_1}} dl_1 - \int\limits_{l_1'} \frac{1}{D_{l_1 l_1'}} dl_1' \right) dl_1} \tag{11.7}$$

$$L_{l_1 h} = \frac{\mu \cos\theta_2}{4\pi \int\limits_{l_1} \left(\int\limits_{h} \frac{1}{D_{l_1 h}} dh - \int\limits_{h'} \frac{1}{D_{l_1 h'}} dh' \right) dl_1} \tag{11.8}$$

$$L_{l1/2} = \frac{\mu \cos\theta_3}{4\pi \int\limits_{l_1} \left(\int\limits_{l_2} \frac{1}{D_{l_1 l_2}} dl_2 - \int\limits_{l_2'} \frac{1}{D_{l_1 l_2}} dl_2' \right) dl_1} \tag{11.9}$$

In Eqs. (11.7)–(11.9), θ_1, θ_2, θ_3 respectively is the angle between the blade and blade, blade and mirror tower, the blade and blade. μ is the spatial permeability.

From the above analysis, the total capacitance and total inductance of a single blade can be obtained. According to the definition of wave impedance, the wave impedance of the blade can be obtained, as shown in Eq. (11.10):

$$Z_b = \sqrt{\frac{L_g + 2L_{l1/2} + L_{lh}}{C_g + 2C_{l1/2} + C_{lh}}} \tag{11.10}$$

In this paper, the blade wave impedance model is compared with the conventional fan blade wave impedance model. Not only the blade length and the height of the drum are taken into account, but also the rotation angle of the blade is taken into account.

11.1.3 Sensitivity Analysis

The single pile offshore wind turbine is used as the standard model in the simulation (parameters shown in Table 11.1), and the transient potential of the offshore wind turbine nacelle is analyzed. The internal equipment circuit adopts separate grounding system, the transient potential of the internal equipment is much less than the transient potential of the nacelle, and the potential difference between the equipment and the nacelle can be approximately equal to the nacelle potential. Therefore, the transient potential analysis in the nacelle is the main factor.

After simulation calculation, the rotation angle has less influence on the lightning transient response of the offshore wind turbine, and the influence of rotation angle can be neglected in the simulation process. Therefore, this section is not taking into account the impact of the rotation angle.

(a) **Seawater Depth Influence on Transient Response of WT Lightning Impulse**

For an offshore WT, the depth of the sea layer is an important factor affecting the grounding parameters. According to Eq. (11.3), with the decrease of the sea water layer, the impulse grounding resistance of the offshore wind turbine will increase. When the sea water layer is 0 m, it can be approximately regarded as a land WT. According to the different soil layer resistivity, the grounding resistance changes in the range of 0–20 Ω. At different ocean depths, the transient potential of the wind turbine nacelle is obtained from the integrated model, as shown in Fig. 11.4.

It can be seen from Fig. 11.4 that the transient overvoltage of the lightning impact of the offshore WT and the onshore WT is different, the transient voltage of the land WT is obviously larger than that of the offshore WT, so it is very necessary to model and analyze the offshore WT separately.

It can be seen that the maximum value of transient potential in the nacelle is related to the depth of the seawater layer, and the maximum value appears in the first oscillation period.

The complex environment of offshore WTs makes them very different from onshore WTs. The greater the depth of the sea, the smaller the impulse grounding resistance, the peak value of the transient voltage of the nacelle decrease. Under the premise of ensuring the normal operation of the equipment, the deeper the sea water, the more conducive to lightning protection design. This has guiding significance for

Table 11.1 Standard model parameters in simulation

Wind power unit parameters (Unit)	Parameter selection
Height of tower body H (m)	120
The average radius of the tower r_t (m)	2.3
The average thickness of the tower wall d_t (m)	0.0257
The length of the blade L (m)	70
Referral radius r_b (m)	0.001
The depth of the sea h_1 (m)	15
Seawater resistivity ρ_1 (Ωm)	5
Sand soil resistivity ρ_2 (Ωm)	1500
Ground base radius r_g (m)	3
Extend the length of the seabed h_2 (m)	10
Length of grounding l_g (m)	25

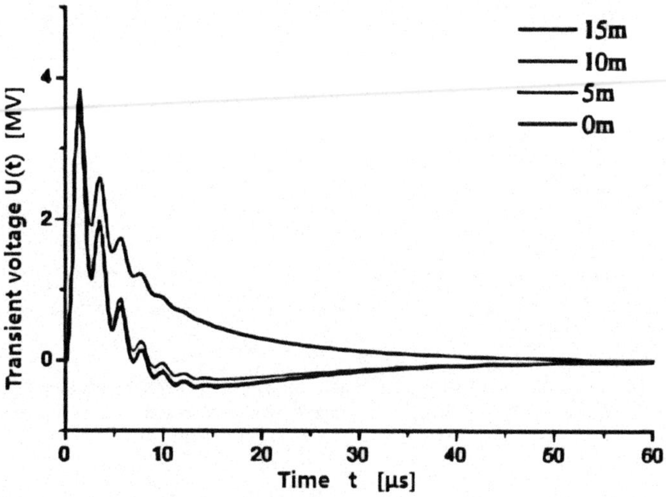

Fig. 11.4 Effect of ocean depth on the transient potential

the site selection of offshore WT. In full consideration of other conditions, the WT should be selected in deep seawater.

(b) **Influence of Blade Length and Tower Height on WT Lightning Impulse Transient Response**

The other parameters are kept unchanged, and the length of the blades ($L = 40$ m, $L = 70$ m, and $L = 100$ m) and the height of the tower ($H = 90$ m, $H = 120$ m, and $H = 150$ m) are simulated. The influence of the blade length and the tower height on the transient potential of the nacelle (Fig. 11.5) and the oscillation frequency (as shown in Table 11.2) are analyzed. The transient voltage peak of the nacelle can be approximated in Eq. (11.11):

$$U_{\mathrm{p}} = U_{\mathrm{h}} + U_{\mathrm{g}} \tag{11.11}$$

where, U_p is the voltage peak in the nacelle, U_h and U_g are the voltage drops on the tower and the ground.

It can be seen that as the tower height increases, the transient potential of the WT nacelle shows an upward trend. With the increase of blade length, the transient potential of the WT nacelle is decreasing. When the WT is struck by lightning, a high tower and short blades will increase the probability of damage to the nacelle.

The simulation results have guiding significance for selecting the height of the tower and the length of the blade when constructing an offshore WT. Short towers and long blades are good for WT lightning protection.

(c) **Lightning Parameters Influence on WT Lightning Impulse Transient Response**

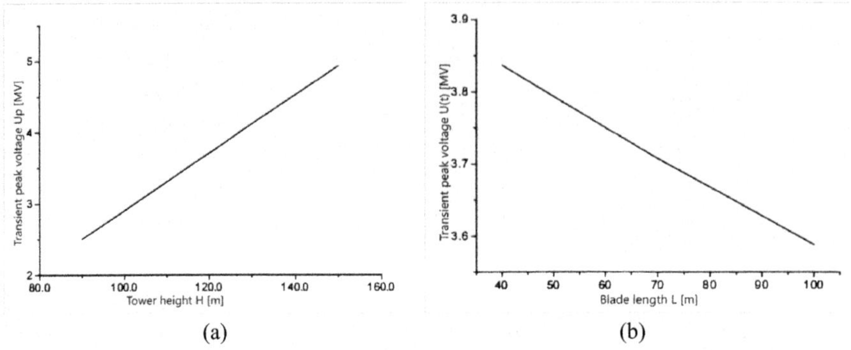

Fig. 11.5 Effect of blade length and tower height on transient potential peak. **a** Effect of tower height on transient potential peak. **b** Effect of blade length on transient potential peak

Table 11.2 Effect of blade length and tower height on transient potential oscillation frequency

Leaf blade length L (m)	Tower height H (m)	Oscillation frequency (MHz)
40	90	2.5936
70	90	2.4593
100	90	2.3584
40	120	3.8370
70	120	3.7073
100	120	3.5882
40	150	5.0767
70	150	4.9643
100	150	4.8425

The time of the wave front and the time of the wave tail are the two main parameters of the lightning current waveform. Figure 11.6 is a schematic diagram of three kinds of lightning current: 4/20, 8/20, and 8/50 μs. Using the three different wave fronts and wave times, the lightning current waveform is simulated to get the transient potential at the nacelle, as shown in Table 11.3.

The overall impedance of the wind turbine is perceptive, so the voltage value in the nacelle is available in Eq. (11.12):

$$U = L\frac{di}{dt} + iR \tag{11.12}$$

where, L is the wind turbine equivalent inductance, R is the wind turbine equivalent resistance, i is the lightning current. di/dt is the sum of the changes of the lightning flow itself and the variation of the current reflection.

It can be seen that the wave front time has a great influence on the transient process of the lightning stroke offshore wind turbine. Since the wave front time is

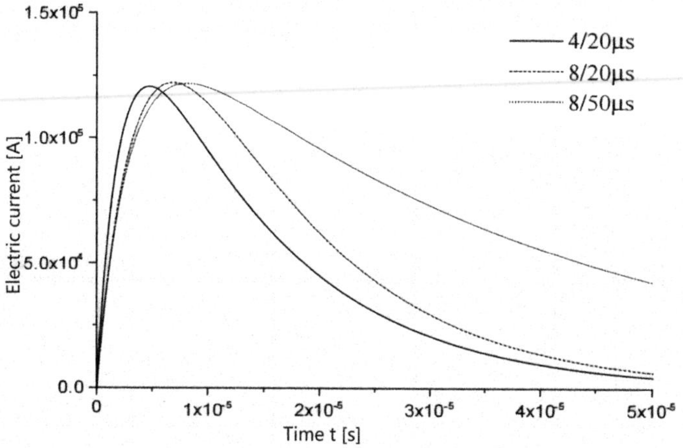

Fig. 11.6 Comparison of different lightning current waveform

Table 11.3 Transient potential of different lightning parameters at WT nacelle

Lightning parameters (μs)	Peak transient voltage at the engine room (MV)
4/20	4.9
8/20	3.7
8/50	3.6

much smaller than the wave wake time, when the wave front time decreases, the high-frequency component obviously increases, and the *di*/*dt* value becomes larger, then the transient potential peak value is higher obviously. The lightning current wave steepness in front of the wave is very big. *U* is mainly decided by the first item in Eq. (11.12). The gradient of the double exponential current wave in front of the wave is the higher, the overvoltage maximum usually occurs in the first period.

Lightning currents tail gently, the wave tail time change has little effect on the *di*/*dt*, then the impact on the nacelle transient potential is also smaller.

There is no fixed waveform in the nature of lightning, but through this section analysis, it can be concluded that when other parameters are the same, the smaller the time of the wave front, the greater the possibility of damage to the WT. In the design of lightning protection, priority should be given to the lightning wave with less front time.

(d) **Lightning Point Influence on WT Lightning Impulse Transient Response**

The offshore WT is struck by lightning, and the lightning current may be injected from the tip of the wind turbine blade tip or the rear of the engine room. Therefore, the different lightning point positions will lead to the change of lightning current on the wind turbine, and then have an impact on the transient process of lightning

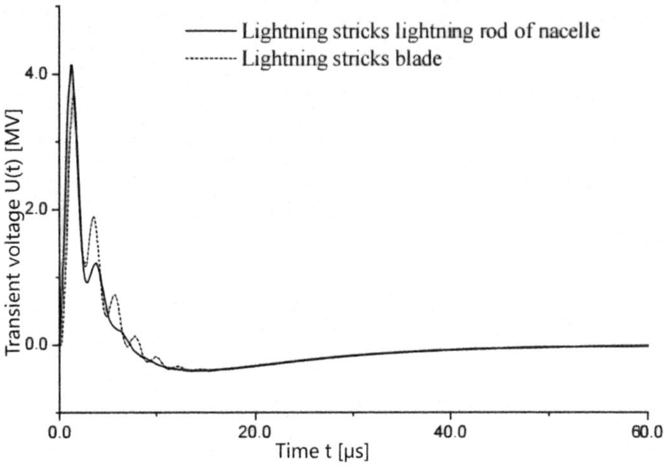

Fig. 11.7 Effect of lightning strike point on the transient potential

stroke [7]. In this paper, the transient potential response of the wind turbine nacelle is studied by analyzing the two different lightning strike points.

As shown in Fig. 11.7, due to the change of the current path, when lightning strikes the lightning rod of the nacelle, the transient potential peak is about 11.6% higher than when lightning strikes blades. Therefore, emphasis should be placed on the nacelle lightning protection. The nacelle is the focus of lightning protection, in order to prevent lightning damage to all kinds of equipment in the cabin, causing major accidents.

11.2 High-frequency Electromagnetic Field Distribution for Offshore Wind Turbines

11.2.1 Transient Electromagnetic Field Calculation Model

(a) **Numerical Analysis of Transient Electromagnetic Field of Offshore Wind Turbine**

The main object of this section is the electromagnetic field distribution around the wind turbine (especially inside the tower). The spatial electromagnetic field distribution is mainly determined by the discharge path of the lightning current on the surface of the wind turbine tower and the specific current distribution. In response to this problem, many scholars have divided the tower into a multi-conductor system, as shown in Fig. 11.8d. Some scholars regard the whole tower as a long cylinder with equal area of upper and lower bottoms [8], as shown in Fig. 11.8b, or divide it into several truncated cones to form a series of unequal truncated cone models, as shown in Fig. 11.8c.

Each corresponding sub-part is represented by RLC coupled π-type circuit units (as shown in Fig. 11.9), forming a transient equivalent calculation electrical network. Under the action of an external lightning current source, the transient lightning current distribution on the surface of the offshore wind turbine tower is obtained [9].

Based on the distribution of lightning current on the surface of the tower, the dipole method is used to calculate the electromagnetic field. The dipole method is also known as the Lorentz condition method, and the solved electromagnetic fields are represented by currents.

Any electromagnetic field satisfies the basic equation (Maxwell's equation):

$$\nabla \cdot E = \rho/\varepsilon_0 \tag{11.13}$$

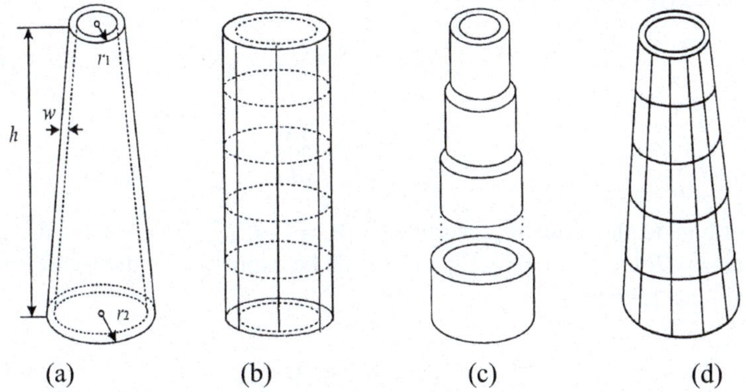

Fig. 11.8 Discrete grid subdivision of offshore wind turbine tower. **a** Schematic diagram of the tower. **b** Cylinder model. **c** Continuum truncated cone model. **d** The division of tower body line network

Fig. 11.9 Coupled π-type circuit unit

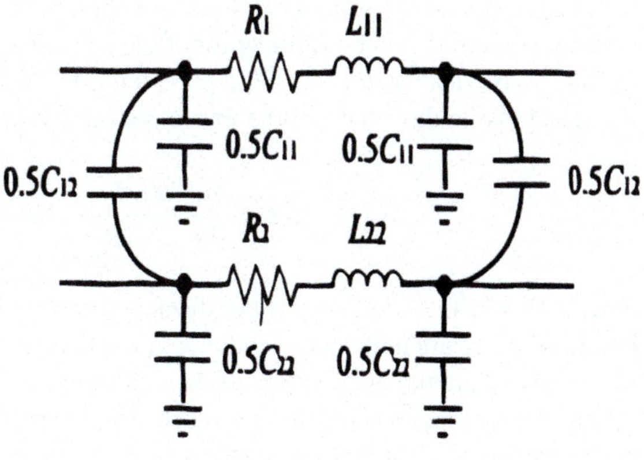

$$\nabla \cdot B = 0 \tag{11.14}$$

$$\nabla \times E = -\frac{\partial B}{\partial t} \tag{11.15}$$

$$\nabla \times B = \mu_0 J + \frac{1}{c^2}\frac{\partial E}{\partial t} \tag{11.16}$$

where, ρ is the charge density; J is the current density; ε_0 is the vacuum dielectric constant; μ_0 is the vacuum permeability; c is the speed of light ($c = 1/\sqrt{\varepsilon_0\mu_0}$). In the air medium, the wave velocity is almost equal to the speed of light, which is 2.998×10^8 m/s.

According to Maxwell's equation, the lightning spatial electromagnetic field of wind turbine tower is represented by vector magnetic potential A and scalar potential Φ, respectively, as follows:

$$B = \nabla \times A \tag{11.17}$$

$$E = -\nabla\Phi - \frac{\partial A}{\partial t} \tag{11.18}$$

According to the Lorentz condition, $\nabla\Phi = -c^2 \int_{\infty}^{t} \nabla(\nabla \cdot A)dt$, the electric field intensity E represented by Eq. (11.6) can be expressed by the vector magnetic potential A, as follows:

$$E = -\frac{\partial A}{\partial t} + c^2 \int_{\infty}^{t} \nabla(\nabla \cdot A)dt \tag{11.19}$$

The current of the lightning current path on the surface of the wind turbine tower can be divided into multiple current micro-elements, each of which is regarded as a current-carrying electric dipole as the field source of the electromagnetic field. In order to obtain the spatial electromagnetic field intensity at a certain moment, a current element dz' in the Z direction of the coordinate system is selected, as shown in Fig. 11.10. The vector magnetic potential dA generated by any point P(r, θ, z) in the space inside the wind turbine tower can be expressed as follows:

$$dA = \frac{\mu_0}{4\pi}\frac{i(z', t - \frac{R}{c})}{R} dz' k_0 \tag{11.20}$$

where, $R = \sqrt{(z - z')^2 + r^2}$ is the distance between the source point and the field point; x_0, y_0, z_0 are unit vectors in the x, y, z axis directions of the coordinate system; r_0, θ_0 and k_0 are the unit vectors of the cylindrical coordinate system.

The vector magnetic potential A is substituted into the electric field strength E and the magnetic field strength B, and the spatial derivative is represented by the time

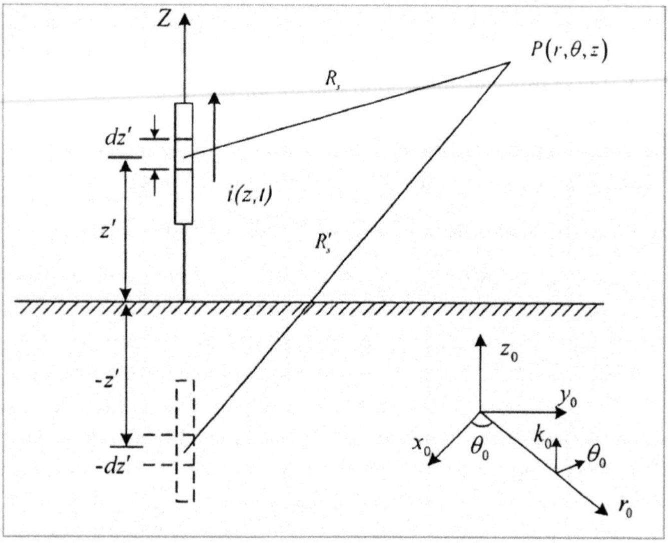

Fig. 11.10 Schematic diagram of space conductor

derivative, which is as follows:

$$\frac{\partial i\left(z', t - \frac{R}{c}\right)}{\partial z} = -\frac{z - z'}{cR}\frac{\partial i\left(z', t - \frac{R}{c}\right)}{\partial t} \tag{11.21}$$

$$\frac{\partial i\left(z', t - \frac{R}{c}\right))}{\partial r} = -\frac{r}{cR}\frac{\partial i\left(z', t - \frac{R}{c}\right)}{\partial t} \tag{11.22}$$

Based on the distribution of lightning current on the surface of the tower, the spatial electromagnetic field generated by any infinitesimal current element can be solved through numerical calculation.

$$\begin{aligned} dE = {} & \frac{dz'}{4\pi\varepsilon_0}\left[\frac{3r(z - z')}{R^5}\int_0^t i\left(z', \tau - \frac{R}{c}\right)d\tau + i\left(z', t - \frac{R}{c}\right)\right. \\ & \left. + \frac{r(z - z')}{c^2R^3}\frac{\partial i\left(z', t - \frac{R}{c}\right)}{\partial t}\right] r_0 + \frac{dz'}{4\pi\varepsilon_0}\left[\frac{2r(z - z') - r^2}{R^5}\int_0^t i\left(z', \tau - \frac{R}{c}\right)d\tau\right. \\ & \left. + \frac{2r(z - z') - r^2}{cR^4} i\left(z', t - \frac{R}{c}\right) - \frac{r(z - z')}{c^2R^3}\frac{\partial i\left(z', t - \frac{R}{c}\right)}{\partial t}\right] k_0 \end{aligned} \tag{11.23}$$

$$dB = \frac{\mu_0 dz'}{4\pi}\left[\frac{r}{R^3} i\left(z', -\frac{R}{c}\right) + \frac{r}{cR^2}\frac{\partial i\left(z', t - \frac{R}{c}\right)}{\partial t}\right]\theta_0 \tag{11.24}$$

Using the above formula to calculate the electric and magnetic fields of all conductor microelements divided on the tower at a point in space, the magnetic and electric fields at this specified point can be obtained by using vector superposition, and the electromagnetic field distribution in the space around the wind turbine can be derived by repeated solving [10]. However, this method requires complex numerical calculations, and the results are not intuitive enough.

(b) **Finite Element Simulation Model of Offshore Wind Turbine**

Anosoft's Maxwell software is a powerful, accurate and easy-to-use two-dimensional/three-dimensional (2D/3D) electromagnetic field finite element analysis software. At present, Maxwell has become an indispensable tool for engineering designers and researchers in the design process of electronic products. With the continuous update of Maxwell's version, its application field is also expanding. At present, it is widely used in motor, electronics, power electronics, AC/DC transmission, power supply, power system and other fields.

Anosoft Maxwell uses computer numerical simulation instead of model test. The solution module includes electrostatic field, static magnetic field, time-varying electric field, time-varying magnetic field, eddy current field, transient field, etc. It can be used to analyze the characteristics of static, transient, steady state, normal and fault conditions of various equipment. The basic principle is based on Maxwell's differential equations, using the finite element discrete form, the engineering electromagnetic field calculation problem is transformed into a matrix solution problem. Among them, the transient field solver can easily design arbitrary waveform voltage and current, or use the circuit diagram plotter and embedded simulator to co-simulate with the external circuit to analyze the transient electromagnetic characteristics of the analysis object.

Based on the engineering data of offshore wind turbines (Table 11.4), a three-dimensional (3D) model is built in Anosoft Maxwell software. Among them, the tower part adopts steel_1008 in Maxwell's property, and the air model is established inside and outside the tower to facilitate the study of the spatial electromagnetic field around the tower part during the transient process of lightning strike on the offshore wind turbine [11].

As shown in the grid subdivision diagram (as shown in Fig. 11.11), the fan tower part is the focus of analysis, and the grid subdivision is meticulous, while the surrounding space (i.e., the air part) is slightly sparse, and the grid between the two gradually fits from small to large.

Table 11.4 Offshore wind turbine tower parameters

Parameter name (unit)	Value
Tower height H (m)	120
Top surface radius r_{top} (m)	2
Bottom radius r_{bottom} (m)	2.4
Shell wall thickness (m)	0.0257

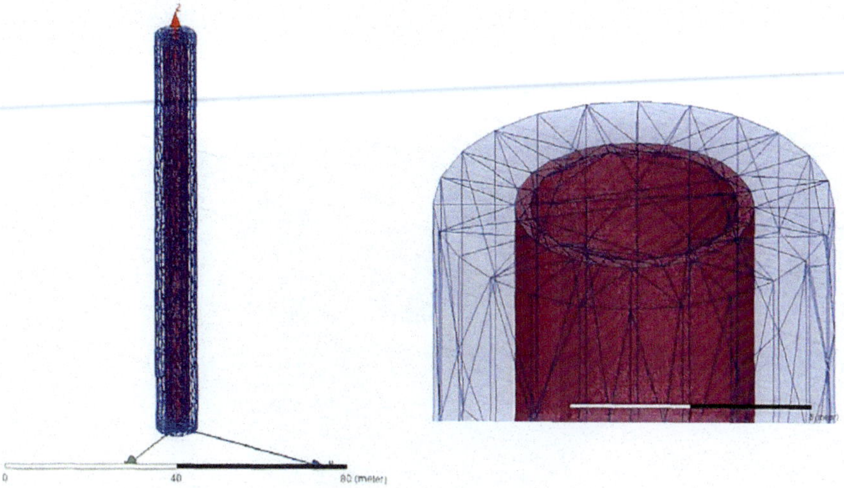

Fig. 11.11 Grid subdivision schematic diagram

In terms of boundary conditions, this paper applies the first-type homogeneous boundary condition (i.e., $A_z = 0$ boundary condition) to the air model, and sets the outer edge of the tower as the second-type homogeneous boundary condition where the rate of change of magnetic potential along the normal direction of the boundary line is zero.

11.2.2 Factors Influencing Electromagnetic Field Intensity

After completing the modeling of the tower model, a standard lightning current of 4/20 μs with an amplitude of 120 kA was applied to a node on the top of the tower. In the actual analysis of electromagnetic fields, transient analysis is required when the voltage or current changes irregularly (i.e., when the applied excitation is a function of time, position, or velocity). Given that we need to conduct a lightning transient analysis, the transient solver was selected for solving and analysis.

When t = 4e − 6 s, the spatial magnetic field distribution of the tower bottom sitting on the plane is shown in Fig. 11.12.

Based on the above results, it can be seen that the magnetic field intensity at a certain point in the space around the wind turbine tower is inversely related to the distance from the tower wall. The closer to the tower wall, the greater the magnetic induction intensity will be, and the minimum magnetic induction intensity is roughly located in the center of the tower. The magnetic field intensity outside the wind turbine tower is significantly higher than that inside the tower, indicating that the lightning

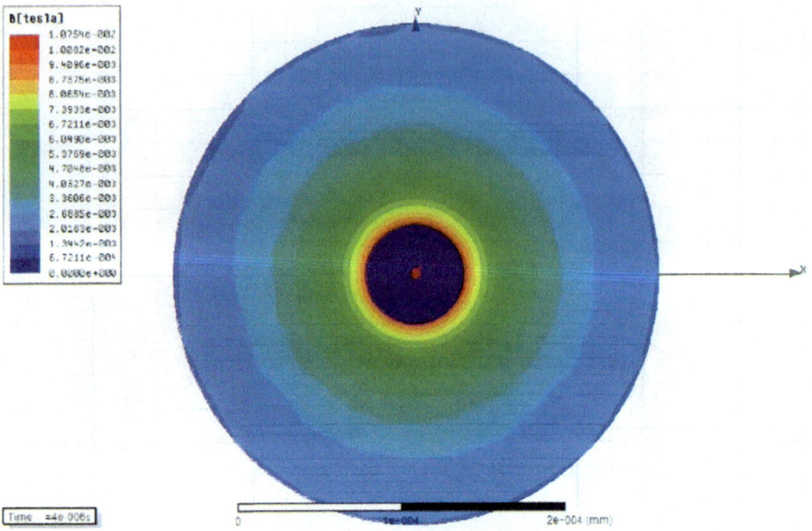

Fig. 11.12 The spatial magnetic field distribution map of the tower bottom sitting on the plane when t = 4e − 6 s

current shows a certain degree of skin effect in the process of downward conduction along the tower wall.

In order to further accurately describe the corresponding relationship between the electromagnetic field distribution in the internal space of the tower and the position of each node, Fig. 11.13 is the horizontal plane where the top of the tower is located. The coordinate system is established, and the leftmost point on the X-axis is set as the corresponding point at the link between the blade and the tower (that is, the node applying the lightning current). Figures 11.14 and 11.15 are the distribution diagrams of the maximum magnetic induction intensity at the top of the tower and the bottom of the tower.

Based on the above results, it can be seen that the spatial magnetic field distribution around the offshore wind turbine exhibits symmetry, with mutual symmetry on both sides of the connection between the blades, nacelle, and tower. As the spatial magnetic field around the offshore wind turbine gradually flattens from top to bottom, it also reflects that the lightning current gradually becomes evenly distributed on the surface of the tower as it discharges from the top down.

Taking the magnetic field distribution on the planes where the top and bottom of the tower are located as an example, the maximum magnetic flux density in some areas approaches 10^{-2} T. In such cases, some electronic devices may be unable to operate normally due to electromagnetic interference exceeding tolerable levels, which in turn has a certain impact on the safe and stable operation of the wind turbine. In the installation design of internal equipment and components of offshore wind turbines, it is advisable to install more sensitive electronic equipment in positions with the

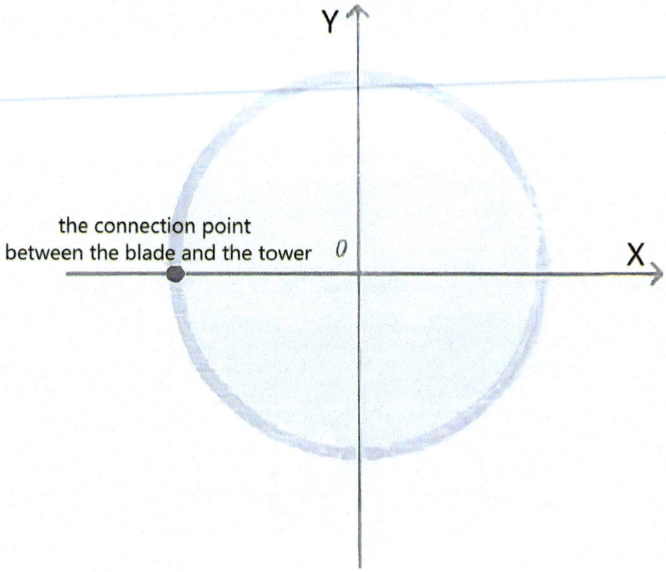

Fig. 11.13 Position corresponding relationship diagram

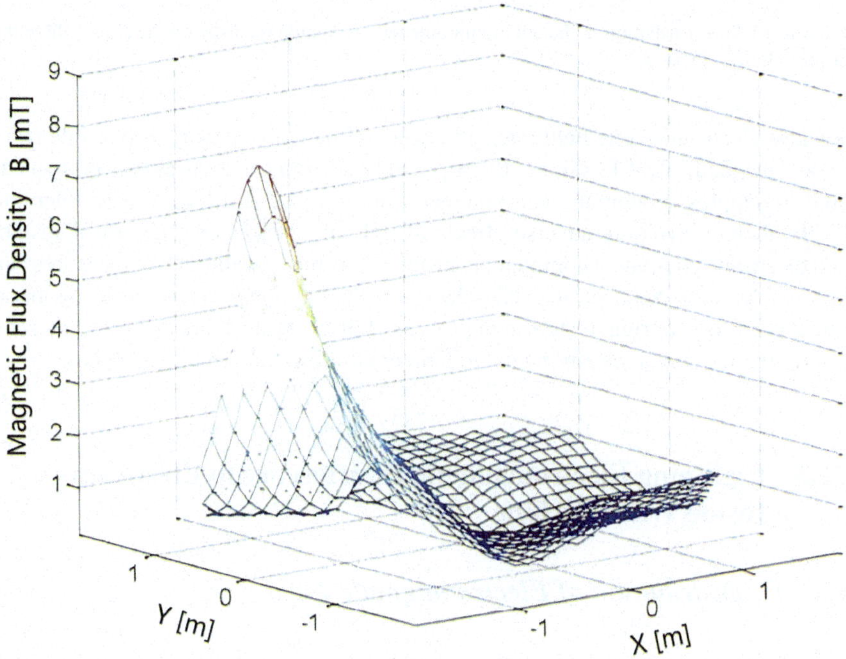

Fig. 11.14 The distribution of the maximum magnetic induction intensity on the plane at the top of the tower

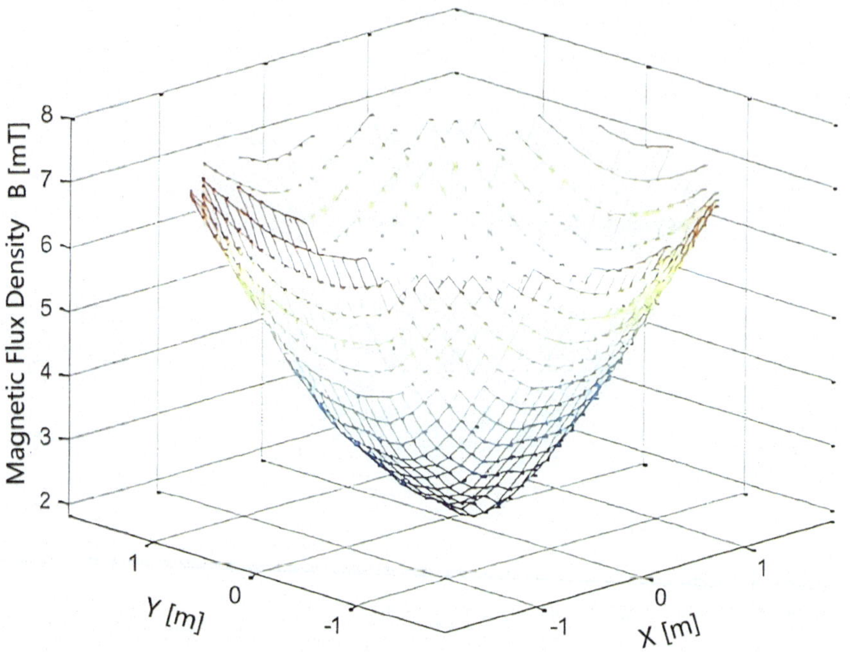

Fig. 11.15 The distribution of the maximum magnetic induction intensity on the plane where the tower bottom is located

weakest electromagnetic field intensity, keeping them as far away as possible from strong magnetic field locations. Even possible shielding measures can be considered to suppress potential electromagnetic interference and reduce it to tolerable levels, thereby avoiding adverse effects of lightning strikes on the control system and electronic equipment components inside the turbine. Additionally, measures can be considered on the tower to evenly distribute the lightning current, ensuring that it remains uniform during the transient process of discharging along the tower, thereby improving the spatial electromagnetic field environment around the turbine.

11.3 Lightning Electromagnetic Interference to Electronics Inside Wind Turbine

11.3.1 Distribution of Electromagnetic Field

A large number of complex numerical calculations have brought huge workload to the analysis of the problem. Therefore, this section uses efficient electromagnetic field simulation analysis software to analyze the electromagnetic field distribution in the

Table 11.5 Parameters of wind turbine [12]

Name	Parameters (m)
Blade length	45
Nacelle size	13 × 4.7 × 4.5
Hub diameter	4
Tower height	95
Top diameter of tower	2.376
Bottom diameter of tower	4.2
Tower thickness	0.034

internal space of the tower when lightning strikes, and uses the efficient and intuitive characteristics of the electromagnetic simulation analysis software to display and analyze the results more intuitively and effectively.

Based on the parameters of the 2 MW wind turbine shown in Table 11.5, a three-dimensional simulation model of the wind turbine was established in an electromagnetic simulation software. In the model, the materials of the nacelle, hub, and tower were set as cast iron, while the interior and exterior of the tower were set as air models. The calculation step size Δt was set to 0.01 μs. A 10/350 μs lightning current with a peak amplitude of 100 kA was injected from the tip of the blade, and the electromagnetic fields at different heights and positions inside the tower during a lightning strike were discussed and analyzed.

Probes were placed on the axis at $z = 85$ m and $z = 70$ m of the tower respectively to conduct simulation analysis on the electric field intensity at different axial positions. The simulation results are shown in Fig. 11.16. It can be seen from the figure that the electric field intensity near the top of the tower is significantly higher than that at the bottom of the tower, indicating that the electric field intensity at the top of the tower is directly proportional to the height of the tower.

The peak results of electric field intensity at different heights at the axis of the tower are shown in Fig. 11.17. By analyzing the peak curve, it can be seen that as the lightning current flows into the earth along the tower, the electric field intensity inside the tower gradually weakens from top to bottom. However, at the bottom of the tower, due to the high potential counterattack of lightning, the instantaneous high potential is formed on the ground, which increases the electric field intensity. Therefore, the bottom of the wind turbine tower still has a high field strength when lightning strikes.

Simulation analysis was conducted on the electric field intensity at the axis and near the inner wall of the tower at distances of 10 m and 30 m from the ground at the bottom of the tower using probes. The simulation results are shown in Fig. 11.18. It can be observed from the simulation results that at the plane $z = 10$ m from the ground at the bottom of the tower, the electric field intensity near the inner wall of the tower is higher than that at the axis, with a larger fluctuation amplitude compared to the axis. However, the overall amplitude is lower than the electric field intensity

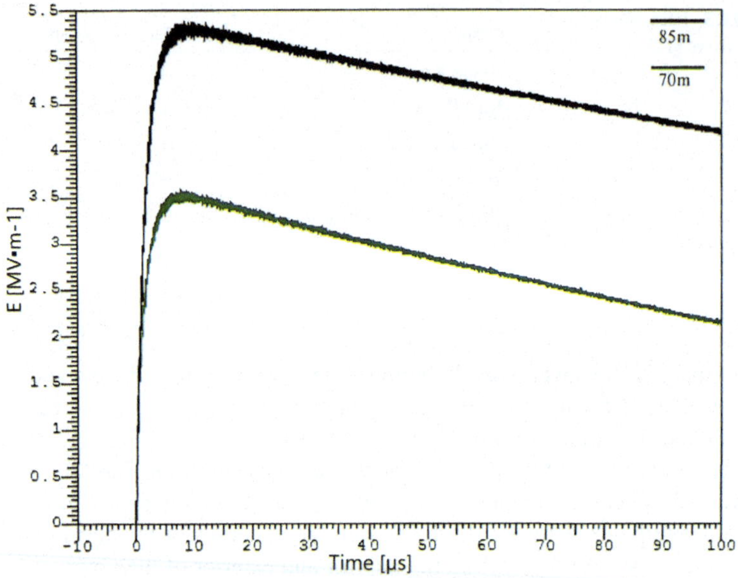

Fig. 11.16 Comparison of electric field intensity at different heights at the top axis of the tower

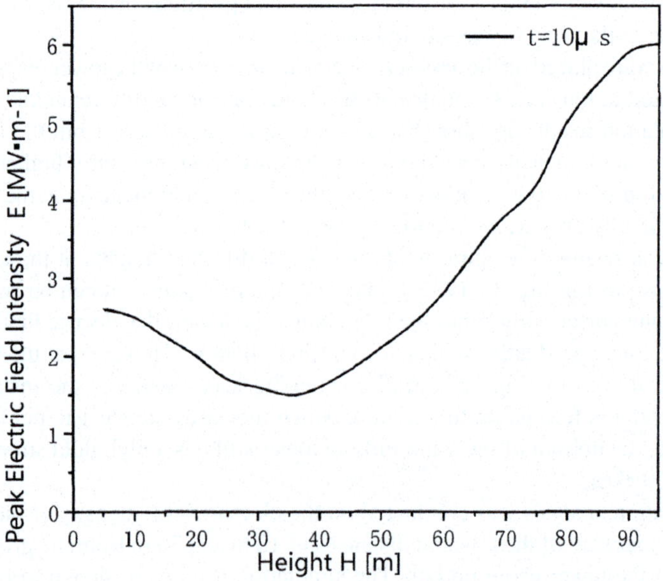

Fig. 11.17 The peak value of electric field intensity at different heights at the axis of the tower cylinder

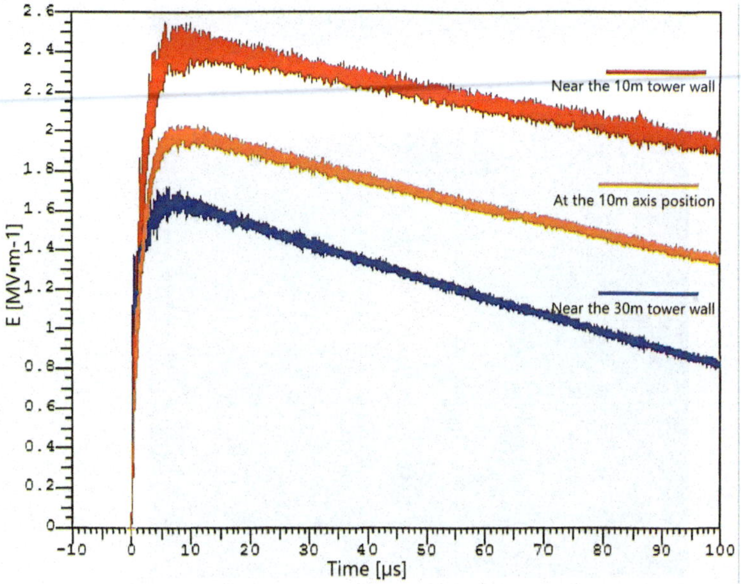

Fig. 11.18 Comparison of electric field intensity at different locations and heights at the bottom of the tower

at the axis at z = 70 m from the ground. At the plane z = 30 m from the ground, the electric field intensity near the inner wall of the tower is lower than that at z = 10 m.

Figure 11.19 shows the distribution of electric field intensity in the plane space where the tower is 10 m away from the ground when t = 10 μs. It can be found from the analysis of the results that the closer to the inner wall of the tower, the higher the electric field intensity. In engineering applications, the impulse grounding resistance is affected by the soil resistivity. During the propagation of the lightning current along the tower, the electromagnetic field inside the tower will oscillate due to the refraction and reflection of the lightning current, which increases the field strength at the bottom of the tower.

Figures 11.20 and 11.21 are the magnetic field intensity at the axis of the tower from the ground $z = 10$ m and the magnetic field intensity distribution in the plane space when $t = 10$ μs, respectively. The simulation results in the figure show that the magnetic field intensity at the axis of the tower is the lowest. With the shortening of the distance from the inner wall of the tower, the magnetic field intensity gradually increases, and the magnetic field intensity in the outer space of the tower is much higher than that in the inner space of the tower. This is due to the skin effect of the cast steel tower wall during the propagation of lightning current along the tower.

In practical engineering applications, when the magnetic induction intensity of the electromagnetic pulse field generated by lightning strikes exceeds 0.07×10^{-4} T, the microcomputer equipment without shielding measures will cause misoperation of the equipment. When it exceeds 2.4×10^{-4} T, unshielded integrated electronic

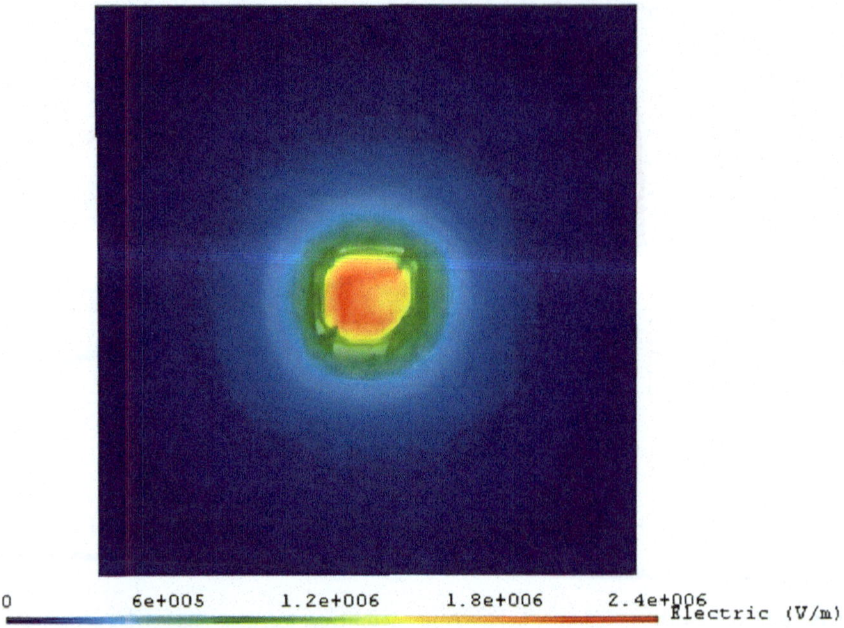

Fig. 11.19 The spatial electric field distribution at the bottom of the tower

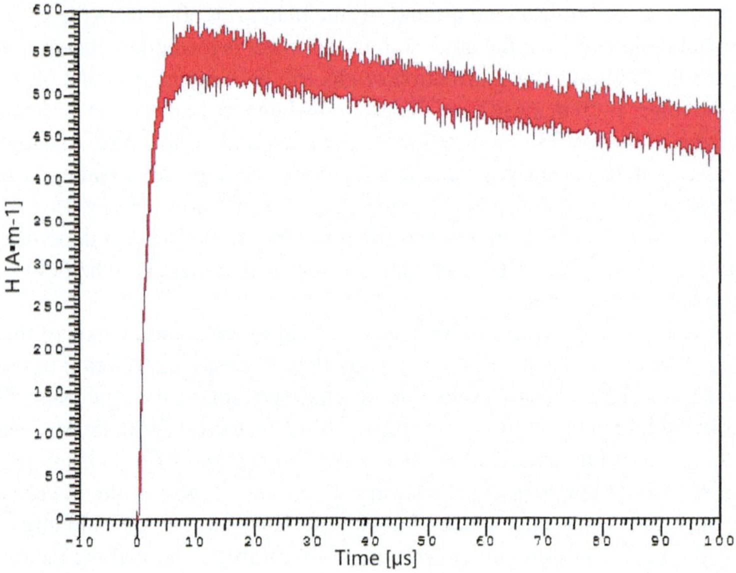

Fig. 11.20 Magnetic field at the bottom axis of the tower

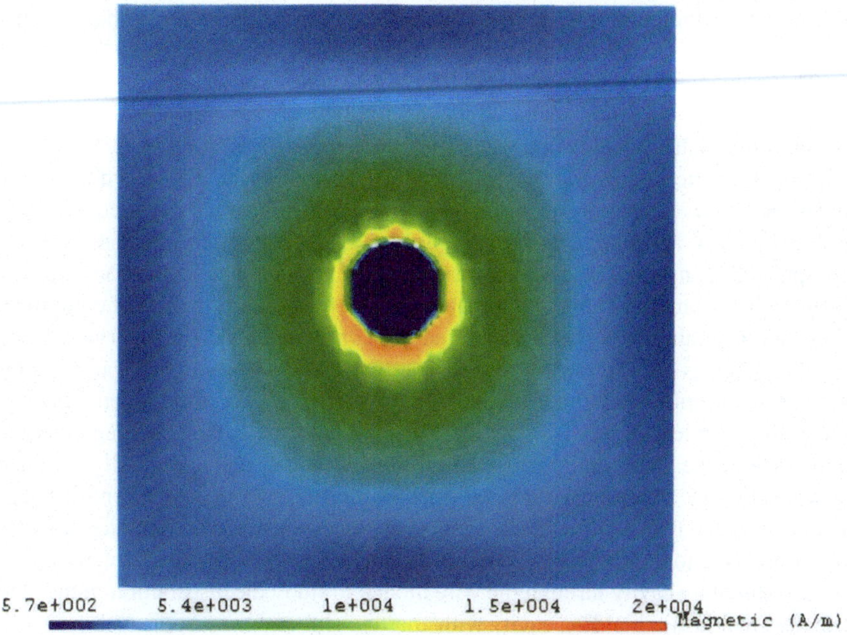

Fig. 11.21 Spatial magnetic field distribution at the bottom of the tower

components will be broken down [13]. Therefore, with the continuous improvement of the integration of electronic components, the lightning electromagnetic pulse field will have a serious impact on the normal operation of electronic equipment.

11.3.2 Optimized Arrangement of Electronic Equipment

The coupling of electromagnetic pulse to electronic communication equipment is mainly divided into two coupling modes: front-door coupling and back-door coupling. Front door coupling refers to the electromagnetic pulse coupled to the signal transceiver system through the antenna, etc., destroying the front-end electronic equipment; backdoor coupling means that the interference source enters through the aperture and interferes with the normal operation of the electronic equipment or circuit [14]. For the electronic equipment in the shielding body, the electromagnetic pulse is coupled into its interior through the holes and slits on the shielding body, which seriously weakens the shielding effect and affects the performance of the internal electronic components. For the wire through the shielding box, the external electromagnetic pulse can be directly coupled on the through wire, and the coupled energy is introduced into the shielding box through the through wire in the way of conduction coupling and radiation coupling. This section mainly discusses the aperture coupling in the electromagnetic pulse coupling path of the communication system and the crosstalk coupling of the through wire.

The main control cabinet of the wind turbine tower foundation is the core device for the reliable operation of the wind turbine. It is responsible for the collection of important data, signal processing and the determination and execution of logic control instructions, and realizes communication with the cabinet. In order to avoid the interference of external electromagnetic pulses, the control cabinet is equivalent to a shielding box, and a large number of integrated control components and electronic equipment are installed in the tower-based control cabinet. Due to the connection of lines such as communication cables and the need for heat dissipation of equipment, there are inevitably wire penetrations and gaps on the cabinet. The external electromagnetic pulses will be coupled to the interior through holes and through cables, so that the internal electronic equipment is in a complex electromagnetic environment. Strong interference signals will be generated on the cable shielding layer and core wire in the shielding cavity, and even electronic components will be broken down, seriously affecting the normal operation of the weak current control equipment connected to the cable. In this section, the electromagnetic simulation software is mainly used to build the shielding cavity with aperture slits and its internal cable model, shielding cavity through the wire model, to study the distribution of the electric field inside the shielding cavity with aperture slits and the response of the cables inside the cavity under the lightning EMP, to discuss the influence of the aperture slits' shapes, sizes, layout, and aperture slits' areas on the coupling of EMP, and to analyze the factors affecting the crosstalk coupling of the through the wire.

(a) **Field-Line Coupling of Shielded Cables in Cavities with Perforated Seams**

This book uses a three-dimensional full-wave electromagnetic simulation software based on the finite integral time domain method (FIT) and the transmission line matrix method (3D-TLM). The cable studio in the software has a rich cable library, including single wire, shielding wire and twisted pair. And the user can customize the multi-core cable. The software also supports the transient and frequency domain co-simulation of the three-dimensional electric field and the circuit, with adaptive meshing. The solution of the cable field-line coupling problem in a complex electromagnetic pulse environment is a combination of field and circuit, and the solution process is shown in Fig. 11.22.

(1) **Analysis of the Internal Electric Field Within a Cavity with Apertures and Slots**

Firstly, a three-dimensional model of the shielding cavity is built in the microwave studio of the electromagnetic simulation software. Figure 11.23 is a schematic diagram of the shielding cavity structure model with aperture under plane wave irradiation. The size of the shielding cavity is set to 2 m $\times$ 1.6 cm $\times$ 1.2 cm, the thickness is 2 mm, and the gap is set at 50 cm from the central axis of the upper surface of the cavity. According to the analysis of the electromagnetic field inside the tower during lightning strikes in the previous section, the field strength simulation results of the P point at the bottom of the tower at 10 m from the ground are extracted

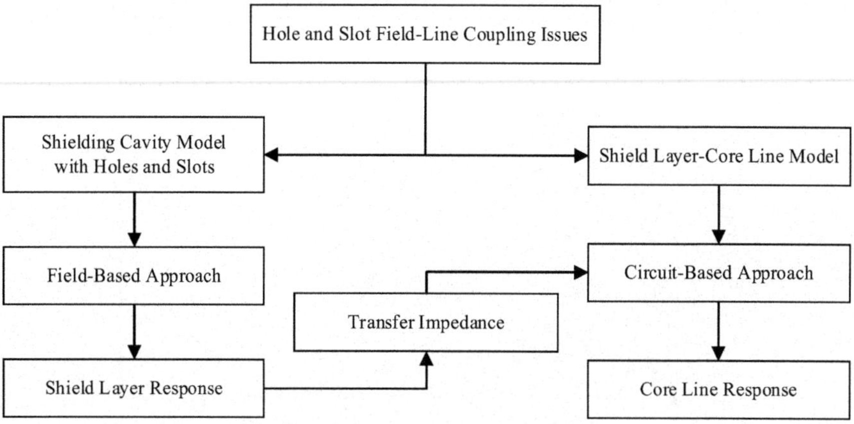

Fig. 11.22 Shielded cavity cable coupling problem with perforated slit

and used as the pulse excitation source. The pulse incident angle is $\theta = 180°$, $\varphi = 90°$, and the polarization angle is $\alpha = 180°$. The electric field intensity coupled to the cavity through the aperture of the shielding cavity is simulated by the simulation software, and the influence of the shape, size, arrangement and area of the aperture on the electromagnetic pulse coupling is discussed and analyzed.

The slit of the shielding cavity is set at the top of the cavity, 40 cm away from the central axis. The dimensions are set as a rectangular hole of 2 cm × 50 cm, a square hole of 10 cm × 10 cm, and a circular hole with a diameter of 11.8 cm. The electric field probe is set at the center of the shielding cavity, and the electric field intensity comparison is shown in Fig. 11.24.

From the comparison results in the figure, it can be found that when the aperture area is the same, if the aperture opened in the shielded cavity is rectangular, the field

Fig. 11.23 Shielding cavity model with apertures

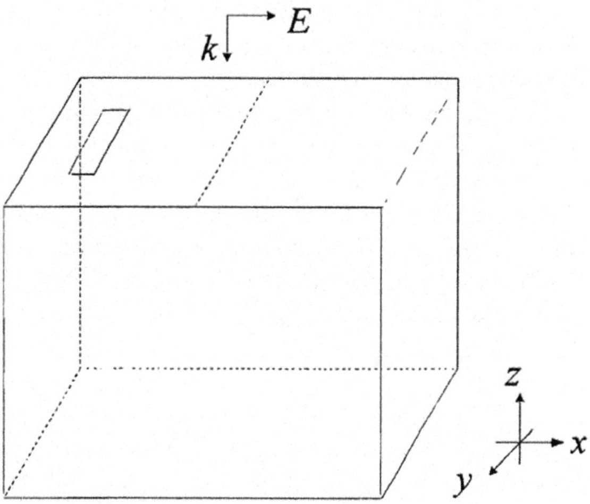

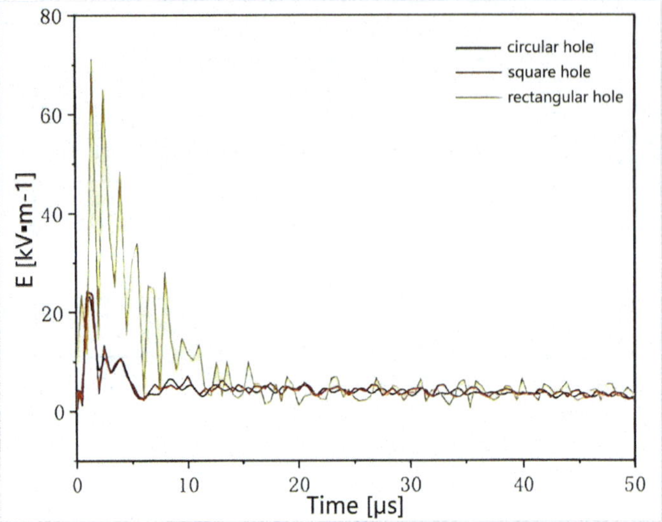

Fig. 11.24 Comparison of electric field intensity at the center of cavity with different shapes of apertures

intensity coupled to the interior through the aperture is the highest, which is about 3 times that of square and circular apertures of the same area; when the aperture is circular and square, the field intensity coupled to the interior of the shielded cavity through the circular aperture is slightly lower than that through the square aperture, and the difference between the two is not very obvious.

For rectangular apertures with the same area, when the length-to-width ratio changes from 1:4 to 1:2.5, the electric field intensity at the center of the shielded cavity decreases. The comparison result is shown in Fig. 11.25. This indicates that not only the shape of the aperture of the shielded cavity affects the energy coupled into the cavity, but also the length-to-width ratio of rectangular apertures, which have the most significant effect on coupling, has a greater impact. The greater the ratio, the more pronounced the effect.

Considering the influence of the length–width ratio of the rectangular aperture on the electromagnetic pulse coupling, the rectangular aperture of 5 cm × 20 cm is decomposed into four 5 cm × 5 cm square aperture arrays with a spacing of 2 cm. Under the same conditions, the electric field intensity at the center of the shielding cavity is simulated. The results are shown in Fig. 11.26a. Compared with the electric field intensity coupled at the center of the shielding cavity under the rectangular aperture with the same area, the electric field intensity coupled into the shielding cavity under the square aperture array is obviously lower, but the fluctuation range of the electric field intensity increases. This is because the mutual coupling between the apertures in the aperture array will affect the energy coupled into the shielding cavity. When the spacing of the square aperture array increases, the mutual coupling between the apertures weakens. As shown in Fig. 11.26b, the interval of the square

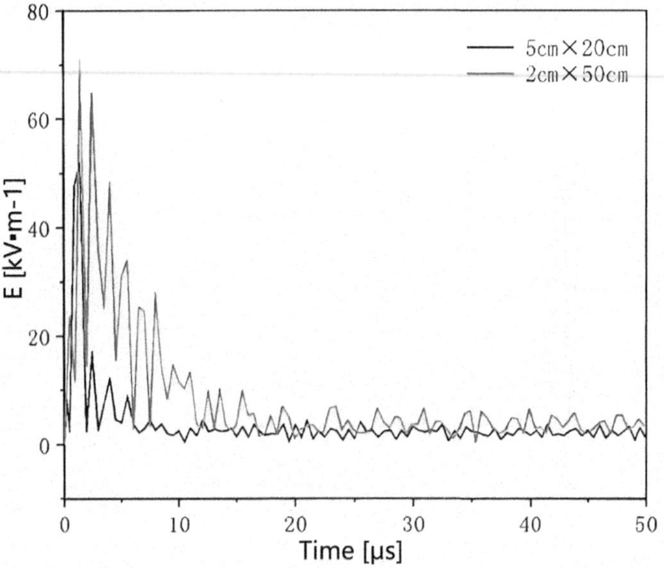

Fig. 11.25 Comparison of electric field strengths at the center of rectangular aperture slit cavities with different aspect ratios

aperture array increases to 5 cm, and the electric field intensity coupled into the shielding cavity also decreases.

After analyzing the influence of the shape of the aperture in the shielded cavity on the voltage coupled into the cavity, the effect of aperture area on coupling is discussed. Based on the above analysis, a square aperture with minimal coupling effect is selected as the research object. Under the same conditions, the electric field intensity at the center of the shielded cavity with 5 cm × 5 cm and 3 cm × 3 cm

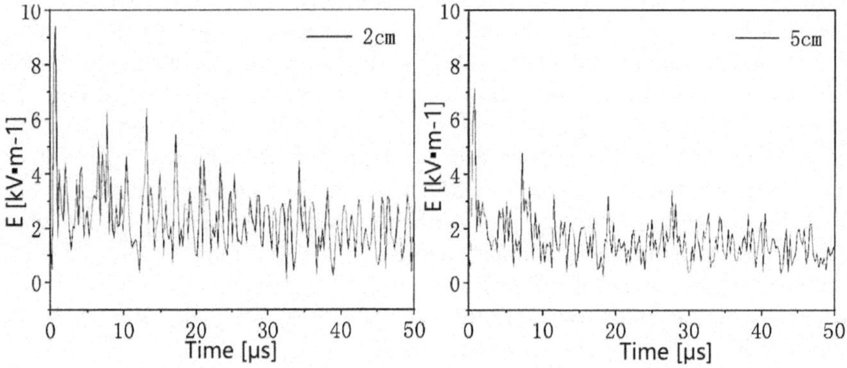

Fig. 11.26 Electric field strength at the center of square aperture array cavity with different spacing. **a** Hole array spacing 2 cm. **b** Hole array spacing 5 cm

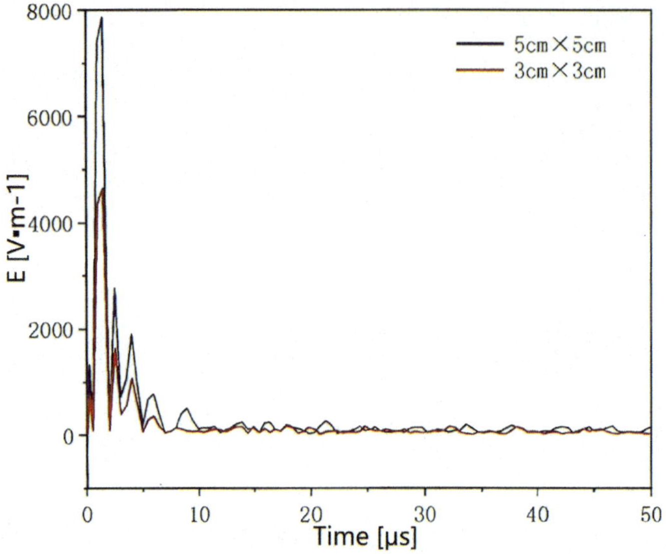

Fig. 11.27 Electric field intensity at the center of cavity with square apertures of different areas

square apertures is simulated, and the results are shown in Fig. 11.27. By comparison, it can be found that the electric field intensity coupled to the center of the shielded cavity through a 5 cm × 5 cm square aperture is 1.7 times that of a 3 cm × 3 cm square aperture. Thus, it can be concluded that the area of the aperture opened on the shielded cavity is proportional to the energy coupled into the cavity through the aperture, which is proportional to the electric field intensity inside the cavity.

(2) **Response of the Cable Inside the Chamber**

Figure 11.28a shows a schematic diagram of a single transmission line model in a shielded cavity with apertures under plane wave irradiation. The internal transmission line of the cavity is a 50 cm shielded twisted pair cable located 10 cm above the bottom. The structural model of shielded twisted pair cable is shown in Fig. 11.28b, with a core wire diameter of 1 mm, a shielding layer thickness of 0.2 mm, and an outer insulation layer diameter of 5 mm. The transient voltage response of the transmission line's shielding layer inside the shielded cavity with a 2 cm × 50 cm rectangular aperture is simulated for both grounded and ungrounded cases at both ends. The wiring situation of the cable in the DS studio is shown in Fig. 11.29, which depicts two grounding methods: ungrounded at both ends of the shielding layer and single-ended grounding of the shielding layer. The simulation results of induced voltage on the shielding layer and core wire under these two grounding methods are shown in Figs. 11.30 and 11.31, respectively.

Through comparison of simulation results, it is found that when both ends of the twisted-pair shielding layer are not grounded, the induced voltage on the shielding layer and the core wire is higher than that when it is grounded at one end. Therefore,

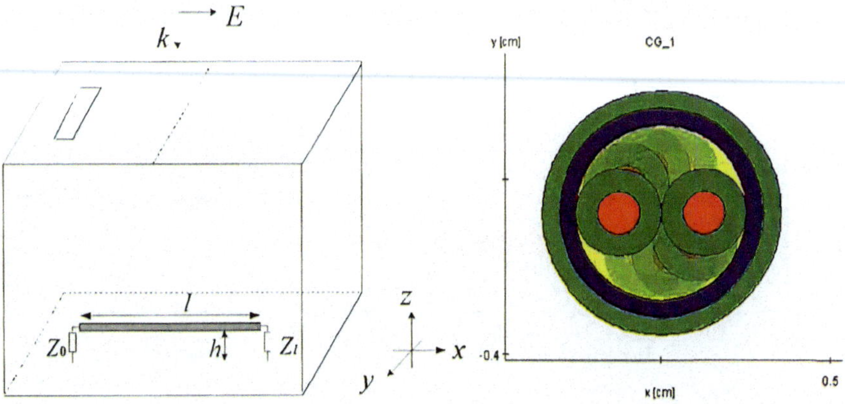

Fig. 11.28 Shielded twisted pair coupling model with perforated slit shielded cavity. **a** Model of a single transmission line in a shielded cavity with perforated slits. **b** Shielded twisted pair model

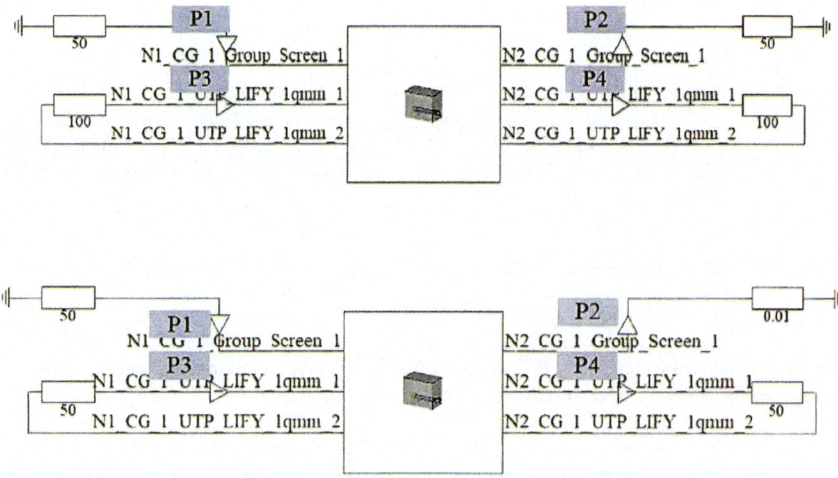

Fig. 11.29 DS Studio twisted pair shield grounding method. **a** Both ends of the shielding layer are not grounded. **b** Single-ended grounding of the shielding layer

the single-ended grounding method of the shielding layer can effectively suppress electromagnetic pulse interference.

(b) **Coupling of the Continuous Conductor Inside the Tower's Internal Cavity**

Crosstalk refers to electromagnetic coupling between closely spaced cables, where the signal in one cable is coupled into an adjacent cable, generating interference signals. This is a near-field coupling problem. Coupling interference between cables, devices, and cables and devices within a system is a common electromagnetic compatibility issue, and the interference caused by coupling between cables is often

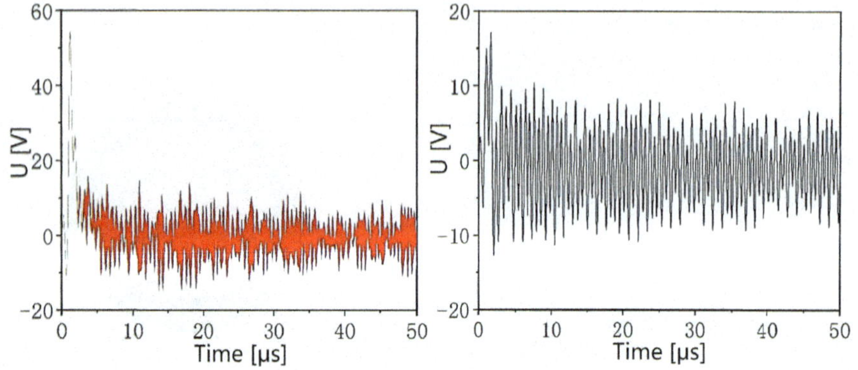

Fig. 11.30 Induced voltage when both ends of the twisted-pair shielding layer are not grounded. **a** Induced voltage on the shielding layer. **b** Induced voltage on the core wire

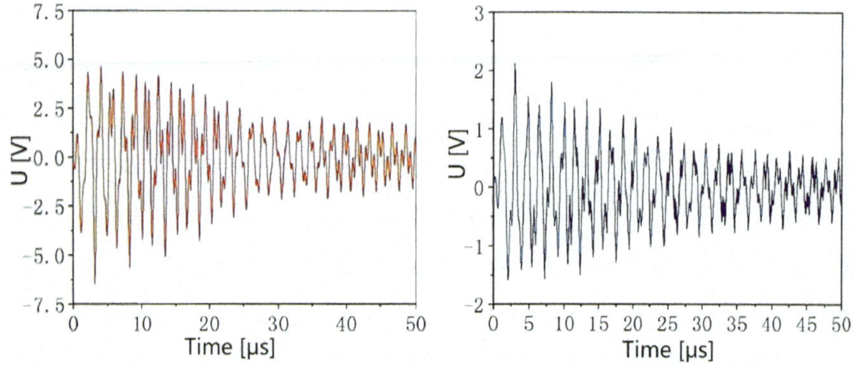

Fig. 11.31 Induced voltage when the twisted-pair shielding layer is grounded at one end. **a** Induced voltage on the shielding layer. **b** Induced voltage on the core wire

overlooked. When a wind turbine is struck by lightning, the lightning current flows into the ground along the tower base. During this period, there may be lightning current invading the internal circuitry, causing interference to other internal cables. Therefore, analyzing the crosstalk of the continuous conductor on its internal cables provides valuable reference for studying the electromagnetic compatibility of wind turbine control systems. By building a model of a shielded cavity with a continuous conductor, we can discuss the effects of the type of interfered cable, the length of the interfering cable, and the distance between cables on crosstalk coupling.

(1) **Principle of Crosstalk Coupling Between Cables**

As shown in Fig. 11.32a, there are two cables: the interfering line *ab* and the interfered line *cd*. Both cables are at a height of *h* from the ground, with a diameter of *d*, and a distance of *D* between them. The inductive coupling of crosstalk, also known as electromagnetic coupling, refers to the process where the magnetic field generated when current flows through a cable induces a voltage in an adjacent cable through mutual inductance. Figure 11.32b illustrates the coupling principle.

The formula for calculating the inductance L of the conductor is:

$$L = 0.14\log\left(\frac{4h}{d}\right) \quad (\mu H/ft) \tag{11.23}$$

The mutual inductance M between the two cables is given by,

$$M = 0.07\log\left[1 + \left(\frac{2h}{D}\right)^2\right] \quad (\mu H/ft) \tag{11.24}$$

As shown in Fig. 11.32b, R_d is the disturbed cable load, and the starting induced voltage V_d for is calculated as

$$|V_d| = 2\pi f M i_1 \frac{R_d}{R_c + R_d}\frac{1}{\sqrt{1 + \frac{\omega^2 L_2^2}{(R_c+R_d)^2}}} \tag{11.25}$$

When lightning current invades signal cables and propagates along transmission lines, electromagnetic coupling can induce voltage or current interference on surrounding lines. Severe crosstalk interference may lead to incorrect signal transmission, which in turn causes the control system and related electronic equipment to malfunction, seriously affecting their normal operation. Figure 11.33 shows a cavity measuring 2 m × 1.6 m × 1.2 m with a thickness of 2 mm. A circular aperture with a radius of 2 cm is located at (0.5, 0.5, 4) on the top of the cavity, through which a 6 m coaxial cable RG58 passes. The section inside the cavity measures 70 cm. Inside the cavity, there is a 50 cm single wire placed parallel to the coaxial cable, with a distance of 5 cm between them. Taking the injection of 1 kA, 1.2/50 μs lightning

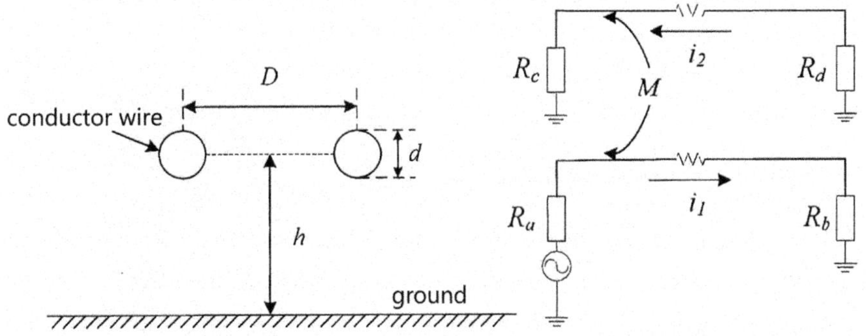

Fig. 11.32 Dual cable crosstalk structure schematic. **a** Dual cable sectional diagram. **b** Dual cable coupling schematic

current into the coaxial cable as an example, the induced voltage of the single cable inside the cavity is analyzed.

Figure 11.34 shows the crosstalk voltage of a single cable within the cavity. Under the interference of coaxial cables, the induced voltage amplitude of the single wire in the cavity reaches nearly 80 V. It can be seen from this that the crosstalk caused by the continuous conductor within the shielded cavity to its internal cables is a non-negligible electromagnetic compatibility issue. If the cable shielding measures within the cavity are inadequate, it can cause severe electromagnetic interference issues and damage to connected equipment.

(2) **The Influence of the Type of Interfered Cable on Crosstalk**

For the factors affecting the crosstalk between the cables, the main considerations are the type of the interfered cable, the length of the interfering cable penetrating into the cavity, and the distance between the interfered cable and the penetrating interfering cable. The response of crosstalk coupling between shielded twisted pair cables and coaxial cables is analyzed by first comparing a single line without a shield.

The 50 cm long shielded twisted-pair and coaxial cable are placed in parallel on the top and bottom of the coaxial cable through the shielding cavity at a distance of 5 cm. The parameters of the shielded twisted pair are the same as those mentioned earlier, and the coaxial cable used is RG58. The cable shielding layer adopts a two-end ungrounded approach. The wiring diagram for DS Studio is shown in Fig. 11.35. Probes are placed on the shielding layer and the core wire respectively to simulate their induced voltage. The results are shown in Figs. 11.36 and 11.37, respectively.

By comparing the induced voltage with that of a single wire, it is found that the cable's shielding layer can effectively shield crosstalk coupling. The amplitude of the induced voltage on the shielded twisted pair core wire decreased from nearly 80 V for a single wire to 2 V, while the induced voltage on the coaxial cable's core wire is even lower, at around 1.5 V. This suggests that improving the shielding effectiveness of cables inside the cavity can effectively suppress crosstalk coupling between cables.

(3) **The Influence of the Length of the Interference Cable on Crosstalk**

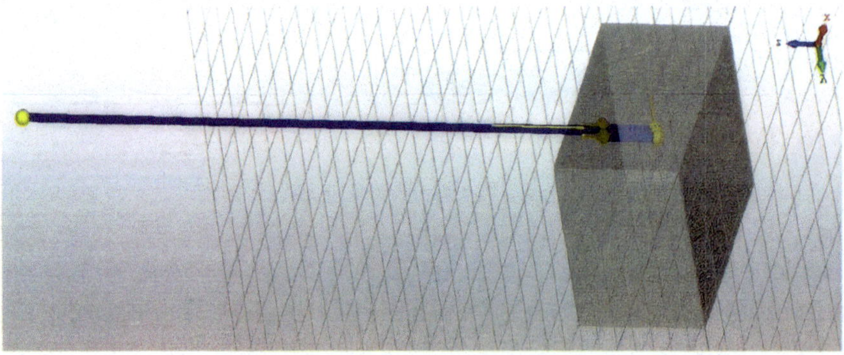

Fig. 11.33 Cavity through-conductor crosstalk modeling

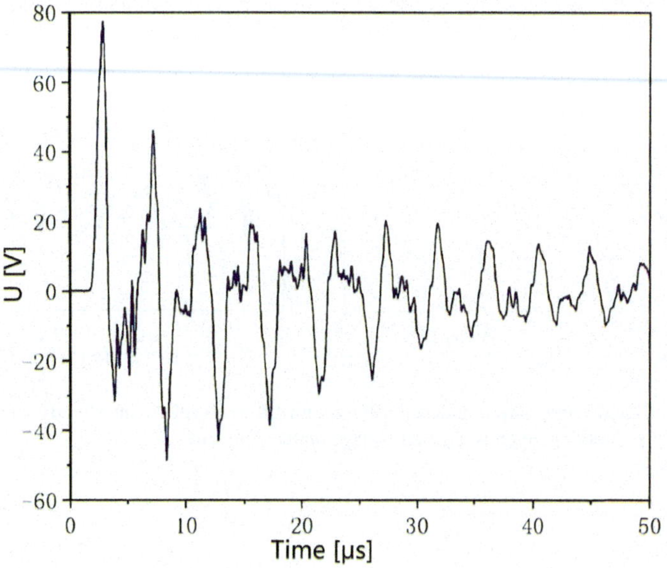

Fig. 11.34 Single cable crosstalk voltage in cavity

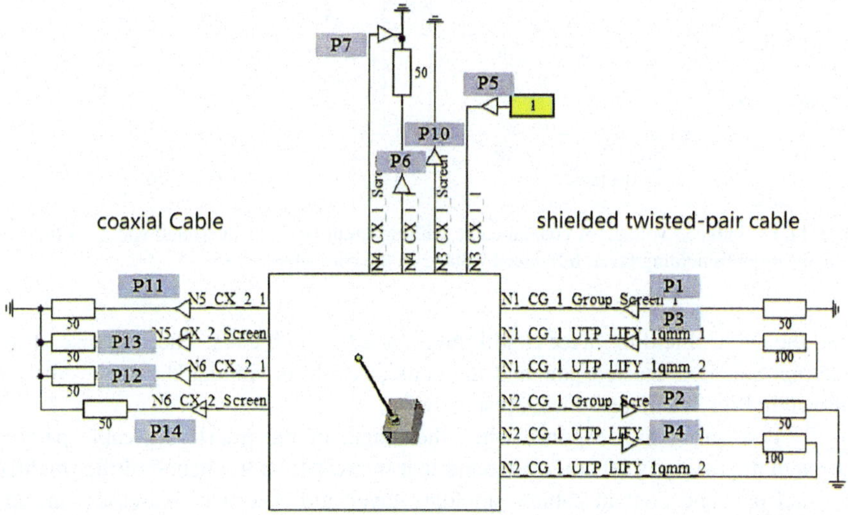

Fig. 11.35 DS Studio coaxial cable and shielded twisted pair wiring

The length of the interference cable penetrating into the cavity is changed from 70 to 20 cm. The 50 cm long shielded twisted pair and coaxial cable are placed directly above and below the coaxial cable penetrating the shielding cavity at a distance of 5 cm. The parameters of the shielded twisted pair and the coaxial cable are unchanged,

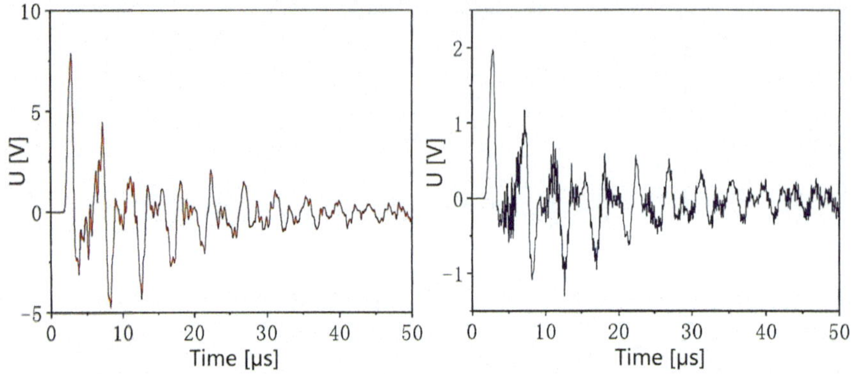

Fig. 11.36 Shielded twisted pair induced voltage with different types of interfered cables. **a** Induced voltage on the shielding layer. **b** Induced voltage on the core wire

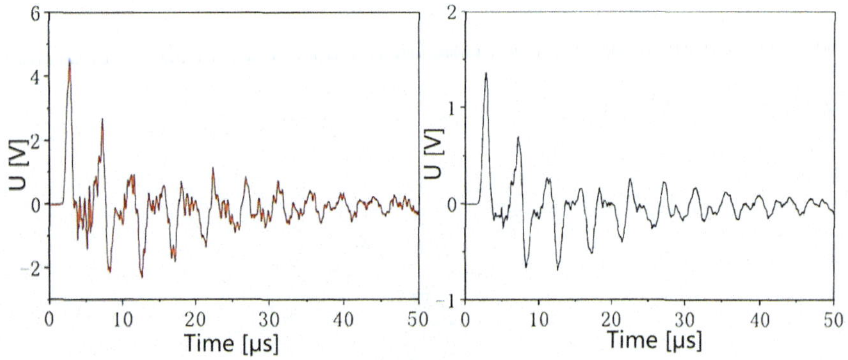

Fig. 11.37 Induced voltage of coaxial cable with different types of interfered cables. **a** Induced voltage on the shielding layer. **b** Induced voltage on the core wire

and the cable shielding layer is still ungrounded at both ends. The response of the internal shielded twisted pair and the coaxial cable is simulated. The results are shown in Figs. 11.38 and 11.39, respectively.

By comparing the response when the length of the interfering cable passing through the cavity is 70 cm, it is found that the response frequency of the shielded twisted pair and coaxial cable's shielding layer and core wire increases, and the response amplitude of the shielding layer decreases significantly to 20 and 30% of the original value, respectively. This suggests that reducing the length of the interfering cable entering the through-cavity can effectively suppress crosstalk coupling between cables.

(4) **The Influence of the Distance Between Cables on Crosstalk**

Considering the influence of the distance between the interfered cable and the interfering cable passing through the cavity on crosstalk coupling, the distance between

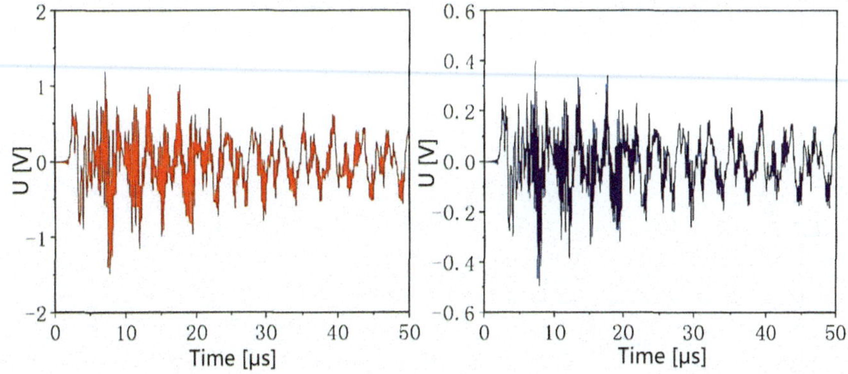

Fig. 11.38 Shielded twisted pair induced voltage with different interference cable lengths. **a** Induced voltage on the shielding layer. **b** Induced voltage on the core wire

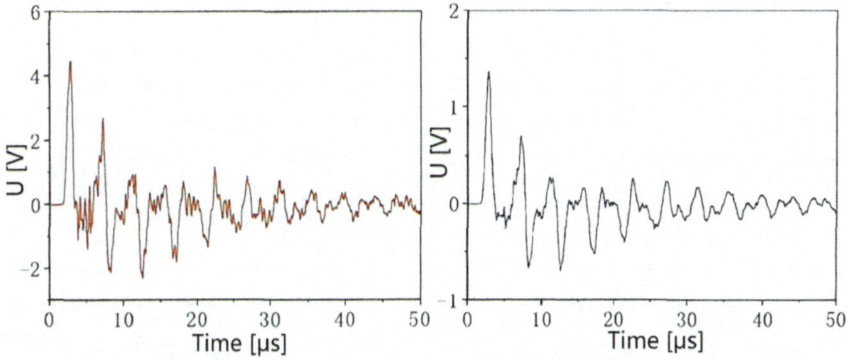

Fig. 11.39 Induced voltage of coaxial cable with different interference cable lengths. **a** Induced voltage on the shielding layer. **b** Induced voltage on the core wire

the two is increased from the original 5–10 cm. The coaxial cable of the continuous conductor passing through the cavity remains 70 cm, and all other conditions remain unchanged. The responses of the internal shielded twisted pair and coaxial cable are simulated, and the results are shown in Figs. 11.40 and 11.41, respectively.

By comparing the results with those in Figs. 11.36 and 11.37, it can be found that compared to the response when the interfering cable and the interfered cable are 5 cm apart, the responses of the shielded twisted pair and the coaxial cable shielding layer decrease by approximately 25 and 27%, respectively. This indicates that increasing the distance from the interfering cable can effectively reduce the impact of crosstalk coupling between cables.

(5) **The Influence of Cable Arrangement Position on Crosstalk**

Assuming that the portion of the coaxial cable through the conductor that passes into the cavity is still 70 cm, a 50 cm long shielded twisted pair of wire and coaxial

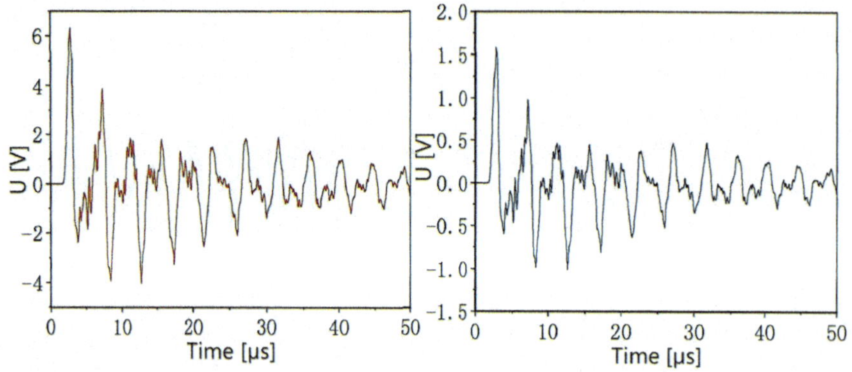

Fig. 11.40 Shielded twisted pair induced voltage with different distances between cables. **a** Induced voltage on the shielding layer. **b** Induced voltage on the core wire

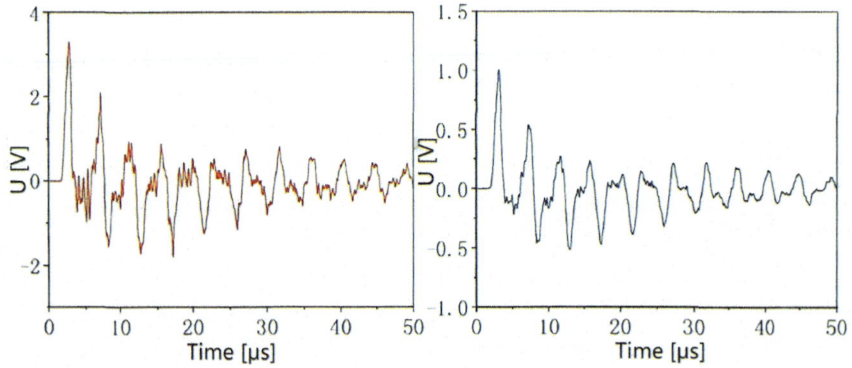

Fig. 11.41 Induced voltage of coaxial cable with different distances between cables. **a** Induced voltage on the shielding layer. **b** Induced voltage on the core wire

cable are placed 5 cm apart from the top and bottom of the coaxial cable that passes through the shielded cavity, respectively. The parameters of shielded twisted pair and coaxial cable are unchanged, and the cable shielding layer is still ungrounded at both ends. When the interfered cable and the interfered cable are in a vertical position, the response of the internal shielded twisted pair and the coaxial cable is simulated. The results are shown in Figs. 11.42 and 11.43, respectively.

Comparing the induced voltage results in Figs. 11.36 and 11.37, where the shielded twisted pair and coaxial cable are placed parallel to the interfering cable, it is found that when the interfered cable and the interfering cable are in a perpendicular position, the response fluctuations of the shielded twisted pair and coaxial cable increase significantly. However, the overall amplitude of the shielded layer's response decreases by approximately 30 and 27%, respectively, which is a noticeable decrease. This

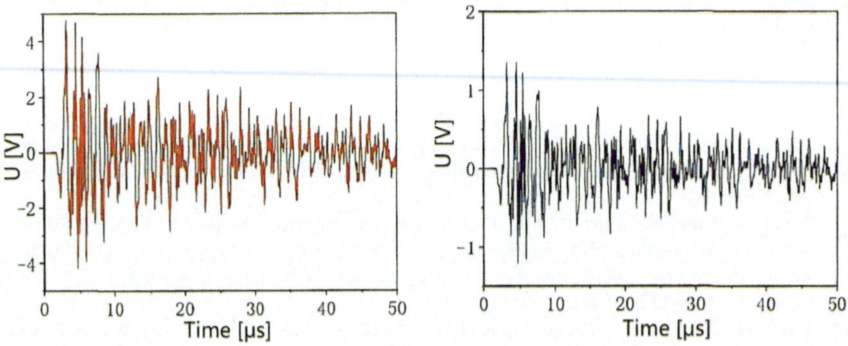

Fig. 11.42 Shielded twisted pair induced voltage with different cable layout positions. **a** Induced voltage on the shielding layer. **b** Induced voltage on the core wire

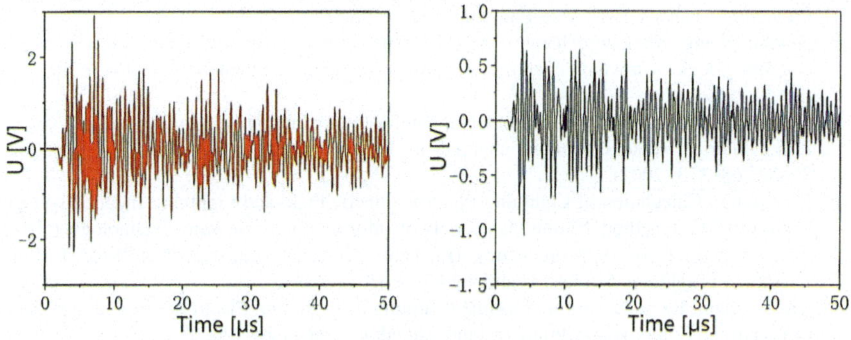

Fig. 11.43 Induced voltage of coaxial cable with different cable layout positions. **a** Induced voltage on the shielding layer. **b** Induced voltage on the core wire

indicates that placing the interfered cable perpendicular to the interfering cable can effectively reduce the impact of crosstalk coupling.

Through an analysis of several factors that affect crosstalk between cables, including the type of interfered cable, the length of the interfering cable passing through the cavity, the distance between the interfered cable and the interfering cable passing through the cavity, and the arrangement position of the interfered cable, the following conclusions are drawn: Shielded twisted pair cables and coaxial cables have a significant inhibitory effect on crosstalk coupling compared to unshielded single-wire pairs, and coaxial cables have a slightly better inhibitory effect; minimizing the length of the interfering cable passing through the shielded cavity can effectively reduce the impact of crosstalk coupling; increasing the distance between the interfered cable and the interfering cable passing through the cavity, and trying to keep the interfered cable in a perpendicular position to the interfering cable passing through the cavity, can also have a certain inhibitory effect on crosstalk coupling between cables.

References

1. Wu, Z.L., Wang, F.W.: Blower Foundation Type in Marine Wind Electric Field and Calculation Method. Port Waterw. Eng. 249–258 (2008).
2. Ma, J., Dawalibi, P., Southey, R.D.: On the equivalence of uniform and two-layer soils to multilayer soils in the analysis of grounding systems. IEE Proc. Gener. Transm. Distrib. 143, 49–55 (1996).
3. Pan, W.X., Liu, T.C., Wang, B., Min, D.: Grounding Computation Method for Layered-Soil with Arbitrary Massive Texture Foundation. Trans. China Electrotech. Soc. 31, 145–151 (2016).
4. Zhang, X.Q.: The Impulse Resistance and Coefficient of Vertical Grounding Rod. J. Electr. Power. 15, 155–157 (2000).
5. Liu, C.H., Zhang, X.Q., Xue, J., Chen, X.W., Liu, X.P.: A Calculation Method of the Electrical Parameters of Sloping Conductors in Lightning Protection Systems. Insul. Surge Arresters. 135, 81–85 (2012).
6. Wang, X.H.: Investigations on the Electromagnetic Effects in Wind Turbines Struck by Lightning; Beijing Jiao Tong University: Beijing, China, 2010.
7. Zhang, Y.J., Sima, W.X., Zhang, Z.J.: Summary of the Study of Tower Models for Lightning Protection Analysis. High Volt. Eng. 2006, 32(2), 93–97.
8. Xiaohui Wang, Xiaoqing Zhang.: Analysis of the lightning electromagnetic interference to electronic devices in wind generation systems. High Voltage Engineering. 35(8): 2019–2023 (2009).
9. Buccella C, Orlandi A.: An efficient technique for the evaluation of lightning-induced voltage in a cylindrical vessel containing charged oil. IEEE Transactions on Industry Applications. 39(2): 368–373 (2003).
10. Xiao Tang.: Calculation of Lightning Electromagnetic Field and Lightning-Induced voltages based on FDTD method. Nanjing University of Information Science and Technology. (2014).
11. Wu, Hao.: Study of High-Frequency Transient Electromagnetic Effect of Wind Turbines Offshore in Lightning Strokes. Shandong University, (2016).
12. Gai, Xinnan.: Research on Field-to-line Coupling of Lightning Electromagnetic Interference of Electronic Equipment in Wind Turbine. Shandong University, (2020).
13. Xiaoqing Zhang.: Lightning protection arrangement for electronic equipment in buildings. Application of Electronic Technique. (07): 44–45 (1998).
14. Fuchang Lin.: Research on the induction interference of indoor secondary cables when lightning strikes buildings. Huazhong University of Science and Technology, (1996).

The manufacturer's authorised representative in the EU is Springer Nature Customer Service Centre GmbH, Europaplatz 3, 69115 Heidelberg, Germany. If you have any concerns regarding our products, please contact ProductSafety@springernature.com

Printed and bound by CPI Group (UK) Ltd, Croydon, CR0 4YY
01/10/2026
02225907-0004